Small Molecule DNA and RNA Binders

M. Demeunynck, C. Bailly, W. D. Wilson (Eds.)

Related Titles from WILEY-VCH

Anneliese E. Barron (Ed.)

DNA Sequencing

2002, ca. 500 pages.
Hardcover. ISBN 3-527-30599-8

Susanne Brakmann and Kai Johnsson (Eds.)

Directed Molecular Evolution of Proteins

2002, 368 pages.
Hardcover. ISBN 3-527-30423-1

O. Zerbe, R. Mannhold, H. Kubinyi and G. Folkers
(Eds.)

BioNMR in Drug Research

2002, ca. 350 pages.
Hardcover. ISBN 3-527-30465-7

Douglas T. Gjerde, Christopher P. Hanna and David D.
Hornby

DNA Chromatography

2002, 243 pages.
Hardcover. ISBN 3-527-30244-1

Small Molecule DNA and RNA Binders

From Synthesis to Nucleic Acid Complexes

Volume 1

M. Demeunynck, C. Bailly, W. D. Wilson (Eds.)

Dr. Martine Demeunynck
Université Joseph Fourier
BP53
38041 Grenoble cedex
France

Dr. Christian Bailly
Institut de Recherches sur le Cancer
INSERM Unité 124
Place de Verdun
59045 Lille cedex
France

Prof. Dr. W. David Wilson
Department of Chemistry
Georgia State University
University Plaza
Atlanta GA 30303-3083
USA

Cover design
Christian Coulombeau

■ This book was carefully produced. Nevertheless, editors, authors and publisher do not warrant the information contained therein to be free of errors. Readers are advised to keep in mind that statements, data, illustrations, procedural details or other items may inadvertently be inaccurate.

Library of Congress Card No.: applied for
A catalogue record for this book is available from the British Library.
Bibliographic information published by Die Deutsche Bibliothek
Die Deutsche Bibliothek lists this publication in the Deutsche Nationalbibliografie; detailed bibliographic data is available in the Internet at http://dnb.ddb.de.

Printed in the Federal Republic of Germany.
Printed on acid-free paper.

Typesetting Asco Typesetters, Hong Kong
Printing betz-druck gmbH, Darmstadt
Bookbinding Litgas & Dopf Buchbinderei GmbH, Heppenheim

ISBN 3-527-30595-5

Contents

Preface

The ultimate goal of most organic-medicinal chemists is to see the small molecule that they have synthesized become a useful drug for the treatment of human diseases. Unfortunately, even with modern technology this is an extremely rare event. In most cases, the compounds designed and synthesized (generally with pain and passion) have a brief existence that does not exceed the first biological activity assay. The valley between chemistry and therapeutics is deep and difficult to cross but nevertheless the two disciplines are intimately associated. It is our goal to help construct a bridge between the makers of the small molecules and the users. Over the past two decades, a relatively large number of useful anticancer and antiparasitic drugs have been discovered or rationally designed based on the principle of nucleic acids recognition. A better understanding of the molecular rules that govern interactions between small molecules and the many sequences and structures of DNA and RNA is pivotal to the development of novel drug candidates. How does the drug adapt to the nucleic acid target (and *vice versa*)? How do nucleic acid structures affect ligand binding? How do small molecules read the genetic information? These types of questions continue to excite our scientific curiosity and the quest for better DNA/RNA binders drives modern researchers much as the search for the Holy Grail did the ancients.

Design and development of nucleic acid targeted drugs is a challenging enterprise but real breakthroughs have been made in recent years and many are reported here. This volume is intended to give the reader an up-to-date view of the current status and expected developments in research involving ligand-nucleic acid recognition. This book was built on a discussion among the three of us on how chemistry, biophysical chemistry and pharmacology serve our field to help design new drugs. Our different but complementary view angles on the subject prompted us to edit this volume focussed on DNA/RNA recognition by a variety of small molecules: peptides, intercalators, groove binders, metal complexes. The various DNA structures that can be targeted by drugs are also considered and the field of natural products is partially covered. Altogether, the 25 chapters of this volume survey most of the drug categories that bind, bond or cleave nucleic acids. The reader will notice the diversity of small molecules mentioned here, from marine products to platinum complexes, from G4-binders to RNA cleaving agents, from abasic site selective agents to aptamers, as well as the panel of biophysical and biochemical approaches routinely used to investigate the structures and dynamics of drug-nucleic acids complexes. The portraits of specific drug families (anthracyclines, indolocarbazoles, bleomycins, ...) are also thoroughly presented. The amalgam was deliberately chosen to cross ideas of organic chemists and biophysicists and those more interested in the therapeutic end point of the research.

The volume starts with a general introduction (magisterially presented in a British style) and then it flies over the world, from several countries in Europe (Spain, Italy, France, Czech Republic, UK) to the USA, via Japan and New Zealand, illustrating the essential international character of the research (and the friendly atmosphere of the edition). Inevitably we have neglected (mostly for consideration of space) a number of interesting areas that should have been cited here, such as clinical applications. But the gallery of molecules presented in this volume must be considered as a live exhibit to explore and to use for further drug design. Come on in, and like us, become fascinated by the "small molecules" that bind or bite the genetic material in its many forms. We hope you will also find examining this volume an enriching experience.

The enterprise was very exciting and proceeded smoothly (with no delay!) thanks to the enthusiastic contribution of all the authors. We are grateful to everyone for delivering their manuscript on time (and in some cases even well before the deadline!) and for making our task as editors such a memorable one. We also thank our "artist" Christian Coulombeau who kindly drew the front cover, sort of a railway to the future. Finally, we shall dedicate this volume to our colleagues who left the world too quickly to contribute (Marc Leng, David S. Sigman, and Peter A. Kollman in particular).

M.D., C.B., W.D.W.
Grenoble, Lille and Atlanta
September 2002

Contributors

Dr. Christian Bailly
Institut de Recherches sur le Cancer de Lille
INSERM U-524
Place de Verdun
59045 Lille Cedex
France

Prof. Dr Jacqueline K. Barton
Division of Chemistry and Chemical
    Engineering
California Institute of Technology
Pasadena, CA 91125
USA

Dr. David W. Boykin
Department of Chemistry
Georgia State University
Atlanta, Georgia 30303
USA

Dr. Viktor Brabec
Institute of Biophysics
Academy of Sciences of the Czech Republic
Kralovopolska 135
61265 Brno
Czech Republic

Dr. Jonathan B. Chaires
Department of Biochemistry
University of Mississippi Medical Center
2500 North State Street
Jackson, Mississippi 39216-4505
USA

Dr. Jean-François Constant
LEDSS
Université Joseph Fourier
BP 53
38041 Grenoble cedex 9
France

Dr. Peter C. Dedon
Division of Bioengineering and Environmental
    Health
Massachusetts Institute of Technology
Cambridge, MA 02139
USA

Dr. William A. Denny
Auckland Cancer Society Research Centre
Faculty of Medicine and Health Science
The University of Auckland, Private Bag 92019
Auckland

Prof. Dr. Keith R. Fox
Division of Biochemistry & Molecular Biology
School of Biological Sciences
University of Southampton
Bassett Crescent East
Southampton SO16 7PX
UK

Prof. Dr. Sidney M. Hecht
Department of Chemistry
University of Virginia
Charlottesville, Virginia 22901
USA

Dr. Laurence H. Hurley
Howard J. Schaeffer Endowed Chair in
    Pharmaceutical Sciences
Arizona Cancer Center
1515 N. Campbell Ave.
Room 4949
Tucson, AZ 85724
USA

Dr. Moses Lee
Department of Chemistry
Furman University
Greenville, SC 29613
USA

Dr. Eric Long
Department of Chemistry
Indiana University/Purdue University
Indianapolis, IN 46202-3274
USA

Dr. Jean-Louis Mergny
Laboratoire de Biophysique
Museum National d'Histoire Naturelle
INSERM U 201 CNRS UMR 8646
43 rue Cuvier
75231 Paris cedex 05
France

Prof. Dr. Stephen Neidle
Cancer Research UK Biomolecular Structure
   Group
The School of Pharmacy
University of London, 29–39
Brunswick Square
London WC1N 1AX
UK

Dr. Manlio Palumbo
Professor of Medicinal Chemistry
Department of Pharmaceutical Sciences
Via Marzolo, 5
35131 Padova
Italy

Dr. Laurence H. Patterson
Department of Pharmaceutical and Biological
   Chemistry
School of Pharmacy
University of London
Brunswick Square
London, WC1N 1AX
UK

Dr. Daniel Pilch
Department of Pharmacology
University of Medicine and Dentistry of New
   Jersey-Robert Wood Johnson Medical School
Piscataway, NJ 08854
USA

Dr. Tariq M. Rana
Department of Pharmacology
Robert Wood Johnson Medical School
675 Hoes Lane, Piscataway, NJ 08854
USA

Dr. Steve Rokita
Department of Chemistry and Biochemistry
University of Maryland
College Park, Maryland 20742
USA

Dr. Mark Searcey
Department of Pharmaceutical and Biological
   Chemistry
The School of Pharmacy
University of London
29/39 Brunswick Square
London WC1N 1AX
UK

Dr. Shigeori Takenaka
Department of Applied Chemistry
Faculty of Engineering
Kyushu University
Fukuoka 812-8581
Japan

Dr. David E. Thurston
Professor of Anticancer Drug Design
The School of Pharmacy
University of London
Brunswick Square
London
UK

Prof. Dr. Yitzhak Tor
Department of Chemistry & Biochemistry
University of California, San Diego
9500 Gilman Drive
San Diego, CA 92093-0358
USA

Dr. Jean-Pierre Vigneron
Collège de France, Chimie des interactions
   moléculaires
11 pl Marcelin Berthelot
75231 Paris cedex 5
France

Prof. M.J. Waring
Department of Pharmacology
University of Cambridge
Tennis Court Road
Cambridge CB2 1QJ
UK

Prof. Dr. W. David Wilson
Department of Chemistry
Georgia State University
University Plaza
Atlanta, GA 30303-3083
USA

1
Forty Years On

Michael J. Waring and L. P. G. Wakelin

1.1
Early Experiments Prior to Molecular Modeling

The quest to understand specific interactions between drugs and nucleic acids dates back a long time – more than 40 years. Even though the concept of gene targeting could not be explicitly formulated until much later, there were early realizations that DNA could provide a fine receptor for drugs. A major turning point in the history of drug binding to DNA, the publication in 1961 of the intercalation hypothesis by Leonard Lerman [1], in many people's estimation represents the true birth of the subject, but it would be wrong to neglect mention of the contributions of earlier workers. These workers knew they were dealing with drug–nucleic acid interactions and must have had some inkling of the future importance of the topic. Among them were the histologists who employed dyes such as aminoacridines to stain cells and tissue sections, particularly the fluorescent dye acridine orange, whose capacity to cause nuclei to fluoresce bright green while the cytoplasm fluoresced red was a valuable tool in histology and cell biology. Indeed in the researches of these pioneers can be found the first evidence that particular dyes can react differently with different kinds of nucleic acid-containing structures and therefore that the small molecules must be capable of some form of discrimination based upon what we would today call molecular recognition. From the variable and sometimes capricious performance of substances such as acridines employed as stains it could also be surmised that depending upon the solvent conditions a given dye might react in more than one way with its 'receptor,' foreshadowing the concept of heterogeneity in binding that was later to occupy the attention of biophysicists.

At the same time, thanks to the seminal work of Paul Ehrlich half a century earlier, the usefulness of dyes – particularly aminoacridines – as antiseptics and antimalarials was widely recognized so that the connection between cell staining and useful biological activity was more than implicit. Thus it happened at a critical moment that the potency of proflavine as a mutagen was recognized. This led to the brilliant experimental work of Crick, Brenner and colleagues [2] showing that exposure of bacteriophage-infected bacteria to proflavine produced frameshift mutations – a phenomenon that enabled them to deduce the triplet nature of the genetic code. Meanwhile, the careful experiments of Peacocke and Skerrett [3] on the interaction of proflavine with purified DNA were under way and the first truly quantitative measurements of a reversible drug–DNA binding reaction became available, complete with a proper description of the metachromatic shift in the absorption spectrum, application of spectrophotometry and equilibrium dialysis to determine genuine binding constants, and clear evidence of the occurrence of secondary binding after saturation of the strong primary binding sites had been accomplished.

Now all the elements were in place for Lerman, at that time working in the Cambridge MRC Laboratory of Molecular Biology with Crick and Brenner, to get to work on the intercalation hypothesis. Stone and Bradley [4] disposed of the secondary interaction of acridine orange with nucleic acids by attributing it to the formation of stacked aggregates of dye bound externally to the polyanion.

Two other pre-intercalation areas of endeavor must be mentioned, the first of which is the action of the antibiotic actinomycin D. Actinomycin had been discovered in the 1940s and was the first antibiotic found to be highly active against certain tumors – indeed, through the 1950s and early 1960s it was reckoned to be the most potent anticancer agent available in the arsenal of chemotherapy. The antibiotic was known to be capable of inhibiting nucleic acid synthesis in susceptible cells, a process that was consequently identified as a prime target for anticancer chemotherapy. The discovery of mRNA and the process of gene transcription owes much to the earnest work of early cell biologists who showed that actinomycin was an exquisitely selective inhibitor of transcription by virtue of its specific inhibitory action on the newly discovered enzyme RNA polymerase; that in turn was attributable to tight but reversible binding of actinomycin to the double-helical DNA template [5]. These discoveries firmly established the business of ligand–DNA interaction as a matter of concern to biologists, clinicians, and a breed of pharmacologists who later emerged as key players in founding what was to become the illustrious discipline of molecular interactions.

The second area of endeavor, though it had little influence on the development of ideas about reversible ligand–nucleic acid interactions, is the remarkable work of people like Kohn, Brooks, and Lawley on nitrogen mustards and comparable alkylating agents used for cancer chemotherapy [6, 7]. We should recall that nitrogen mustards were the very first chemicals used to treat cancer, prompted by unhappy events that occurred during the Second World War; it is indeed salutary that so evidently worthy a purpose as the alleviation of suffering from one of humanity's most dreaded diseases should have come about in such an inauspicious manner. The determined attentions of a few medically minded individuals capable of grappling with rather complicated and messy chemistry did a lot to clarify the mechanisms of action of these highly reactive substances, and again the critical target turned out to be DNA. Painstaking analysis of the products formed *in vitro* and *in vivo* when cells were exposed to mustards eventually identified the N7 position of the guanine ring as the most reactive (i.e. nucleophilic) site for alkylation of DNA, and the perceived correlation of anticancer activity with the possession of two alkylating centers spaced some five atoms apart led to the concept that bifunctional reactivity must be crucial for therapeutic effect.

1.2
Formulation of Molecular Models and Mechanisms of Binding to DNA

Before we return to the historic turning point at which the intercalation hypothesis was born, it is logical to finish consideration of the early alkylating agents by referring to their identification as crosslinking agents capable of covalently linking the complementary strands of the DNA double helix. At first it was thought that this action would adequately explain their cytotoxic activity through inhibiting the progress of the replication fork, but more recently the possible contribution of *intra*-strand crosslinks and DNA–protein crosslinks has brought this assumption into question [8, 9]. Meanwhile other types of powerful alkylating agents have been discovered that do not necessarily form interstrand crosslinks but are endowed with excellent biological activity. Moreover, an early twist to the tale of covalent reaction with DNA came with the finding that the antibiotic mitomycin C must be activated by reduction prior to forming inter-strand DNA crosslinks; this discovery added impetus to the idea of bioreductive activation of pro-drugs, particularly for cancer treatment, which has become an important focus for the efforts of several groups of drug designers (see Chapter 9). There is also a complex relation between bond-forming and bond-breaking interactions with nucleic acids that can be seen with several DNA-binding compounds described elsewhere in this volume (see Chapters 3 and 23).

A unifying theme that runs through these lines of work, and indeed throughout the volumes of this publication, is the extraordinarily sophisticated chemistry that attends the reaction of many compounds with nucleic acids, not to mention the amazing biosynthetic capabilities of the organisms that produce those substances that are of natural origin. Neither should we belittle the remarkable inventiveness and achievements of the organic chemists who increasingly are succeeding in their efforts to design strategies to come up with novel DNA-reactive compounds for chemotherapy as well as other purposes.

Returning to the historical thread, we go back to the year 1961, which was when the first reasonably explicit model for binding of a drug to the double helix – the intercalation hypothesis – was proposed for the interaction of aminoacridines like proflavine with DNA [1]. It is no secret, though not often appreciated outside laboratories of molecular biology, that the notion of frameshift mutation furnished a degree of inspiration for the model. However, the idea of intercalation was not universally acclaimed: indeed it was greeted with profound skepticism in certain quarters. Lerman's original evidence, drawn from observations of changes in viscosity, sedimentation coefficient, and X-ray diffraction from oriented fibers of DNA, was perfectly reasonable so far as it went. But that was not far enough to satisfy many of the "real" structure solvers, who made it clear that they were not going to believe the postulate unless and until it had been verified by their own favorite "direct" technique as opposed to the admittedly rather indirect evidence adduced by Lerman. One of the present authors remembers conversations including such phrases as "do you believe in intercalation?", as if it were an article of faith akin to religion. In due course, experiments were devised to verify or disprove the hypothesis, eventually to the satisfaction (or conversion) of the most hardened skeptics.

One early experiment was the circular DNA unwinding test, based upon the generally (but not quite universally) agreed expectation that intercalation must locally unwind the double helix [10]. It worked, and confirmed ethidium together with aminoacridines and several other interesting ligands, including actinomycin, as intercalators [11]. By the same token, antibiotics like netropsin and distamycin were identified as something else: minor groove binders as we now know [11]. It also became clear that different drugs unwind the helix by different angles when they intercalate, and the test even unearthed certain ligands that could unwind the double helix somewhat without apparently intercalating in the usual sense: steroidal diamines and triphenylmethane dyes [12, 13]. Questions still remain to be answered about these ligands. Of course a legacy of this early work is the detailed understanding of higher order structure, especially circularity, of DNA which studies on drug interactions have helped to elucidate. Thirty-five years after it was first shown to unwind circular DNA, ethidium is still routinely used to isolate plasmids.

Perhaps because of the seminal contributions of physical (bio)chemists during the early years of probing mechanisms of drug–nucleic acid interaction, the study of reaction kinetics soon emerged as a powerful tool for throwing light on the forces involved [14]. Don Crothers, an influential advocate of the kinetic approach, used to remark that the study of kinetics was uniquely valuable, if only because it added a new dimension – time – to the analysis of the phenomena. He was absolutely right. A highlight was the discovery that some ligands which bound well but not outrageously tightly to DNA could be characterized by on-rates and off-rates many orders of magnitude slower than ostensibly comparable substances. The anthracycline antibiotic nogalamycin is a good case in point; its slow association and dissociation kinetics are attributable to the disruption of base pairing needed to "thread" its bulky sugar substituents through the double helix [15]. Slow dissociation kinetics have been correlated with improved biological activity, and underlie the success of Phillips' relatively recent assay for transcription termination at particular drug-binding sites on DNA [16]. Direct ligand transfer between binding sites on DNA without involving complete dissociation from the polymer was evidenced many years ago and has given rise to the "shuffling" concept whereby ligands are supposed to migrate one-dimensionally along a DNA molecule in search of better (tighter) binding sites [17, 18].

1.3
Specificity of Nucleotide Sequence Recognition

Although the value of DNA and, to a lesser extent, RNA as a target for selective drug action had been evident from the outset, it also quickly became apparent that few known drugs showed much, if any, selectivity for binding to particular nucleotide sequences. Yet the holy grail of selectively suppressing gene expression was conceived early on, together with the realization that to attain this end it would be necessary to recognize moderately long stretches of base pairs. Eventually it was calculated that one might need to recognize a sequence composed of a number of base pairs in the high teens in order to identify a single targeted site in the human genome. The first experiments aimed at examining drug-binding preferences were crude and laborious to say the least, consisting of little more than attempts to detect different levels of binding to nucleic acids from different sources. Scatchard plots were employed to determine affinity constants, together with the frequency of binding sites, initially by simple and inappropriate means that were eventually much improved by better theoretical treatments like those of McGhee and von Hippel [19]. Sometimes the available methods (spectroscopy, equilibrium dialysis, etc.) were simply inapplicable because of the poor aqueous solubility of the ligands under investigation and alternative techniques had to be devised, such as solvent partition analysis used for the quinoxaline antibiotics [20]. Much effort was required just to establish a preference for, say, GC-rich DNA. Then the steady development of chemical methods for polynucleotide synthesis began to extend the range of synthetic, defined sequences available to the investigator and furnished substrates that could be used to examine whether or not a particular sequence would support interaction with a drug of interest.

A quantum leap occurred in the early 1980s with the invention of footprinting methodology in several laboratories at much the same time, using enzymes or Dervan's cleverly designed synthetic reagent MPE-Fe(II) to cut a cloned radiolabeled DNA fragment [21–23]. At a stroke it became possible to identify exactly where the preferred binding sites for a ligand were on a substrate that amounted to a real gene or a chosen fragment of a known gene. Although only semiquantitative at first, methods were quickly developed to adapt the technology to provide passable binding constants so that a true thermodynamic comparison of ligand affinity and capacity to discriminate between different sites could be gained in a single experiment or series of experiments. The power of the footprinting technique can be gauged from the reports of sequence-selectivity to be found in several chapters of this book. With its application, a substantial database of binding affinities for different sequences has been amassed, so that it is now becoming possible to enquire about general mechanisms that underlie the recognition of particular base-pair sequences, such as whether binding occurs predominantly in the major groove, the minor groove (much the most common with small molecules), or occasionally both.

Some workers have focused attention on the distinction between "digital" and "analog" readouts of sequence information, based on the notion that microstructural variation in the exact parameters of the double helix (groove width, for example) can sensitively reflect nucleotide sequence heterogeneity and therefore afford a means of sequence recognition that is independent of direct, specific contacts with the base pairs themselves. One of the techniques that can throw light on such questions involves looking at the behavior of DNA molecules containing unnatural nucleotide substitutions, which have the effect of shifting, removing, or adding specific base substituents. Such experiments have amply confirmed the dominant role of the 2-amino group of guanine in directing many ligands to their preferred binding sites, and have also thrown light upon related questions like the role of base pair substituents in modulating groove width, reactivity towards alkylating agents, helix curvature or flexibility, and the sequence-dependent winding of DNA around the histone octamer in nucleosome core particles [24].

While footprinting and related gel methodology continues to play a major role in studies of this sort it has recently been joined by the elegant but simple method of competition dialysis, whereby the relative binding of a test ligand to many different types of nucleic acids can be assessed at the same time [25]. This method is of particular interest for investigating structure-specific binding of drugs to nucleic acids or indeed other polymers, whether natural or synthetic.

1.4
Details at the Atomic and Molecular Levels

Insight into the structure and dynamics of intercalation complexes has progressed by a close synergy between theoretical and experimental approaches, which today has developed to the point at which it is now possible to give a complete molecular description of what a drug–DNA complex looks like at the atomic level, and how its constituent atoms move in solution. The techniques that have proved invaluable in this quest are X-ray crystallography, NMR spectroscopy, quantum chemistry, molecular mechanics, and molecular dynamics. Lerman himself used X-ray fiber diffraction data from proflavine–DNA complexes as part of the initial evidence he marshaled for the intercalation hypothesis [1], and Fuller and Waring adopted the technique to produce the first molecular model of the ethidium–DNA complex [26].

These early attempts at model building took an important step forward at the beginning of the 1970s when Sobell and colleagues solved the crystal structure of a 2:1 actinomycin–deoxyguanosine complex, which enabled them to construct a fairly precise intercalation model based upon purely geometrical constraints [27]. Later in the 1970s the commercial availability of DNA and RNA dinucleoside monophosphates made possible crystallographic and NMR studies of intercalated mini-duplexes of ethidium and aminoacridines such as 9-aminoacridine, proflavine, and acridine orange. The crystallographic studies by Sobell, Neidle, Rich, and their colleagues provided the first truly atomic description of intercalation complexes, and unequivocally proved that the DNA duplex could indeed stretch so as to sandwich acridine and phenanthridine chromophores between two base pairs [28–30].

These were seminal studies that not only provided insight into the fine details of individual drug–DNA complexes, but also revealed modifications to the geometry of the sugar–phosphate backbone generally required to open the intercalation cavity. Armed with the latter information, Neidle and others were able to construct molecular models of suitably modified B- and A-DNA duplexes containing stereochemically sound intercalation cavities [31, 32]. This provided the means for many investigators, Neidle, Pullman, and Hopfinger prominent amongst them, to use molecular mechanics and related methods to propose plausible structures for intercalation complexes of individual drugs [33–35].

Meanwhile, developments by Pullman and others in the application of quantum mechanical methods to determine the molecular electrostatic potential surrounding DNA, led to the important realization that the electrostatic field is strongly dependent on helical geometry, and that for B-type duplexes the field strength is strongest along the floor of the major and minor grooves [36, 37]. Moreover, it transpired that the intensity of the field in the grooves is sequence-dependent, so that for some sequences the potential is greatest in the minor groove, whereas for others, the global maximum is to be found in the major groove [36, 37]. Naturally, these sites of maximum potential provide "hotspots" for binding of cations, and positively charged, or highly dipolar, ligands. Interestingly, these studies also made clear that there are strong end effects for short oligonucleotides, so that the field strength is greatest in the middle of a short DNA fragment [36, 37]: a phenomenon that provides a potential motive force for driving the movement of DNA-bound ligands along the helix. These studies, in common with molecular mechanics-based modeling, emphasized the theoretical importance of electrostatics in determining the mode and strength of DNA–ligand interactions, and provided general insights into how intercalating agents and minor groove binders might interact with DNA.

In parallel work, Manning, Record, and their colleagues applied polyelectrolyte theory to DNA in order to establish a theoretical framework for interpreting the experimental observation that the DNA affinity of positively charged ligands is strongly ionic strength-dependent [38, 39]. Polyelectrolyte theory provides the fundamental insight that the electrostatic field of DNA causes the condensation of mobile cations in the region of its surface, so that for a B-type duplex the local concentration is about 1 M. It is the displacement of these mobile cations into the bulk solvent of lower ionic strength by fixed positively charged ligands that provides the entropic source that drives the observed dependency of affinity on salt concentration. In this way, polyelectrolyte theory links DNA molecular electrostatic potential and thermodynamics to enable a comprehensive description of the major forces that hold DNA–ligand complexes together: a union that Wilson, Chaires, Graves, and many others have used to great effect [40–42].

Structural studies made another quantum leap in the 1980s, courtesy of organic chemists, who provided the means to synthesize longer DNA fragments, initially in solution, and later by solid phase methods. The availability of self-complementary hexa-, octa-, and dodecanucleotides led to a truly wonderful explosion of crystallographic and NMR studies of DNA complexes containing intercalators and minor groove binders. At this time we were treated to crystal structures of the complexes of the quinoxaline bisintercalators echinomycin and triostin A, complete with a switch from Watson–Crick to Hoogsteen pairing in the base pairs immediately flanking the intercalated chromophores [43, 44], a multitude of complexes of anthracyclines related to adriamycin and daunomycin [e.g. 45], and to the complexes of the threading anthracycline nogalamycin [46]. A little later came the actinomycin–DNA structures [47]. It was at this time that NMR spectroscopy began to reveal its true power in providing near-atomic resolution structures in solution, a result of the union of NOESY data with molecular dynamics simulations, and we saw consonance with the crystallographic structures for the quinoxaline antibiotics nogalamycin, actinomycin, and the anthracyclines [48–52].

NMR even began to steal a march on crystallography by yielding solution structures of the luzopeptin, chromomycin, and pluramycin DNA complexes [53–55]. We still await the crystal structures. Simultaneously, much progress was made in the crystal and solution structures of DNA complexes of a multitude of minor groove-binding ligands encompassing oligopyrrole ligands such as netropsin and distamycin, bisbenzimidazoles related to Hoechst 33258, bisquaternary ligands and the bisamidine ligands related to pentamidine [56–58]. The NMR studies of these systems also provided information concerning the internal dynamics of the complexes, such as base pair breathing rates, and on occasions could define binding kinetics as well [59–61].

As the 1980s moved into the 1990s, developments in quantum mechanics and molecular dynamics began to lift traditional molecular modeling, based primarily on molecular mechanics calculations or rudimentary molecular dynamics performed *in vacuo*, out of the realm of the purely theoretical into simulations that can provide realistic descriptions of the complete drug:DNA:solvent:cation system. These calculations have progressed in the hands of the NMR spectroscopists and enlightened molecular modelers to the point where we can now determine atomic charge distributions by *ab initio* quantum mechanical methods, and formulate molecular dynamics simulations using more reliable force-fields that include the DNA, the drug, water, and cations [62–66]. In addition to the obvious advantages that these new developments bring by including solvent and counterion to deal properly with electrostatics and solvation, most significantly they intrinsically include hydrophobic bonding and entropy effects. We often forget the importance of these latter phenomena, because most DNA-binding ligands are charged, but should remember that they frequently dominate the energetics of drug–receptor complexes in the wider world of pharmacology, and are particularly significant in determining the DNA affinity of neutral ligands such as actinomycin, echinomycin, and luzopeptin.

Much of the fundamental work in this area was pioneered by Peter Kollman, who provided the community with AMBER, one of the most important computation tools [67]. His recent sudden death in his prime is a sad loss for us all, and he will be sorely missed by colleagues grateful for his contribution.

So, today, we stand in the happy position of being able to recruit crystallography, NMR spectroscopy, and molecular dynamics to provide a true atomic description for practically any DNA–ligand system. The limitations on the experiments are that the complex must crystallize and diffract, or, for NMR, the lifetime of the complex must be sufficiently long to allow build-up of NOE signals before the ensemble dissociates. The limitations on molecular dynamics are now reduced to concerns about computational power: a problem that is rapidly vanishing for most laboratories. Recent achievements of note are the crystal structures of a bisanthracycline complex [68], a DNA–PNA triplex [69], a PNA duplex [70], complexes with the sequence-selective hairpin polyamide ligands of Dervan [71, 72], and complexes of the acridinecarboxamide topoisomerase poisons [73, 74]. These works afford eloquent testimony to the current sophistication of our field, and have provided profound insights into the molecular determinants of drug–DNA complexes that confer upon the ligands their important biological properties.

One of the greatest surprises of recent times is the curious structure formed by the acridinecarboxamides, and similar ligands, with d(CGTACG)$_2$ in which four duplexes come together to exchange, by strand invasion, their terminal cytosine bases and to provide an intercalation cavity that accommodates two drug molecules [75, 76]. This structure is surely beyond our wildest imaginings and one can only wonder at the marvels of the solid state. However, it serves to remind us that we should be cognisant of the possibility of unexpected structures in circumstances where concentrations are high, water activity low, and mobility restricted: the cellular nucleus, for example.

A newcomer to the scene is atomic (scanning) force microscopy which promises to revolutionize the study of ligand–receptor interactions in general, including drug–nucleic acid complexes. Although barely out of its infancy, AFM has already produced excellent images of single DNA molecules with or without bound drugs. These studies have convincingly illustrated the changes in contour length and altered supercoiling of circular DNA caused by intercalative binding [77]. The singular advantages of AFM lie in its ability to acquire images in vacuum, under liquid, or in air at ambient, elevated, or cryogenic temperatures – not to mention the extreme economy in its requirements for materials when single molecules are the objects of examination. As the technology evolves to produce better images at higher resolution we can expect to admire genuine pictures of drugs attached to nucleic acid molecules under "wet" conditions in an aqueous environment. Perhaps what readers of this book have been seeing in their mind's eye for many years.

1.5
Identification of Motifs for Drug Design

As we noted at the beginning of this chapter, medicinal chemists have constantly sought pointers to mechanisms that might serve as a springboard for drug design, and latterly have been keeping a watchful eye on motifs arising from structural work that might provide a rational basis for designing new nucleic acid-binding ligands. Since the middle of the twentieth century numerous pointers or "leads" have emerged from the study of nucleic acids and their interactions with small molecules. At first the intercalation hypothesis itself attracted attention to the peculiar susceptibility of helical structures to deformation by binding of planar aromatic molecules, and the manifest existence of several useful drugs acting in this way encouraged the design and synthesis of many more. The endeavor received something of a boost with the discovery 13 or 14 years later of bis-intercalation, by echinomycin [78] and diacridines [79, 80] in particular, that clearly indicated the potential advantages of tighter binding and sequence-selectivity to be gained by synthesizing novel bis-intercalators. This was a task comfortably within the capabilities of organic chemists, and also afforded opportunities for early computer modelers to flex their muscles. The need to respect the limitations of the neighbor exclusion hypothesis [81] was recognized along with a whole new area of structure–activity relations. Sadly, despite considerable effort few synthetic bis-intercalators ever reached the clinic.

In parallel, attempts were made to modify classical minor groove binders so as to endow them with the capacity to recognize GC as well as AT base pairs, termed lexitropsins, but still no new drugs emerged [82]. Next came combilexins, in which intercalators were combined with other DNA-interactive moieties such as minor groove binders, alkylators and free radical generators or other mediators of poly-nucleotide strand cleavage [83–85]. Notwithstanding the failure of these programs to deliver quickly a new generation of wonder drugs (a notion which appears un-realistically optimistic in retrospect), they helped to build up a strong corpus of knowledge and experience which will have a lasting influence on the business of drug design.

As the concept of specific gene targeting gained momentum in the last years of the 1990s, four fresh approaches to the recognition of biologically meaningful se-quences of base pairs made their appearance. The first was triple helix formation, dating back to a very early observation from Rich's laboratory in 1957, in which a third polynucleotide strand sequence specifically occupies the major groove of the double helix. Its most common manifestation relies upon Hoogsteen pairing with the intact Watson–Crick base pairs and is restricted to polypurine tracts in the DNA target, but other motifs are possible and drug stabilization of such structures is important (see Chapter 14). The second approach, invented by Ole Buchardt in Copenhagen and now championed by Peter Nielsen [86] uses synthetic polyamide (or peptide) mimics of polynucleotides, termed PNA, to form double-stranded, triple-stranded, or strand-displacement complexes with a DNA target. The third approach, with its origins in the zinc-finger motif identified by Aaron Klug and his colleagues as underlying the sequence-specific recognition of DNA-response ele-ments by many proteins including hormone receptors and transcription factors, relies upon exploiting the natural motif(s) employed by peptides and proteins to recognize DNA [87]. More can be read about this topic in Chapter 16.

The fourth approach, arising from a chance observation that the minor groove binder distamycin could form a 2:1 complex with double-helical DNA by squeezing two antibiotic molecules side-by-side into the minor groove [88] was immediately recognized by Peter Dervan as a new motif endowed with tremendous potential for generating sequence readout by application of a new twist to the old lexitrop-sin concept. Dervan synthesized dimeric derivatives of *N*-methylpyrrole and *N*-methylimidazole ring-containing polyamides linked like a hairpin, showed that they could recognize and bind to targeted DNA sequences with subnanomolar affinity, and formulated a set of pairing rules governing the recognition of all four Watson–Crick base pairs in the minor groove [89]. Recently it has been found that a similar motif is applicable to synthetic diamidine minor groove binders as recounted in Chapter ▌.

1.6
Actions on Nucleoproteins, Chromatin, and Enzymes

Much of the most recent research on small molecule DNA and RNA binders has, quite rightly, been devoted to studies on more biologically relevant systems – building upon the detailed chemistry and physico-chemical principles established in studies on interactions with purified nucleic acids occupying the last four decades of the twentieth century. During the latter part of that period advances in biochemistry and cell biology rendered it possible to begin to make serious efforts aimed at understanding how drugs might affect nucleic acids in their native state – particularly the nucleoproteins of the cell nucleus. Early experiments on chromatin–drug interactions had generally been inconclusive to say the least, but with the rapid advances that followed the isolation of nucleosome core particles, together with better understanding of chromatin structure, it became feasible to design experiments with nucleic acid-binding drugs that might yield useful, concrete information.

Of course, binding of many ligands to the DNA in chromatin or core particles often does little more than disrupt the DNA–protein association so that the integrity of the structure is gradually impaired and eventually lost, releasing the DNA (presumably with lots of drug bound) from the histones and any other chromosomal proteins. However, some drugs do seem to have a peculiar effect on the winding of DNA around the histone octamer at moderate concentrations, provoking a change in the rotational orientation as judged by hydroxyl radical or enzyme attack. The altered rotational orientation may be as much as 180 degrees, and a similar effect can be produced by incorporating unnatural nucleotides into the DNA which alter its flexibility [90, 91]. Needless to say, the local unwinding of the double helix associated with intercalation could play a significant part in this phenomenon, though some minor groove binders can produce it too.

More recently the involvement of specific enzymes in the correct functioning, packaging, and maintenance of chromosome structure has been clarified to the extent that novel drug actions of great therapeutic importance and promise can be envisaged. We have already referred to the transcription stop assay that can be used to determine drug-binding sites where RNA polymerase activity is terminated most effectively [16]. Another enzyme of crucial importance for the functioning of chromosomes in replication and transcription is topoisomerase, whose two principal forms (topo I and topo IIα and β) are both targets for drug attack via inhibition or poisoning. Small molecules that act on these enzymes, several of them high on the list of newly developed anticancer agents, are described in detail in Chapters 19, 20, and 21. But perhaps the most interesting and remarkable enzyme to have emerged as a novel drug target is telomerase, the enzyme responsible for the maintenance of chromosome ends in eukaryotes, which is a reverse transcriptase that operates to extend the telomeric structures at the ends of chromosomes. Those structures are characterized by multiple repeats of a short DNA sequence rich in guanine nucleotides which predisposes them to form peculiar quadruplex structures that represent a potential target for attachment of drugs, thereby inhibiting telomerase activity and the preservation of chromosome ends. Since telomere maintenance appears to be correlated with malignancy, the G-quadruplex structures could offer a marvelous target for selective suppression of cancer cells as described in Chapters 12 and 13, where drug design strategies are described using molecular modeling based upon the accumulated experience recounted throughout this book.

References

1 L. S. LERMAN. Structural considerations in the interaction of deoxyribonucleic acid and acridines. *J. Mol. Biol.* **1961**, *3*, 18–30.

2 S. BRENNER, L. BARNETT, F. H. C. CRICK, A. ORGEL. The theory of mutagenesis. *J. Mol. Biol.* **1961**, *3*, 121–124.

3 A. R. PEACOCKE, J. N. H. SKERRETT. The interaction of aminoacridines with nucleic acids. *Trans. Faraday Soc.* **1956**, *52*, 261–279.

4 A. L. STONE, D. F. BRADLEY. Aggregation of acridine orange to polyanions: the stacking tendency of deoxyribonucleic acids. *J. Am. Chem. Soc.* **1961**, *83*, 3627–3634.

5 E. REICH, I. H. GOLDBERG. Actinomycin and nucleic acid function. *Prog. Nucleic Acids Res. Mol. Biol.* **1964**, *3*, 183–234.

6 K. W. KOHN, N. H. STEIGBIGEL, C. L. SPEARS. Crosslinking and repair of DNA in sensitive and resistant strains of *E. coli* treated with nitrogen mustard. *Proc. Natl Acad. Sci. USA* **1965**, *53*, 1154–1161.

7 P. D. LAWLEY, P. BROOKES. Molecular mechanism of the cytotoxic action of difunctional alkylating agents and of resistance to this action. *Nature* **1965**, *206*, 480–483.

8 K. W. KOHN. DNA filter elution: a window on DNA damage in mammalian cells. *Bioessays* **1996**, *18*, 505–513.

9 G. B. BAUER, L. F. POVIRK, Specificity and kinetics of interstrand and intrastrand bifunctional alkylation by nitrogen mustards at a GGC sequence. *Nucleic Acids Res.* **1997**, *25*, 1211–1218.

10 L. V. CRAWFORD, M. J. WARING. Supercoiling of polyoma virus DNA measured by its interaction with ethidium bromide. *J. Mol. Biol.* **1967**, *25*, 23–30.

11 M. J. WARING. Variation in the supercoiling of the DNA double helix as evidence for intercalation. *J. Mol. Biol.* **1970**, *54*, 247–279.

12 M. J. WARING, J. W. CHISHOLM. Uncoiling of bacteriophage PM2 DNA by binding of steroidal diamines. *Biochim. Biophys. Acta* **1972**, *262*, 18–23.

13 L. P. G. WAKELIN, A. ADAMS, C. HUNTER, M. J. WARING. Interaction of crystal violet with nucleic acids. *Biochemistry* **1981**, *20*, 5779–5787.

14 H. J. LI, D. M. CROTHERS. Relaxation studies of the proflavine–DNA complex: the kinetics of an intercalation reaction. *J. Mol. Biol.* **1969**, *39*, 461–477.

15 K. R. FOX, C. BRASSETT, M. J. WARING. Kinetics of dissociation of nogalamycin from DNA: comparison with other anthracycline antibiotics. *Biochim. Biophys. Acta* **1985**, *840*, 383–92.

16 D. R. PHILLIPS, C. CULLINANE. Transcriptional footprinting of drug-DNA interactions. *Methods Mol. Biol.* **1997**, *90*, 127–145.

17 J. L. BRESLOFF, D. M. CROTHERS. DNA-ethidium reaction kinetics:

demonstration of direct ligand transfer between DNA binding sites. *J. Mol. Biol.* **1975**, *95*, 103–123.

18 K. R. Fox, M. J. Waring. Footprinting reveals that nogalamycin and actinomycin shuffle between DNA binding sites. *Nucleic Acids Res.* **1986**, *14*, 2001–2014.

19 J. D. McGhee, P. H. von Hippel. Theoretical aspects of DNA-protein interactions: co-operative and non-co-operative binding of large ligands to a one-dimensional homogeneous lattice. *J. Mol. Biol.* **1974**, *86*, 469–489.

20 M. J. Waring, L. P. G. Wakelin, J. S. Lee. A solvent-partition method for measuring the binding of drugs to DNA. Application to the quinoxaline antibiotics echinomycin and triostin A. *Biochim. Biophys. Acta* **1975**, *407*, 200–212.

21 C. M. Low, H. R. Drew, M. J. Waring. Sequence-specific binding of echinomycin to DNA: evidence for conformational changes affecting flanking sequences. *Nucleic Acids Res.* **1984**, *12*, 4865–4879.

22 M. W. Van Dyke, R. P. Hertzberg, P. B. Dervan. Map of distamycin, netropsin, and actinomycin binding sites on heterogeneous DNA:DNA cleavage-inhibition patterns with methidiumpropyl-EDTA.Fe(II). *Proc. Natl Acad. Sci. USA* **1982**, *79*, 5470–5474.

23 M. J. Lane, J. C. Dabrowiak, J. N. Vournakis. Sequence specificity of actinomycin D and Netropsin binding to pBR322 DNA analyzed by protection from DNase I. *Proc. Natl Acad. Sci. USA* **1983**, *80*, 3260–3264.

24 C. Bailly, M. J. Waring. Use of DNA molecules substituted with unnatural nucleotides to probe specific drug–DNA interactions. *Methods Enzymol.* **2001**, *340*, 485–502.

25 J. Ren, J. B. Chaires. Rapid screening of structurally selective ligand binding to nucleic acids. *Methods Enzymol.* **2001**, *340*, 99–108.

26 W. Fuller, M. J. Waring. Molecular model for the interaction of ethidium bromide with DNA. *Ber. Bunsenges. Phys. Chem.* **1964**, *68*, 805–808.

27 H. M. Sobell, S. C. Jain. Stereochemistry of actinomycin binding to DNA. II. Detailed molecular model of actinomycin-DNA complex and its implications. *J. Mol. Biol.* **1972**, *68*, 21–34.

28 C. C. Tsai, S. C. Jain, H. M. Sobell. X-ray crystallographic visualization of drug-nucleic acid intercalative binding: structure of an ethidium-dinucleoside monophosphate crystalline complex, ethidium: 5-iodouridylyl (3′-5′) adenosine. *Proc. Natl Acad. Sci. USA* **1975**, *72*, 628–632.

29 H. S. Shieh, H. M. Berman, M. Dabrow, S. Neidle. The structure of drug-deoxydinucleoside phosphate complex; generalized conformational behavior of intercalation complexes with RNA and DNA fragments. *Nucleic Acids Res.* **1980**, *8*, 85–97.

30 A. H. Wang, J. Nathans, G. van der Marel, J. H. van Boom, A. Rich. Molecular structure of a double helical DNA fragment intercalator complex between deoxy CpG and a terpyridine platinum compound. *Nature* **1978**, *276*, 471–474.

31 H. M. Berman, S. Neidle, R. K. Stodola. Drug-nucleic acid interactions: conformational flexibility at the intercalation site. *Proc. Natl Acad. Sci. USA* **1978**, *75*, 828–832.

32 K. J. Miller, J.F. Pycior. Interaction of molecules with nucleic acids. II. Two pairs of families of intercalation sites, unwinding angles, and the neighbor-exclusion principle. *Biopolymers* **1979**, *18*, 2683–2719.

33 S. A. Islam, S. Neidle, B. M. Gandecha, M. Partridge, L. H. Patterson, J. R. Brown. Comparative computer graphics and solution studies of the DNA interaction of substituted anthraquinones based on doxorubicin and mitoxantrone. *J. Med. Chem.* **1985**, *28*, 857–864.

34 K. X. Chen, N. Gresh, B. Pullman. A theoretical investigation on the sequence selective binding of daunomycin to double-stranded polynucleotides. *J. Biomol. Struct. Dyn.*, **1985**, *3*, 445–466.

35 Y. NAKATA, A. J. HOPFINGER. Predicted mode of intercalation of doxorubicin with dinucleotide dimers. *Biochem. Biophys. Res. Commun.* **1980**, *95*, 583–588.

36 A. PULLMAN, B. PULLMAN. Molecular electrostatic potential of nucleic acids. *Q. Rev. Biophys.* **1981**, *14*, 289–380.

37 P. M. DEAN, L. P. G. WAKELIN. Electrostatic compounds of drug-receptor recognition. I. Structural and sequence analogues of DNA poly-nucleotides. *Proc. R. Soc. Lond.* **1980**, *B209*, 453–471.

38 R. A. FRIEDMAN, G. S. MANNING. Polyelectrolyte effects on site-binding equilibria with application to intercalation of drugs with DNA. *Biopolymers* **1984**, *23*, 2671–2714.

39 M. T. RECORD, JR., C. F. ANDERSON, T. M. LOHMAN. Thermodynamic analysis of ion effects on the binding and conformational equilibria of proteins and nucleic acids: the roles of ion association or release, screening, and ion effects on water activity. *Q. Rev. Biophys.* **1978**, *11*, 103–178.

40 F. A. TANIOUS, S. F. YEN, W. D. WILSON. Kinetic and equilibrium analysis of a threading intercalation mode: DNA sequence and ion effects. *Biochemistry* **1991**, *30*, 1813–1819.

41 J. B. CHAIRES, S. SATYANARAYANA, D. SUH, I. FOKT, T. PRZEWLOKA, W. PRIEBE. Parsing the free energy of anthracycline antibiotic binding to DNA. *Biochemistry* **1996**, *35*, 2047–2053.

42 R. M. WADKINS, D. E. GRAVES. Thermodynamics of the interactions of m-AMSA and o-AMSA with nucleic acids: influence of ionic strength and DNA base composition. *Nucleic Acids Res.* **1989**, *17*, 9933–9946.

43 A. H. WANG, G. UGHETTO, G. J. QUIGLEY, T. HAKOSHIMA, G. A. VAN DER MAREL, J. H. VAN BOOM, A. RICH, The molecular structure of a DNA-triostin A complex. *Science*, **1984**, *225*, 1115–1121.

44 G. UGHETTO, A. H. WANG, G. J. QUIGLEY, G. A. VAN DER MAREL, J. H. VAN BOOM, A. RICH. A comparison of the structure of echinomycin and triostin A complexed to a DNA fragment. *Nucleic Acids Res.* **1985**, *13*, 2305–2323.

45 G. J. QUIGLEY, A. H. WANG, G. UGHETTO, G. A. VAN DER MAREL, J. H. VAN BOOM, A. RICH. Molecular structure of an anticancer drug-DNA complex: daunomycin plus d(CpGpTpApCpG). *Proc. Natl Acad. Sci. USA* **1980**, *77*, 7204–7208.

46 L. D. WILLIAMS, M. EGLI, G. QI, P. BASH, G. A. VAN DER MAREL, J. H. VAN BOOM, A. RICH, C. A. FREDERICK. Structure of nogalamycin bound to a DNA hexamer. *Proc. Natl Acad. Sci. USA* **1990**, *87*, 2225–2229.

47 S. KAMITORI, F. TAKUSAGAWA, Crystal structure of the 2:1 complex between d(GAAGCTTC) and the anticancer drug actinomycin D. *J. Mol. Biol.* **1992**, *225*, 445–456.

48 X. L. GAO, D. J. PATEL. NMR studies of echinomycin bisintercalation complexes with d(A1-C2-G3-T4) and d(T1-C2-G3-A4) duplexes in aqueous solution: sequence-dependent formation of Hoogsteen A1.T4 and Watson–Crick T1.A4 base pairs flanking the bisintercalation site. *Biochemistry* **1988**, *27*, 1744–1751.

49 D. E. GILBERT, J. FEIGON. Proton NMR study of the [d(ACGTATACGT)]$_2$-2 echinomycin complex: conformational changes between echinomycin binding sites. *Nucleic Acids Res.* **1992**, *20*, 2411–2420.

50 M. S. SEARLE, J. G. HALL, W. A. DENNY, L. P. G. WAKELIN. NMR studies of the interaction of the antibiotic nogalamycin with the hexadeoxyribonucleotide duplex d(5′-GCATGC)$_2$. *Biochemistry* **1988**, *27*, 4340–4349.

51 X. LIU, H. CHEN, D. J. PATEL. Solution structure of actinomycin-DNA complexes: drug intercalation at isolated G-C sites. *J. Biomol. NMR* **1991**, *1*, 323–347.

52 C. ODEFEY, J. WESTENDORF, T. DIECKMANN, H. OSCHKINAT. 2-Dimensional nuclear magnetic resonance studies of an intercalation complex between the novel semisynthetic anthracycline 3′-

deamino-3′-(2-methoxy-4-morpholinyl)-doxorubicin and the hexanucleotide duplex d(CGTACG). *Chem. Biol. Interact.* **1992**, *85*, 117–126.

53 M. S. SEARLE, J. G. HALL, W. A. DENNY, L. P. G. WAKELIN. Interaction of the antitumour antibiotic luzopeptin with the hexanucleotide d(5′-GCATGC)$_2$: one and two-dimensional NMR studies. *Biochem. J.* **1989**, *259*, 433–441.

54 X. L. GAO, D. J. PATEL. Solution structure of the chromomycin-DNA complex. *Biochemistry* **1989**, *28*, 751–762.

55 M. HANSEN, S. YUN, L. H. HURLEY. Hedamycin intercalates the DNA helix and, through carbohydrate-mediated recognition in the minor groove, directs N7-alkylation of guanine in the major groove in a sequence-specific manner. *Chem. Biol.* **1995**, *2*, 229–240.

56 M. L. KOPKA, C. YOON, D. GOODSELL, P. PJURA, R. E. DICKERSON. The molecular origin of DNA-drug specificity in netropsin and distamycin. *Proc. Natl Acad. Sci. USA* **1985**, *82*, 1376–1380.

57 P. PJURA, K. GRZESKOWIAK, R. E. DICKERSON. Binding of Hoechst 33258 to the minor groove of B-DNA. *J. Mol. Biol.* **1987**, *197*, 257–271.

58 T. A. LARSEN, D. S. GOODSELL, D. CASIO, K. GRZESKOWIAK, R. E. DICKERSON. The structure of DAPI bound to DNA. *J. Biomol. Struct. Dyn.* **1989**, *7*, 477–491.

59 D. J. PATEL. Antibiotic–DNA interactions: intermolecular nuclear Overhauser effects in the netropsin-d(C-G-C-G-A-A-T-T-C-G-C-G) complex in solution. *Proc. Natl Acad. Sci. USA* **1982**, *79*, 6424–6428.

60 A. PARDI, K. M. MORDEN, D. J. PATEL, I. TINOCO, JR. Kinetics for exchange of the imino protons of the d(C-G-C-G-A-A-T-T-C-G-C-G) double helix in complexes with the antibiotics netropsin and/or actinomycin. *Biochemistry* **1983**, *22*, 1107–1113.

61 D. J. PATEL, L. SHAPIRO, Sequence-dependent recognition of DNA duplexes. Netropsin complexation to the AATT site of the d(G-G-A-A-T-T-C-C) duplex in aqueous solution. *J. Biol. Chem.* **1986**, *261*, 1230–1240.

62 P. HERZYK, S. NEIDLE, J. M. GOODFELLOW. Conformation and dynamics of drug-DNA intercalation. *J. Biomol. Struct. Dyn.* **1992**, *10*, 97–139.

63 B. DE PASCUAL-TERESA, J. GALLEGO, A. R. ORTIZ, F. GAGO. Molecular dynamics simulations of the bis-intercalated complexes of ditercalinium and Flexi-Di with the hexanucleotide d(GCGCGC)$_2$: theoretical analysis of the interaction and rationale for the sequence binding specificity. *J. Med. Chem.* **1996**, *39*, 4810–4824.

64 H. E. WILLIAMS, M. S. SEARLE. Structure, dynamics and hydration of the nogalamycin-d(ATGCAT)$_2$ complex determined by NMR and molecular dynamics simulations in solution. *J. Mol. Biol.* **1999**, *290*, 699–716.

65 D. L. BEVERIDGE, K. J. MCCONNELL. Nucleic acids: theory and computer simulation, Y2K. *Curr. Opin. Struct. Biol.* **2000**, *10*, 182–196.

66 S. A. HARRIS, E. GRAVATHIOTIS, M. S. SEARLE, M. OROZCO, C. A. LAUGHTON. Cooperativity in drug-DNA recognition: a molecular dynamics study. *J. Am. Chem. Soc.* **2001**, *123*, 12658–12663.

67 T. CHEATHAM III, P. A. KOLLMAN. Molecular dynamics simulation of nucleic acids. *Annu. Rev. Phys. Chem.* **2000**, *51*, 435–471.

68 G. G. HU, X. SHUI, F. LENG, W. PRIEBE, J. B. CHAIRES, L. D. WILLIAMS. Structure of a DNA-bisdaunomycin complex. *Biochemistry* **1997**, *36*, 5940–5946.

69 L. BETTS, J. A. JOSEY, J. M. VEAL, S. R. JORDON. A nucleic acid triple helix formed by a peptide nucleic acid-DNA complex. *Science* **1995**, *270*, 1838–1841.

70 H. RASMUSSEN, J. S. KASTRUP, J. N. NIELSEN, J. M. NIELSEN, P. E. NIELSEN. Crystal structure of a peptide nucleic acid (PNA) duplex at 1.7 A resolution. *Nature Struct. Biol.* **1997**, *4*, 98–101.

71 C. L. KIELKOPF, E. E. BAIRD, P. B. DERVAN, D. C. REES. Structural basis for G.C recognition in the DNA minor groove. *Nature Struct. Biol.* **1998**, *5*, 104–109.

72 C. L. KIELKOPF, S. WHITE, J. W. SZEWCZYK, J. M. TURNER, E. E. BAIRD, P. B. DERVAN, D. C. REES. A structural basis for recognition of A.T and T.A base pairs in the minor groove of B-DNA. *Science* **1998**, *282*, 111–115.

73 A. ADAMS, C. COLLYER, J. M. GUSS, W. A. DENNY, L. P. G. WAKELIN. Crystal structure of the topoisomerase II poison 9-amino-[N-(2-dimethylamino)ethyl]acridine-4-carboxamide bound to the DNA hexanucleotide d(CGTACG)$_2$. *Biochemistry* **1999**, *38*, 9221–9233.

74 A. ADAMS, J. M. GUSS, W. A. DENNY, L. P. G. WAKELIN. Crystal structure of 9-amino[N-(2-morpholino)ethyl]acridine-4-carboxamide bound to d(CGTACG)$_2$: Implications for structure activity relationships of acridinecarboxamide topoisomerase poisons. *Nucleic Acids Res.* **2002**, *30*, 719–725.

75 A. ADAMS, J. M. GUSS, C. A. COLLYER, W. A. DENNY, L. P. G WAKELIN. A novel form of intercalation involving four DNA duplexes in an acridine-4-carboxamide complex of d(CGTACG)$_2$. *Nucleic Acids Res.* **2000**, *28*, 4244–4253.

76 X. L. YANG, H. ROBINSON, Y. G. GAO, A. H. WANG. Binding of a macrocyclic bisacridine and ametantrone to CGTACG involves similar unusual intercalation platforms. *Biochemistry* **2000**, *39*, 10950–10957.

77 P. T. LILLEHEI, L. A. BOTTOMLEY. Scanning force microscopy of nucleic acid complexes. *Methods Enzymol.* **2001**, *340*, 234–251.

78 M. J. WARING, L. P. G. WAKELIN. Echinomycin: a bifunctional intercalating antibiotic. *Nature* **1974**, *252*, 653–657.

79 E. S. CANELLAKIS, Y. H. SHAW, W. E. HANNERS, R. A. SCHWARTZ. Diacridines: bifunctional intercalators. I. Chemistry, physical chemistry and growth inhibitory properties. *Biochim. Biophys. Acta* **1976**, *418*, 277–289.

80 L. P. G. WAKELIN, M. ROMANOS, T. K. CHEN, D. GLAUBIGER, E. S. CANELLAKIS, M. J. WARING. Structural limitations on the bifunctional intercalation of diacridines into DNA. *Biochemistry* **1978**, *17*, 5057–5063.

81 D. M. CROTHERS. Calculation of binding isotherms for heterogenous polymers. *Biopolymers* **1968**, *6*, 575–584.

82 J. W. LOWN, K. KROWICKI, U. G. BAHT, A. SKOROBOGATY, B. WARD, J. C. DABROWIAK. Molecular recognition between oligopeptides and nucleic acids: novel imidazole-containing oligopeptides related to netropsin that exhibit altered DNA sequence specificity. *Biochemistry* **1986**, *25*, 7408–7416.

83 C. BAILLY, N. POMMERY, R. HOUSSIN, J. P. HENICHART. Design, synthesis, DNA binding, and biological activity of a series of DNA minor-groove-binding intercalating drugs. *J. Pharm. Sci.* **1989**, *78*, 910–917.

84 T. A. GOURDIE, K. K. VALU, G. L. GRAVATT, T. J. BORITZKI, B. C. BAGULEY, L. P. G. WAKELIN, W. R. WILSON, P. D. WOODGATE, W. A. DENNY. DNA-directed alkylating agents. 1. Structure-activity relationships for acridine-linked aniline mustards: consequences of varying the reactivity of the mustard. *J. Med. Chem.* **1990**, *33*, 1177–1186.

85 J. W. LOWN, S. M. SONDHI, C. W. ONG, A. SKOROBOGATY, H. KISHIKAWA, J. C. DABROWIAK. Deoxyribonucleic acid cleavage specificity of a series of acridine- and acodazole-iron porphyrins as functional bleomycin models. *Biochemistry* **1986**, *25*, 5111–5117.

86 P. E. NIELSEN, M. EGHOLM. An introduction to peptide nucleic acid. *Curr. Issues Mol. Biol.* **1999**, *1*, 89–104.

87 Y. CHOO, I. SANCHEZ-GARCIA, A. KLUG. In vivo repression by a site-specific DNA-binding protein designed against an oncogenic sequence. *Nature* **1994**, *372*, 642–645.

88 J. G. PELTON, D. E. WEMMER. Structural characterization of a 2:1 distamycin A.d(CGCAAATTGGC) complex by two-dimensional NMR. *Proc. Natl Acad. Sci. USA* **1989**, *86*, 5723–5727.

89 P. B. DERVAN. Molecular recognition of DNA by small molecules. *Bioorg. Med. Chem.*, **2001**, *9*, 2215–2235.

90 C. M. LOW, H. R. DREW, M. J. WARING. Echinomycin and distamycin induce rotation of nucleosome core DNA. *Nucleic Acids Res.* **1986**, *14*, 6785–6801.

91 M. BUTTINELLI, A. MINNOCK, G. PANETTA, M. J. WARING, A. TRAVERS. The exocyclic groups of DNA modulate the affinity and positioning of the histone octamer. *Proc. Natl Acad. Sci. USA* **1998**, *95*, 8544–8549.

2
Targeting HIV RNA with Small Molecules

Nathan W. Luedtke and Yitzhak Tor

2.1
Introduction

The central dogma of biology states that RNA is selectively transcribed from DNA and serves as a messenger to be translated into proteins [1, 2]. Recent discoveries detailing the molecular workings of the cell have inspired new generations of chemists to search for a deeper understanding of the biophysical world. The majority of this book presents important advances in the study of DNA; this chapter is one of the few to explore RNA. We have, therefore, taken the liberty of presenting a general background on why RNA is such a fascinating biomolecule and why RNA small molecule recognition holds great promise for the future.

Far from being a passive carrier of genetic code, RNA is intimately involved in a wide range of biological processes including chemical catalysis and information storage. The vast majority of cellular RNAs form higher order structures with other cellular components to facilitate the exchange of biochemical information. Successful molecular recognition of RNA often precedes catalytic events that are essential to a wide range of cellular activities, including: initiation of DNA replication [3], extension of the telomeric regions of chromosomes [4], splicing of pre-mRNA [5], and iron chelation [6]. In addition, RNA serves as the primary genome of most pathogenic viruses [7].

2.1.1
Translation

Unique patterns of gene expression rely upon an interplay between recognition events and catalytic activities of RNA–protein complexes. Ribosomal RNA accounts for the vast majority of total cellular RNA (80%) and provides both the molecular scaffold and enzymatic activities needed for protein translation [8]. The key step of translation occurs in the ribosome's A-site, where codon–anticodon recognition decodes the mRNA. Upon a correct codon–anticodon match between mRNA and the anticodon loop of tRNA, the ribosome's peptidyltransferase activity catalyzes the formation of a new peptide bond between the amino acid-charged tRNA in the A-site and the growing protein chain on the tRNA in the P-site. Studies have shown that the prokaryotic ribosomes stripped of protein are still capable of limited peptidyltransferase activity [9]. In accordance with this result, recent crystal structures show that the peptidyltransferase active site is composed entirely of rRNA [10]. A single, unusually basic adenosine may be the key player in the mechanism of peptidyl transfer [11, 12].

Transfer RNAs, at 15% of total cellular RNA, are the most common type of "soluble" RNA (i.e. lacking associated protein). The binding of tRNA to the ribosomal A-site is mediated by extensive RNA–RNA interactions (including rRNA–tRNA, and mRNA–tRNA binding). Through these and other important interactions, the ribosome can distinguish the relatively small energetic differences between cognate and non-cognate codon–anticodon pairing to achieve an astounding fidelity of translation (over 99.9% accuracy) [13].

The transport, translation efficiency, and stability of individual messenger RNAs are controlled by numerous protein–RNA, ribonucleoprotein–RNA, and RNA–RNA interactions [14–16]. Upon transcription from DNA, ribonucleoprotein complexes called splisosomes excise the introns from pre-mRNA and from other heterogeneous RNAs. Some organisms are capable of intron excision (splicing) without protein assistance, and provided the first examples of RNA enzymes (or ribozymes) [17]. The translation efficiency of individual mRNAs is regulated at many levels, including the binding of the 5′ and 3′ untranslated regions (UTRs) of the mRNA by proteins [18], microRNAs [19], and even small molecules [20].

2.1.2
RNA Viruses

Viral epidemics have accounted for more human deaths than all known wars and famine combined. Over 65% of the known families of viruses use RNA for a primary genome and cause many of the modern-day plagues, including: AIDS, cancer, hepatitis, smallpox, ebola, and influenza [7]. Most viruses are, however, benign. Approximately 42% of the human genome is composed of transposable elements that multiply by reverse transcription, using an RNA intermediate similar to that of a retrovirus [21]. In general, reverse transcription is a highly error-prone process allowing viral elements to evolve rapidly under selective pressures (such as antiviral drugs). About 8% of the human genome is made up of repetitive genomic elements known as "retrovirus-like elements" [21]. Their structures closely resemble those of retroviruses, carrying the open reading frames common to all retroviruses (Gag, Pol, Env) flanked by 5′ and 3′ long terminal repeats. Overall, the human genome is composed of approximately 50% self-repeating, parasitic sequences. Compare this with the unique (non-repeated) genes, encoding for only ∼5% of the human genome!

2.2
Small Molecules that Modulate RNA Activity

The ability of RNA to facilitate the essential biochemical activities needed for information storage, signal transduction, replication, and enzymatic catalysis has

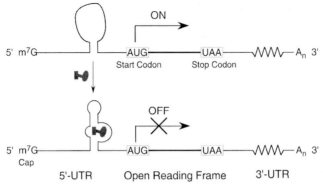

Fig. 2.1. The mature mRNA of an artificial gene construct is actively translated in the absence of small-molecule binding. Upon binding of the aptamer by its cognate small molecule, the translation of a reporter gene is deactivated.

distinguished it as a good candidate for being the central biomolecule in a prebiotic world [22]. If such an "RNA world" did exist, then small molecule–RNA interactions most likely played a key role in the regulation of RNA replication and in modulating RNA enzymatic activity [23].

An impressive conceptual proof demonstrating the ability of small organic molecules to regulate gene expression *in vivo* has recently been illustrated in the context of an artificial gene construct [20]. An RNA aptamer located in the 5′ UTR region of an mRNA, has been shown to inactivate translation of a downstream open reading frame upon the binding of its cognate small molecule (Fig. 2.1). (Note: Aptamers are RNA sequences that are selected from large libraries of RNA sequences for affinity to a predetermined small molecule. Through multiple rounds of mutation, selection, and amplification, the evolution of an RNA sequence to fit onto a predetermined small molecule can be effected. Selected sequences often demonstrate truly astonishing affinity and specificity to their cognate small molecule.) The mechanism proposed for translation inactivation involves the structural rearrangement of the aptamer into a rigid RNA–small molecule complex that cannot be correctly scanned by the ribosomal pre-initiation machinery. Natural systems have also provided examples of how small molecule–RNA binding (often accompanied by RNA structural rearrangements) can modulate RNA activity.

2.2.1
Magnesium (II)

Much like proteins, the primary sequence of an RNA directs its folding into a unique three-dimensional structure [24]. Correct RNA folding, however, typically relies upon the binding of divalent metal ions (especially Mg^{2+}). Mg^{2+} exhibits a low to moderate affinity to many unrelated RNAs. Mg^{2+} binding affinities (K_d) range from 0.01 mM through 10 mM in the presence of 0.1–0.2 M of monovalent ions [25, 26]. Given the three-dimensional structure of an RNA, an electrostatic contour map can be calculated, allowing for the theoretical prediction of the higher affinity Mg^{2+}-binding sites [27, 28].

The Mg^{2+}-induced folding of the *Tetrahymena thermophila* group 1 intron has become an important paradigm for RNA folding [29–31]. In the absence of Mg^{2+}, it adopts flexible structures dominated by duplex regions that are interrupted by internal bulges and stem loops. This secondary structure can largely be predicted from its nucleotide sequence using base-pairing and nearest neighbor rules [32–34]. Upon Mg^{2+} binding, it collapses into a more rigid, enzymatically active, tertiary structure with fewer available conformations. In at least one region of the group 1 intron, Mg^{2+} binding induces a rearrangement of the RNA secondary structure itself [35]. Our limited understanding of cation-binding interactions remains a major obstacle in the accurate prediction of a three-dimensional RNA structure given only its sequence.

There are some cases where small molecules other than metal cations can be used to facilitate the folding and enzymatic activity of RNA. Linear polyamines (like spermine) and aminoglycosides, displace Mg^{2+} from RNA, and have been shown to directly facilitate enzymatic activity of the hairpin and hammerhead ribozymes in the absence of divalent metal ions [36, 37].

2.2.2
Aminoglycosides

Aminoglycoside antibiotics are a diverse family of natural products that interfere with prokaryotic protein biosynthesis (Fig. 2.2). Their ability to non-specifically bind to RNA through electrostatic interactions was described over 20 years ago [39]. The aminoglycosides are also capable, however, of site-specific recognition of prokaryotic rRNA. Footprinting experiments indicated that the aminoglycosides bind to discrete locations within the ribosome [40]. Later experiments showed that aminoglycosides increase the affinity of tRNA to the ribosomal A-site [41], thus providing an attractive mechanism to explain their ability to decrease the fidelity of prokaryotic translation [42]. A recent crystal structure of three aminoglycosides (streptomycin, paromomycin, and spectinomycin) bound to the *Thermus thermophilus* 30S ribosomal subunit confirms the location of these distinct binding sites and provides a high-resolution picture of how RNA–aminoglycoside recognition occurs within a ribonucleoprotein complex [43]. This type of structural information will likely prove indispensable for structure-based design of aminoglycoside derivatives that will have an improved "fit" within their ribosomal binding pockets. Structural information cannot, however, answer basic questions related to the energetics involved in the binding of small molecules to their RNA "receptor." Equilibrium binding constants must be measured in order to establish the actual energetic values associated with RNA–small molecule interactions.

Despite the structural details provided by aminoglycoside–RNA complexes, the energetic contributions made by the pendant hydroxyl groups of the aminoglycosides remain unclear. Hydrogen bonding between these groups and RNA are apparent in some structures [44], but the energetic contributions of hydrogen bond-

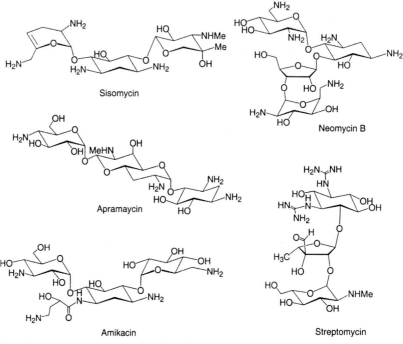

Fig. 2.2. Representative aminoglycosides. Five out the of six amino groups of neomycin B have pK_a values over 7, giving it a highly positive charge under physiological conditions [38].

ing in aqueous media is still debated [45]. In an attempt to determine their role in RNA affinity, the hydroxyl groups of tobramycin were systematically removed (Fig. 2.3). RNA binding was tested by measuring the HH16 ribozyme inhibitory activity of each aminoglycoside derivative [46]. Interestingly, the removal of

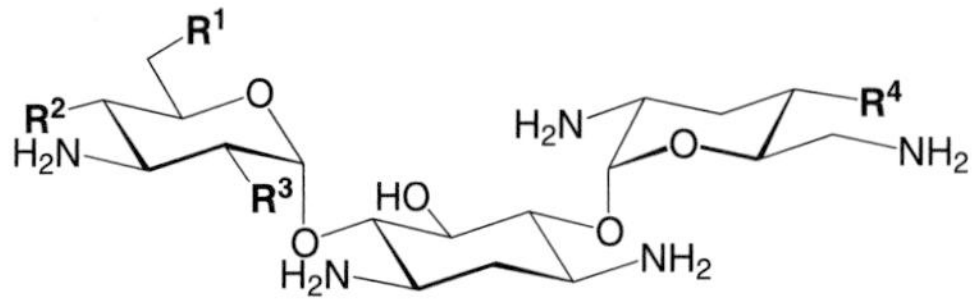

Aminoglycoside	R^1	R^2	R^3	R^4	Relative Ribozyme Cleavage Rate
Tobramycin	OH	OH	OH	OH	1.0
6''-Deoxytobramycin	H	OH	OH	OH	1.4
4''-Deoxytobramycin	OH	H	OH	OH	0.33
2''-Deoxytobramycin	OH	OH	H	OH	0.17
4'-Deoxytobramycin	OH	OH	OH	H	0.33

Fig. 2.3. Summary of hammerhead inhibition by deoxytobramycin derivatives. A lower relative rate suggests better RNA binding.

hydroxyls at positions R^2, R^3, and R^4 (Fig. 2.3) lead to aminoglycosides with *better* RNA binding. The current explanation of this "unexpected" result is that the hydroxyl groups decrease the basicity of neighboring amines. Removing certain hydroxyls, therefore, increases the overall positive charge, and hence RNA affinity of the resulting molecules. These dehydroxylated aminoglycosides have not yet been tested for the binding of other RNAs, so the roles of the hydroxyls in RNA specificity remain unknown.

2.2.3
Ligand Specificity

For the purpose of this discussion, specificity will be defined as the binding affinity (K_{eq}) of a small molecule to a particular RNA site divided by its average affinity to "all" other RNAs.

$$\text{Specificity} = \frac{K_{eq}(\text{interaction of interest})}{\text{average } K_{eq}(\text{other sites of interaction})} \tag{2.1}$$

For practical reasons, specificity is a relative term, where the affinity of the interaction of interest is weighted by the affinity of that same small molecule to "other" nucleic acids. The reported specificity is, therefore, always dependent on the selection of the "other" nucleic acids used for the comparison. In general, larger RNA ligands have a greater potential to exhibit higher RNA specificity. This explains why metal ions typically show a very low specificity for any single RNA.

High specificity is an essential factor for the effective modulation of *in vivo* RNA activity. Specificity is directly related to both bioavailability and receptor occupancy. Aminoglycosides, for example, are not ideal antibiotics. Their promiscuous binding to non-ribosomal RNAs and/or membrane components may be related to the multiple therapeutic side effects exhibited by these compounds [47–49]. Aminoglycosides bind to and inhibit the function of a wide range of unrelated RNAs with moderate activities ($IC_{50} = 0.1$–100 µM) [23, 50]. Aminoglycosides do, however, show excellent specificity for RNA over DNA (Section 2.5.3). Aminoglycosides have, therefore, served as suitable scaffolds for the synthesis of new small molecules targeted to RNA. These derivatives are shown to exhibit dramatically different RNA specificities and altered biological activities as compared to their parent aminoglycosides.

2.2.4
Goals

Notwithstanding the efforts of many groups towards the understanding of how small molecules bind to RNA, there are still no "rules" for the structure-based design of small molecules for a specific RNA tertiary fold. One obstacle is that there are still very few examples of small molecules that bind to natural RNA structures with high specificity. There are other potential reasons as well. For example, RNA is a highly dynamic molecule known to occupy multiple conformations. The structural details of an RNA do not typically entail the potential structural changes it can adopt upon ligand binding. Issues related to induced fit add additional complexity to the problem.

The following fundamental questions remain largely unanswered:
- How do electrostatic interactions affect the RNA specificity of a small ligand?
- How does one design small RNA ligands that exhibit high specificity for a predetermined RNA target?

To help answer these questions, we have addressed a number of goals:
1. To design and synthesize new small molecules that are targeted to a predetermined RNA site, in this case the HIV-1 RRE (Rev Response Element).
2. To use novel fluorescence-based methodologies to rapidly characterize the affinity *and* specificity of the new RRE ligands.
3. To conduct experiments in a systematic fashion so that general questions about RNA–small molecule recognition can be rigorously addressed.

To evaluate the "higher order" biological impacts of RNA binding, we have chosen the HIV-1 Rev–RRE interaction as our model system. This way, new RNA ligands that show promising activities, may eventually prove themselves as future antiviral agents. Our work, along with effort by many other groups, contributes to the growing body of knowledge that aids in the future rational design and use of small molecules directed to RNA [51–53].

2.3
The RRE and HIV Replication

The Rev Response Element (RRE) is an HIV-1 RNA structure essential for viral replication [54–56]. The Rev protein binds to the RRE and facilitates the export of HIV RNA out of the host nucleus, while protecting it from the cell's splicing machinery (Fig. 2.4). The importance of the "underspliced" HIV RNA is two-fold. First, this RNA serves as an open reading frame for proteins that are essential to the construction of new viral particles (including Gag, Pol, and Env). Second, the unspliced HIV RNA is packaged into the outgoing viroids, serving as the primary genome. Successful inhibition of Rev-RRE binding, therefore, prevents the production of new viral particles in two ways: the proteins essential for viroid construction are never translated, and the future HIV genomic RNA becomes highly spliced.

The RRE serves both as the high-affinity Rev-binding site and as part of a highly conserved open reading frame for the fusion domain of the gp41 Env protein (essential for the CD4-dependent viral invasion of the host cell) [57]. This dual function is likely responsible for the RRE's low mutation rate [58]. The RRE is, therefore, a highly attractive target for small molecule therapeutics since mutations that

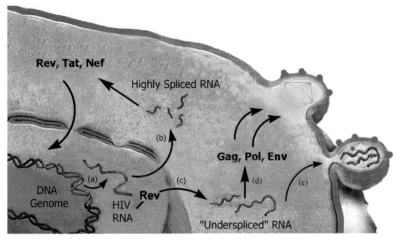

Fig. 2.4. Production of viral particles in an HIV-infected cell. (a) The Tat–TAR interaction facilitates the transcription of viral RNA. (b) In the absence of Rev activity, the HIV RNA becomes highly spliced and the proteins Rev, Tat, and Nef are translated. (c) HIV's "late" replication phase is initiated by the Rev–RRE interaction. (d) From the unspliced and singly spliced HIV RNA, the proteins Gag, Pol, and Env are translated. (e) Unspliced RNA is packaged into outgoing viral particles.

can lead to drug resistance should be impeded. Through the efforts of a number of research groups, many details regarding the sequences, structures, and dynamics of the Rev–RRE interaction have been revealed [54–56]. The 116-amino-acid Rev protein has at least three functional domains, including an arginine-rich motif spanning amino acids 34–50 (Fig. 2.5a). This positively charged domain binds to a single high-affinity site on the RRE (bold in Fig. 2.5b) with high affinity ($K_d = \sim 1$ nM) and high specificity. Additional Rev molecules then polymerize along the length of the RRE by utilizing both protein–protein and protein–RNA interactions. The oligomerization of Rev along the RRE is essential for Rev's ability to transport the HIV genome out of the host's nucleus while protecting it from the host's splicing machinery.

If the Rev protein cannot bind to its high-affinity site on the RRE, subsequent Rev molecules do not bind along the length of RRE and the entire HIV genome becomes highly spliced, thereby breaking the viral replication cycle. For this reason, we have chosen the high-affinity Rev binding site on the RRE (bold in Fig. 2.5b and henceforth referred to as "the" RRE) as our target for small molecules.

2.4
Determination of RRE–Ligand Affinity and Specificity

Numerous methods are available for evaluating a small molecule's ability to inhibit the binding of two macromolecules. Traditional techniques, such as native gel-shift electrophoresis, analytical ultracentrifugation, and equilibrium dialysis are limited

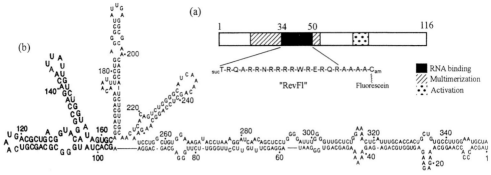

Fig. 2.5. (a) A domain map of Rev protein is shown with the sequence of its RRE-binding domain (amino acids 34–50). The peptide RevFl is used for fluorescence-based assays and is succinylated at its N-terminus, amidated at its C-terminus, and contains a four-alanine spacer to fluorescein. These modifications have been shown to increase the helicity and RRE specificity of this peptide [59]. (b) The RRE secondary structure as defined by Mann [60] spans a total of 351 bases and contains a single high-affinity Rev-binding site (the bold G-rich bulge contained between bases 100 and 160).

in their inability to rapidly screen large numbers of potential inhibitors. Newer methodologies such as mass spectrometry and surface plasmon resonance are powerful tools, but hardware costs can be problematic. Fluorescence-based techniques have proven to be highly sensitive, versatile, and relatively inexpensive for the examination of RNA–ligand interactions [61–63]. Three fluorescence-based techniques are described for evaluation of the RRE affinity and specificity of natural and synthetic Rev–RRE inhibitors.

2.4.1
Fluorescence Anisotropy

Fluorescence anisotropy can be used to measure a small molecule's ability to inhibit the Rev–RRE interaction. The anisotropy value is directly proportional to the tumbling rate of a fluorescein-labeled Rev_{34-50} peptide "RevFl" (Fig. 2.5a). As RRE is added to the fluorescent Rev peptide, their association is observed by an increased anisotropy value (Fig. 2.6a). The increase is due to the slower tumbling rate of the RevFl–RRE complex compared to that of the free RevFl.

Once the RevFl–RRE complex is formed (indicated by high anisotropy values), an inhibitor "X" can be added. Competitive inhibitors will bind to the free RRE, and shift the three-way binding equilbrium towards release of the RevFl peptide from the RevFl–RRE complex (causing a decreased anisotropy of RevFl) (Fig. 2.6b). If the number of inhibitor-binding sites on the RRE is established by an independent method, the absolute affinity of the RRE–"X" binding interaction can be calculated (K_i is equivalent to K_d). In most cases, however, the binding stoichiometry of the inhibitor is not known, so its affinity to the RRE is evaluated by measuring the concentration of inhibitor needed to displace half of the Rev peptide from the RRE (defined as the ligand's IC_{50} value).

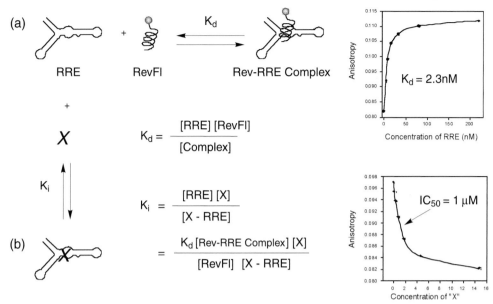

$$K_d = \frac{[RRE]\,[RevFl]}{[Complex]}$$

$$K_i = \frac{[RRE]\,[X]}{[X - RRE]}$$

$$= \frac{K_d\,[Rev\text{-}RRE\ Complex]\,[X]}{[RevFl]\,[X - RRE]}$$

Fig. 2.6. (a) Association of RevFl with the RRE as evident by changes in the fluorescence anisotropy of RevFl. Since this is a simple two-state system, the change in fluorescence anisotropy is directly proportional to the fraction of RevFl bound by the RRE. By non-linear regression, analysis of the association data yields a binding constant for Rev–RRE that is similar to the Rev protein (K_d approx. 2 nM). (b) Upon partial formation of the RevFl–RRE complex (using 10 nM RRE), an inhibitor "X" can be titrated. Binding of the RRE by "X" (in this case neomycin B) is apparent by the concentration-dependent decrease of RevFl's anisotropy signal back to its value when free in solution.

2.4.2
Solid-Phase (Affinity-Displacement) Assay

A novel biophysical approach has been developed to rapidly characterize the relative affinity *and* specificity of a small molecule for the RRE [64]. Following immobilization of a biotinylated RRE transcript onto an insoluble solid support, one equivalent of the fluorescent Rev protein fragment "RevFl" is added. RevFl binds to the immobilized RRE with an affinity the same as measured by fluorescence anisotropy (K_d 2 nM). (Note: Calculating the binding constant of Rev-RRE from solid-phase data is different from in solution, because the volume occupied by the unbound RRE is the same volume occupied by the Rev–RRE complex, so that the actual surface volume does not matter. The K_d(solid phase) = (moles of free RRE)*(concentration of RevFl in solution)/(moles of RevFl–RRE complex).) The fraction of RevFl bound to the immobilized RRE is easily determined by measuring the fluorescence intensity of the RevFl bound to the solid support versus that free in solution (Fig. 2.7a). As in solution, the immobilized Rev–RRE forms a highly specific complex that is not disrupted by addition of a vast excess of other nucleic acids (including DNA, tRNA, and polyA–polyU duplex RNA). (Note: We

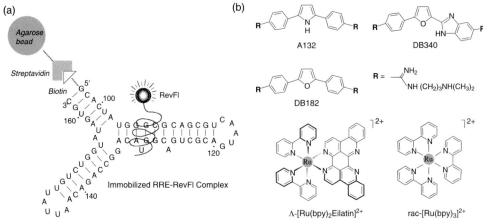

Fig. 2.7. (a) Immobilized Rev–RRE Complex. (b) Some of the compounds used to evaluate the solid-phase assay.

feel that immobilization of the RNA to the solid support by site-specific linkage through biotin–streptavidin provides a high level of environmental homogeneity for each RNA transcript, and that interactions between neighboring RNA transcripts do not occur. The bulky, hydrated matrix components (provided by agarose and streptavidin) are ideal conditions for the study of biologically relevant interactions (as proven by the countless macromolecular binding studies by ELISA and other related techniques).)

Following formation of the immobilized Rev–RRE complex, small molecule inhibitors are titrated in, and the fraction of RevFl free in solution is monitored to determine an IC_{50} value. Most biophysical methods, including fluorescence anisotropy, are not amenable to the analysis of complex mixtures of biomolecules or fluorescent ligands. Fluorescence anisotropy is found to be susceptible to artifacts when used to evaluate ligands that are fluorescent and/or quench the fluorescence of RevFl. The solid-phase assay has been shown to overcome these challenges and to produce displacement data that are consistent with native gel-shift electrophoresis experiments (Tab. 2.1). Another advantage of the solid-phase assay is that the specificity of an RRE ligand can be evaluated by comparing its activity in the

Tab. 2.1. IC_{50} values (µM) for Rev dsiplacement by three methods.

Compound	Fluorescence anisotropy	Solid-phase assay	Gel-shift electrophoresis
Neomycin B	7	7	~7
A132	10	1.2	~1.2
DB182	1.5	0.4	~0.4
DB340	0.3	0.1	<0.1
rac-Ru(bpy)$_3$	170	>10,000	>10,000
Λ-[Ru(bpy)$_2$Eilatin]	1	2	~1

presence and absence of an excess of competing nucleic acids (examples in later sections).

2.4.3
Ethidium Bromide Displacement

The emission intensity of ethidium bromide typically increases upon binding of nucleic acids [65]. Subsequent displacement of ethidium is apparent by a decrease in total fluorescence upon addition of a competitive inhibitor. Displacement experiments can be conducted using a wide range of nucleic acids [62]. Ethidium, however, is known to have variable binding stoichiometries and variable affinities to different nucleic acids [66]. The IC_{50} values for a given inhibitor cannot, therefore, be directly compared between different nucleic acids.

2.5
New RRE Ligands

An important demonstration of small molecules binding to viral RNA was established for the aminoglycoside family of antibiotics. These compounds were reported as having moderate affinities to the RRE (1 µM < IC_{50} < 100 µM) [67]. Adverse side effects and poor anti-HIV activities have prevented the use of aminoglycosides as antiviral agents. In attempts to improve these properties, modifications to aminoglycosides have been made. The resulting compounds show dramatically different RRE affinity, specificity, and anti-HIV activities.

2.5.1
Neomycin–Acridine Conjugates

Both neomycin B and the Rev protein are reported to bind near the purine-rich bulge of the RRE [67]. Flanking this bulge is a single adenosine bulge. Intercalating agents are known to preferentially bind to duplex regions containing a bulged base [68]. To utilize both of these structural features, acridine, a well-studied intercalating agent, was conjugated to the 5″ position of neomycin B (Scheme 2.1). (Note: A "single step" conjugation of acridine to an amine can potentially provide a shorter synthetic route, but the amines are essential for RNA binding). The resulting conjugate "neo–acridine" shows extremely high affinity for the RRE [69]. Both fluorescence anisotropy and the solid-phase assay indicate that neo–acridine binds to the RRE with an affinity comparable to that of the unlabeled Rev peptide itself (Tab. 2.2). At the time of publication, the binding between neo–acridine and the RRE was the highest affinity interaction between a natural RNA and small synthetic ligand reported to date. Enzymatic footprinting experiments showed that neo–acridine binds to the same portion of the RRE as does the Rev peptide (Fig. 2.8).

Neo–acridine was evaluated for nucleic acid specificity using the solid-phase assay (Tab. 2.2). By comparing the IC_{50} values measured in the absence of nucleic

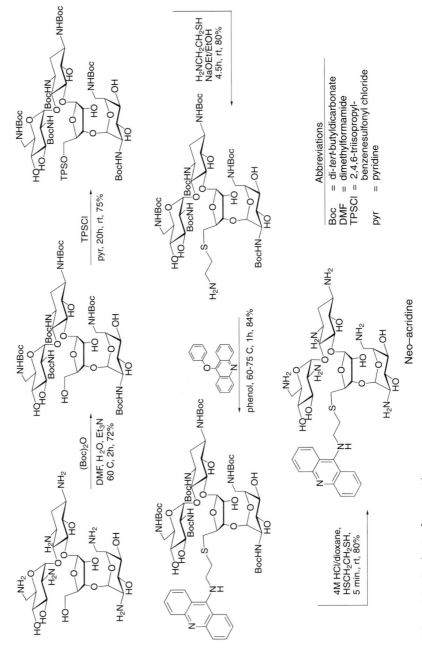

Scheme 2.1. Synthesis of neo–acridine.

Abbreviations

Boc = di-*tert*-butyldicarbonate
DMF = dimethylformamide
TPSCl = 2,4,6-trisopropyl-
benzenesulfonyl chloride
pyr = pyridine

Tab. 2.2. Approximate IC$_{50}$ values (µM) for RevFl displacement by solid phase.

Compound	IC$_{50}$	IC$_{50}$ with DNA	IC$_{50}$ with tRNA
Rev (unlabeled)	0.035	0.080	0.085
Neo–acridine	0.040	0.45	1.2
Neo–N-acridine	0.045	0.15	0.50
Neomycin B	7	8	18
Neo–neo	0.055	0.11	3.5

acid with those measured in an excess of other nucleic acids, the RRE specificity of each compound can be evaluated. Even though both neo–acridine and the Rev peptide bind to the same location of the RRE with similar affinities, neo–acridine has a much lower specificity for the RRE compared with the Rev peptide. Neo–acridine, in fact, binds with high affinity to nearly every RNA and DNA evaluated thus far. The low RRE specificity of neo–acridine can be improved by decreasing the linker length between the two moieties by three atoms (Fig. 2.9). Neo–N-acridine has about the same affinity but a significantly improved RRE specificity compared with neo–acridine (Tab. 2.2). The shorter linker length provided by neo–N-acridine is believed to provide fewer degrees of freedom to the conjugate and hence fewer nucleic acids can bind to it with high affinity.

Ethidium displacement experiments confirm that neo–acridine has a much higher affinity to "non-specific" competitors (like calf thymus DNA) than does neo–N-acridine (Tab. 2.3). Other groups have also synthesized intercalator-glycoside conjugates [70]. Their approach, however, involves incorporation of a variable-length methylene spacer linking paromomycin to either pyrene or thiazole orange. Subsequent evaluation as A-site antagonists reveals that as the number of atoms in the linker increases, the A-site affinity decreases. This is a different trend as observed with our conjugates. It is possible that highly flexible, hydrophobic linkers (including polymethylene) introduce a penalty for RNA binding.

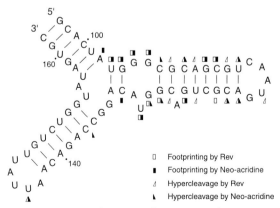

Fig. 2.8. Summary of enzymatic footprinting and hypercleavage observed for Rev$_{34-50}$ and neo–acridine.

Tab. 2.3. Approximate IC_{50} values (μM) for ethidium displacement.

Compound	RRE	Calf thymus DNA
Neo–acridine	0.03	0.2
Neo–N-acridine	0.04	1.3
Neo–neo	0.5	0.45
Neo–N-neo	0.2	0.45

2.5.2
Dimeric Aminoglycosides

The dimerization of aminoglycosides had previously been reported to dramatically increase their ability to inhibit ribozyme activity [71]. Dimerization doubles the total charge of each compound, leading to a dramatically enhanced affinity to many RNAs, including the RRE. The Rev–RRE inhibitory activity of neomycin is enhanced by over 100-fold upon dimerization (Tab. 2.2). The solid-phase assay indicates, however, that the RRE specificity, relative to other RNAs, of neo–neo is even worse than that of neo–acridine (Tab. 2.2). As observed for the neomycin–acridine conjugates, RRE specificity is enhanced by decreasing the linker length between moieties (Fig. 2.9). Ethidium displacement experiments indicate that the dimer neo–N-neo has a higher RRE affinity than neo–neo, and that both compounds have the same affinity to a non-specific nucleic acid (calf thymus DNA) (Tab. 2.3).

2.5.3
Guanidinoglycosides

Effective recognition between Rev and the RRE is mediated, largely, by guanidinium groups present in the Rev's arginine-rich domain [72]. Substitution of arginine at positions 35, 38, 39, or 44 with lysine leads to over a 20-fold loss of *in vivo* RRE binding [73]. We speculated, therefore, that the transformation of the amino groups of the aminoglycosides into guanidinium groups would increase their

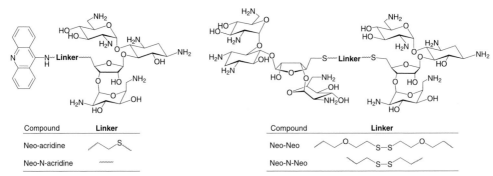

Fig. 2.9. Structures of the neo–acridine and dimeric neomycin conjugates.

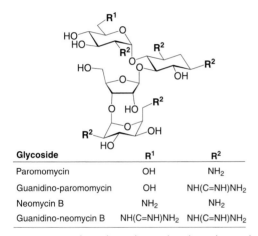

Glycoside	R¹	R²	R³
Kanamycin A	OH	OH	NH_2
Guanidino-kanamycin A	OH	OH	$NH(C=NH)NH_2$
Kanamycin B	NH_2	OH	NH_2
Guanidino-kanamycin B	$NH(C=NH)NH_2$	OH	$NH(C=NH)NH_2$
Tobramycin	NH_2	H	NH_2
Guanidino-tobramycin	$NH(C=NH)NH_2$	H	$NH(C=NH)NH_2$

Glycoside	R¹	R²
Paromomycin	OH	NH_2
Guanidino-paromomycin	OH	$NH(C=NH)NH_2$
Neomycin B	NH_2	NH_2
Guanidino-neomycin B	$NH(C=NH)NH_2$	$NH(C=NH)NH_2$

Fig. 2.10. Aminoglycosides and guanidinoglycosides used in these studies.

affinity and specificity for the RRE. We guanidinylated five aminoglycosides and used fluorescence anisotropy to evaluate the RRE affinity of the resulting "guanidinoglycosides" (Fig. 2.10 and Tab. 2.4) [74]. IC_{50} values were measured at both 10 nM and 100 nM RRE (consistently using 10 nM of RevFl). By taking the ratio of these IC_{50} values (Tab. 2.4), some information regarding the small molecule–RRE binding stoichiometries can be gleaned. A lower ratio suggests that more binding sites for the small molecule are capable of displacing the peptide. Surprisingly, the binding stoichiometries (for peptide displacement) are significantly different for kanamycin A versus kanamycin B, and for paromomycin versus neomycin B. This ratio, however, does *not* change for each compound upon guanidinylation, indicating that the RRE-binding stoichiometries of the guanidinoglycosides are the same as their aminoglycoside precursors. Differences in Rev–RRE IC_{50} values are, therefore, proportional to the differences in RRE affinity. Upon guanidinylation of neomycin B and paromomycin, a five-fold increase in RRE affinity is obtained.

Tab. 2.4. IC$_{50}$ values (μM) for RevFl displacement by anisotropy.

Glycoside	10 nM RRE	100 nM RRE	Ratio (100 nM/10 nM)
Kanamycin A	100	750	7.5
Guanidino-kanamycin A	8	65	8.1
Kanamycin B	15	80	5.3
Guanidino-kanamycin B	0.7	3.5	5.0
Tobramycin	10	45	4.5
Guanidino-tobramycin	0.8	3.8	4.7
Paromomycin	12	65	5.4
Guanidino-paromomycin	3.4	18	5.3
Neomycin B	1	7	7
Guanidino-neomycin B	0.2	1.3	6.5

Upon guanidinylation of the kanamycin family of glycosides, over a 10-fold increase in affinity is gained (Tab. 2.4).

The solid-phase assay indicates that guanidinylation also significantly affects RNA specificity (Tab. 2.5). By dividing the IC$_{50}$ values obtained in the presence of competitor RNAs (in this case duplex poly(A)–poly(U)) with the IC$_{50}$ values obtained in the absence of other nucleic acids, a specificity ratio can be calculated (Tab. 2.5). A lower ratio indicates higher RRE specificity (relative to simple duplex RNA). Upon guanidinylation, this ratio decreases for all the glycosides (except for neomycin B), showing that, in general, guanidinylation increases RRE specificity of the aminoglycosides. Importantly, guanidinylation does *not* affect the RNA over DNA specificity of these glycosides (Tab. 2.5). This is, to our knowledge, the first indication that the identity of the basic group is not an important factor for the ability of aminoglycosides to preferentially bind RNA over DNA.

Analysis of the activity trends apparent for each family of glycosides (Tab. 2.5) suggests some fundamental relationships relating the total charge of a ligand to its

Tab. 2.5. IC$_{50}$ values (μM) for RevFl displacement by solid-phase assay.

Glycoside	IC$_{50}$	IC$_{50}$ with DNA	IC$_{50}$ with RNA	RNA specificity ratio
Kanamycin A	700	750	1300	1.9
Guanidino-kanamycin A	55	55	60	1.1
Kanamycin B	90	90	170	1.9
Guanidino-kanamycin B	3	3	4.7	1.6
Tobramycin	45	50	130	2.9
Guanidino-tobramycin	3	3	4.7	1.6
Paromomycin	55	60	110	2.0
Guanidino-paromomycin	15	18	18	1.2
Neomycin B	7	8	16	2.5
Guanidino-neomycin B	1	1	4.5	4.5

RNA affinity *and* specificity. The total number of basic groups on each compound is roughly proportional to its RRE affinity, and the higher the ligand's affinity, the less likely it is to show specificity for the RRE. These same principles are also reflected in the dimeric aminoglycosides and may, therefore, reflect a general rule.

Other informative activity trends can also be observed (Tab. 2.5). By comparing compounds with the same number of basic groups (like kanamycin B, tobramycin, and paromomycin), additional structure–activity relationships can be inferred. For example, tobramycin is found to have a higher RRE affinity than kanamycin B, even though these two compounds differ only by a single hydroxyl (Fig. 2.10). This finding is consistent with the ability of hydroxyls to decrease the basicity of neighboring amines (giving tobramycin a higher overall charge *and* a lower RRE specificity than kanamycin B) [46]. Upon guanidinylation, the differences between kanamycin B and tobramycin are lost, because the pK_a of guanidinium groups are significantly less variable than those of amines.

Guanidinium groups are highly basic, capable of defined multidirectional hydrogen bonding, and their planar surface may facilitate stacking interactions. The higher basicity of the guanidinoglycosides, as compared to the aminoglycosides, should impart a higher total charge, and therefore higher RNA affinity. Their higher overall positive charge cannot, however, explain their improved RRE specificity (this would oppose the general relationships stated above). To understand why the guanidinoglycosides show improved RRE affinity *and* specificity (relative to the aminoglycosides), a number of urea-containing glycosides were synthesized [75]. Urea is an uncharged structural analog of guanidine. Both urea and guanidinium are capable of denaturing nucleic acids; therefore, the urea glycosides may have bound to RNAs that possess single-stranded character (like the purine-rich bulge of the RRE). The urea glycosides are, however, unable to inhibit Rev–RRE binding through 1 mM. This indicates that electrostatic interactions are critical for the guanidinoglycoside's affinity for RRE. Guanidinoglycoside–RNA binding has been shown (in a different system) to be much less dependent on the total ionic strength of the media compared to the aminoglycosides [76]. This suggests that stacking and/or other hydrophobic interaction(s) also mediate guanidinoglycoside–RNA binding. This may be the origin of the increased RRE specificity of the guanidinoglycosides.

The guanidinoglycosides are found to possess significantly improved RRE affinity, specificity, and anti-HIV activities [77] compared to the aminoglycosides (Fig. 2.11). Other groups have conjugated amino acids to aminoglycosides in order to mimic the Tat protein arginine-rich motif. Conjugation of arginine to kanamycin A is reported to increase its affinity and specificity for the TAR RNA [78]. This synthetic modification effectively doubles the number of basic groups present. Their findings suggest, however, that the guanidinium-containing side chains do not play the dominant role in RNA binding. The α-amines of the amino acids, in close proximity to the "core" glycoside, are reported to be essential for TAR binding. We have made similar observations with tobramycin–arginine conjugates in the Rev–RRE system. A tobramycin–arginine conjugate has the same number of basic groups and a similar RRE affinity as a tobramycin–aminoglycoside dimer (not shown). If

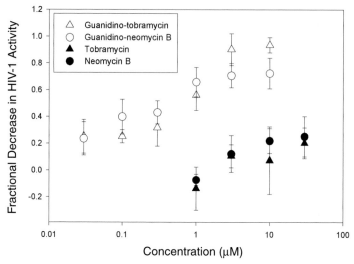

Fig. 2.11. Anti-HIV activities of aminoglycosides and guanidinoglycosides.

the α-amines of the arginine–tobramycin conjugate are blocked by acetylation, the resulting compound has the same number of basic groups as guanidino-tobramycin. Both guanidino-tobramycin and the aceylated tobramycin-arginine conjugate possess five guanidinium groups, and their comparison allows us to establish the contributions made by the flexible, hydrophobic, methylene linkages present in the arginine conjugates. We find that the aceylated tobramycin argine conjugate has over a 20-fold lower RRE-binding activity than guanidino-tobramycin, indicating that a significant energetic penalty is introduced by the methylene linkages of the amino acids.

Neither neo–acridine, nor neo–*N*-acridine possess anti-HIV activity. The guanidino forms of these compounds, however, are active in HIV inhibition. Fluorescence microscopy indicates that mammalian cells show significant cellular uptake of the guanidino forms of these compounds but not of their respective amino forms. As reported for the poly-arginine peptides (including Tat and Rev) the guanidinoglycosides may be actively transported into eukaryotic cells.

2.6
Conclusions

Systematic studies of the binding interactions between small molecules and RNA are essential for deciphering the parameters that govern RNA recognition. As illustrated above, the RRE affinity, specificity, and antiviral activities of aminoglycosides can be dramatically altered through synthetic modifications. The amine groups present on aminoglycosides are essential for RNA binding, but can be substituted by other basic groups including guanidine. The resulting "guanidinoglycosides" have improved affinity and specificity for the RRE, and maintain RNA over DNA specificity. Structure–activity studies in both families of glycosides have revealed trends that suggest general relationships between total charge of small ligands and their RNA affinity *and* specificity. Synthetic modifications that introduce extremely greasy and flexible functionalities onto aminoglycosides are likely to be detrimental to RNA binding. Polycyclic aromatic heterocycles, however, can greatly enhance the RNA-binding affinity, but can decrease the specificity of glycoside conjugates. The nucleic acid specificity of aminoglycoside dimers and of neomycin–acridine conjugates can be tuned by adjusting the length of the linkers that separate each moiety.

RNA plays a pivotal role in the replication of all organisms, including viral and bacterial pathogens. The development of small molecules that can selectively interfere with undesired RNA activity is a promising new direction for drug development. Continuing progress in the understanding of small molecule–RNA recognition may provide future scientists a means for selective targeting of a predetermined RNA regulatory element.

Acknowledgments

We thank the Center For AIDS Research at UCSD for technical assistance and partial support. We are grateful to the National Institute of Health for funding (AI 47673 and GM 58447 to Y.T.). N.W.L. thanks the Universitywide AIDS Research Program for a doctoral fellowship (D00-SD-017).

References

1 F. H. C. Crick. *What Mad Pursuit*, Basic Books, New York, 1988.

2 H. F. Judson. *The Eighth Day of Creation*, Simon & Schuster, New York, 1980.

3 A. Kornberg, T. Baker. *DNA Replication*, W. H. Freeman, New York, 1992.

4 E. H. Blackburn. Telomerase RNA structure and function, in *RNA Structure and Function*, Cold Spring Harbor Laboratory Press, Cold Spring Harbor, 1998, 669–693.

5 T. W. Nilsen. RNA–RNA Interactions in nuclear pre-mRNA splicing, in *RNA Structure and Function*, Cold Spring Harbor Laboratory Press, Cold Spring Harbor, 1998, 279–307.

6 J. B. Hartford, T. A. Rouault. RNA structure and function in cellular iron homeostasis, in *RNA Structure and Function*, Cold Spring Harbor Laboratory Press, Cold Spring Harbor, 1998, 575–602.

7 B. N. Fields, D. M. Knipe, P. M. Howley (eds). *Fundamental Virology*, Lippincott-Raven, Philadelphia, 1996.

8 R. Green, H. F. Noller. Ribosomes and translation. *Annu. Rev. Biochem.* 1997, 66, 679–716.

9 H. F. Noller, V. Hoffarth, L. Zimniak. Unusual resistance of peptidyl transferase to protein extraction procedures. *Science* 1992, 256, 1416–1419.

10 T. R. Cech. The ribosome is a ribozyme. *Science* 2000, 289, 878–879.

11 P. Nissen, J. Hansen, N. Ban, P. B. Moore, T. A. Steitz. The structural basis of ribosome activity in peptide bond synthesis. *Science* 2000, 289, 920–930.

12 G. W. Muth, L. Ortoleva-Donnelly, S. A. Strobel. A single adenosine

with a neutral pKa in the ribosomal peptidyl transferase center. *Science* **2000**, *289*, 947–950.

13 C. G. Kurland. Translational accuracy and the fitness of bacteria. *Annu. Rev. Genet.* **1992**, *26*, 29–50.

14 H. Siomi, G. Dreyfuss. RNA-binding proteins as regulators of gene expression. *Curr. Opin. Gen. Dev.* **1997**, *7*, 345–353.

15 G. Varani, K. Nagai. RNA recognition by RNP proteins during RNA processing. *Annu. Rev. Biophys. Biomol. Struct.* **1998**, *27*, 407–445.

16 N. K. Gray, M. Wickens. Control of translation initiation in animals. *Annu. Rev. Cell. Dev. Biol.* **1998**, *14*, 399–458.

17 T. R. Cech, B. L. Golden. Building a catalytic active site using only RNA, in *The RNA World*, Cold Spring Harbor Laboratory Press, Cold Spring Harbor, 1999, 321–349.

18 M. W. Hentze, L. C. Kuhn. Molecular control of vertebrate iron metabolism – mRNA-based regulatory circuits operated by iron, nitric oxide, and oxidative stress. *Proc. Natl Acad. Sci. USA* **1996**, *93*, 8175–8182.

19 N. C. Lau, L. P. Lim, E. G. Weinstein, D. P. Bartel. An abundant class of tiny RNAs with probable regulatory roles in *Caenorhabditis elegans*. *Science* **2001**, *294*, 858–862.

20 G. Werstuck, M. R. Green. Controlling gene expression in living cells through small molecule-RNA interactions. *Science* **1998**, *282*, 296–298.

21 International Human Genome Sequencing Consortium. Initial sequencing and analysis of the human genome. *Nature* **2001**, *409*, 860–921.

22 R. F. Gesteland, T. R. Cech, J. F. Atkins (eds). *The RNA World*. Cold Spring Harbor Laboratory Press, Cold Spring Harbor, 1999.

23 J. Davies, U. Ahsen, R. Schroeder. Antibiotics and the RNA world: a role for low-molecular-weight effectors in biochemical evolution? in *The RNA World*, Cold Spring Harbor Laboratory Press, Cold Spring Harbor, 1993, 185–204.

24 R. T. Batey, J. A. Doudna. The parallel universe of RNA folding. *Nature Struct. Biol.* **1998**, *5*, 337–340.

25 A. L. Feig, O. C. Uhlenbeck. The role of metal ions in RNA biochemistry, in *The RNA World*, Cold Spring Harbor Laboratory Press, Cold Spring Harbor, 1999, 287–319.

26 V. K. Misra, D. E. Draper. On the role of magnesium ions in RNA stability. *Biopolymers* **1998**, *48*, 113–135.

27 T. Hermann, E. Westhof. Exploration of metal ion binding sites in RNA folds by Brownian-dynamics simulations. *Structure* **1998**, *6*, 1303–1314.

28 V. K. Misra, D. E. Draper. A thermodynamic framework for Mg^{2+} binding to RNA. *Proc. Natl Acad. Sci. USA* **2001**, *98*, 12456–12461.

29 R. Russell, D. Herschlag. Probing the folding landscape of the *Tetrahymena* ribozyme: commitment to form the native conformation is late in the folding pathway. *J. Mol. Biol.* **2001**, *308*, 839–851.

30 S. K. Silverman, M. L. Deras, S. A. Woodson, S. A. Scaringe, T. R. Cech. Multiple folding pathways for the P4-P6 RNA domain. *Biochemistry* **2000**, *39*, 12465–12475.

31 M. S. Rook, D. K. Treiber, J. R. Williamson. An optimal Mg^{2+} concentration for kinetic folding of the *Tetrahymena* ribozyme. *Proc. Natl Acad. Sci. USA* **1999**, *96*, 12471–12476.

32 M. Zuker. On finding all suboptimal foldings of an RNA molecule. *Science* **1989**, *244*, 48–52.

33 C. Gaspin, E. Westhof. An interactive framework for RNA secondary structure prediction with a dynamical treatment of constraints. *J. Mol. Biol.* **1995**, *229*, 1049–1064.

34 A. R. Banerjee, J. A. Jaeger, D. H. Turner. Thermal unfolding of a group I ribozyme – the low-temperature transition is primarily disruption of tertiary structure. *Biochemistry* **1993**, *32*, 153–163.

35 M. Wu, I. Tinoco, Jr. RNA folding causes secondary structure rearrangement. *Proc. Natl Acad. Sci. USA* **1998**, *95*, 11555–11560.

36 D. J. Earnshaw, M. J. Gait. Aminoglycosides and cleavage of the hairpin ribozyme, in *RNA-Binding Antibiotics*. Molecular Biology Intelligence Unit 13, Eurekah.com, 2001, 35–55.

37 B. M. Chowrira, A. Berzal-Herranz, J. M. Burke. Ionic requirements for RNA binding, cleavage and ligation by the hairpin ribozyme. *Biochemistry* **1993**, *32*, 1088–1095.

38 R. E. Botto, B. Coxon. Nitrogen-15 nuclear magnetic resonance spectroscopy of neomycin B and related aminoglycosides. *J. Am. Chem. Soc.* **1983**, *105*, 1021–1028.

39 A. D. Dahlberg, F. Horodyski, P. Keller. Interaction of neomycin with ribosomes and ribosomal ribonucleic acid. *Antimicrob. Agents Chemother.* **1978**, *13*, 331–339.

40 D. Moazed, H. F. Noller. Interaction of antibiotics with functional sites in 16S ribosomal RNA. *Nature* **1987**, *327*, 389–394.

41 T. Pape, W. Wintermeyer, M. V. Rodnina. Conformational switch in the decoding region of 16S rRNA during aminoacyl-tRNA selection on the ribosome. *Nature Struct. Biol.* **2000**, *7*, 104–107.

42 R. Karimi, M. Ehrenberg. Dissociation rate of cognate peptidyl tRNA from the A-site of hyper-accurate and error-prone ribosomes. *Eur. J. Biochem.* **1994**, *226*, 355–360.

43 J. Davies, B. D. Davis. Misreading of ribonucleic acid code words induced by aminoglycoside antibiotics. *J. Biol. Chem.* **1968**, *243*, 3312–3316.

44 A. P. Carter, W. M. Clemons, D. E. Brodersen, R. J. Morgan-Warren, B. T. Wimberly, V. Ramakrishnan. Functional insights from the structure of the 30S ribosomal subunit and its interactions with antibiotics. *Nature* **2000**, *407*, 340–348.

45 M. Hendrix, P. B. Alper, E. S. Priestley, C-H. Wong. Hydroxy-amines as a new motif for the molecular recognition of phosphodiesters: Implications for aminoglycoside-RNA interactions. *Angew. Chem. Intl. Ed.* **1997**, *36*, 95–98.

46 H. Wang, Y. Tor. Electrostatic interactions in RNA aminoglycosides binding. *J. Am. Chem. Soc.* **1997**, *119*, 8734–8735.

47 A. Whelton, H. C. Neu (eds). *The Aminoglycosides: Microbiology, Clinical Use, and Toxicology*, Marcel Dekker, New York, 1982.

48 T. Koeda, K. Umemura, M. Yokota. Toxicology and pharmacology of aminoglycoside antibiotics, in *Aminoglycoside Antibiotics*, Springer-Verlag, Berlin, 1982, 293–356.

49 N. Tanaka. Mechanism of action of aminoglycoside antibiotics, in *Aminoglycoside Antibiotics. Handbook of Experimental Pharmacology*, Vol. 62, Springer-Verlag, Berlin, 1982, 221–266.

50 K. Michael, Y. Tor. Designing novel RNA binders. *Chem. Eur. J.* **1998**, *4*, 2091–2098.

51 W. D. Wilson, K. Li. Targeting RNA with small molecules. *Curr. Med. Chem.* **2000**, *7*, 73–98.

52 S. J. Sucheck, C-H. Wong. RNA as a target for small molecules. *Curr. Opin. Chem. Biol.* **2000**, *4*, 678–686.

53 A. C. Cheng, V. Calabro, A. D. Frankel. Design of RNA-binding proteins and ligands. *Curr. Opin. Struct. Biol.* **2001**, *11*, 478–484.

54 V. W. Pollard, M. H. Malim. The HIV-1 Rev protein. *Annu. Rev. Microbiol.* **1998**, *52*, 491–532.

55 T. J. Hope. The ins and outs of HIV Rev. *Arch. Biochem. Biophys.* **1999**, *365*, 186–191.

56 A. D. Frankel, J. A. T. Young. HIV-1: Fifteen proteins and an RNA. *Annu. Rev. Biochem.* **1998**, *67*, 1–25.

57 H. Schaal, M. Klein, P. Gehrmann, O. Adams, A. Scheid. Requirement of N-terminal amino acid residues of gp41 for human immunodeficiency virus type 1-mediates cell fusion. *J. Virol.* **1995**, *69*, 3308–3314.

58 R. C. Gallo. HIV – The cause of AIDS – An overview of its biology, mechanisms of disease induction, and our attempts to control it. *J. Acquir.*

Immune Defic. Syndr. **1988**, *1*, 521–535.

59 R. Tan, L. Chen, J. A. Buettner, D. Hudson, A. D. Frankel. RNA recognition by an isolated alpha-helix. *Cell* **1993**, *73*, 1031–1040.

60 D. A. Mann, I. Mikaelian, R. W. Zemmel *et al.* A molecular rheostat – cooperative binding to stem-1 of the Rev-response element modulates human immunodeficiency virus type-1 late gene expression. *J. Mol. Biol.* **1994**, *241*, 193–207.

61 K. Hamasaki, R. R. Rando. A high-throughput fluorescence screen to monitor the specific binding of antagonists to RNA targets. *Anal. Biochem.* **1998**, *261*, 183–190.

62 D. L. Boger, B. E. Fink, S. R. Brunette, W. C. Tse, M. P. Hedrick. A simple, high-resolution method for establishing DNA binding affinity and sequence selectivity. *J. Am. Chem. Soc.* **2001**, *123*, 5878–5891.

63 M. Auer, J-M. Seifert, S. Wallace, R. Sleigh. RNA in a miniaturized lead discovery process, in *RNA-Binding Antibiotics*, Molecular Biology Intelligence Unit 13, Eurekah.com, **2001**, 164–176.

64 N. W. Luedtke, Y. Tor. A novel solid-phase assembly for identifying potent and selective RNA ligands. *Angew. Chem. Intl. Ed.* **2000**, *39*, 1788–1790.

65 J-B. LePecq, C. J. Paoletti. A fluorescent complex between ethidium bromide and nucleic acids. Physical-chemical characterization. *J. Mol. Biol.* **1967**, *27*, 87–106.

66 J. L. Bresloff, D. M. Crothers. Equilibrium studies of ethidium-polynucleotide interactions. *Biochemistry* **1981**, *20*, 3547–3553.

67 M. L. Zapp, S. Stern, M. R. Green. Small molecules that selectively block RNA binding of HIV-1 Rev protein inhibit Rev function and viral production. *Cell* **1993**, *74*, 969–978.

68 W. D. Wilson, L. Ratmeyer, M. T. Cegla, *et al.* Bulged-base nucleic acids as potential targets for antiviral drug action. *N. J. Chem.* **1994**, *18*, 419–423.

69 S. R. Kirk, N. W. Luedtke, Y. Tor. Neomycin-acridine conjugate: a potent inhibitor of Rev-RRE binding. *J. Am. Chem. Soc.* **2000**, *122*, 980–981.

70 J. B. H. Tok, J. H. Cho, R. R. Rando. Aminoglycoside hybrids as potent RNA antagonists. *Tetrahedron* **1999**, *55*, 5741–5758.

71 H. Wang, Y. Tor. Dimeric amino-glycosides: design, synthesis and RNA binding. *Bioorg. Med. Chem. Lett.* **1997**, *7*, 1951–1956.

72 J. L. Battiste, H. Mao, N. S. Rao, R. *et al.* Alpha helix-RNA major groove recognition of an HIV-1 Rev peptide RRE RNA complex. *Science* **1996**, *273*, 1547–1551.

73 R. Tan, A. D. Frankel. Costabilization of peptide and RNA structure in an HIV Rev peptide-RRE complex. *Biochemistry* **1994**, *33*, 14579–14585.

74 N. W. Luedtke, T. J. Baker, M. Goodman, Y. Tor. Guanidino-glycosides: a novel family of RNA ligands. *J. Am. Chem. Soc.* **2000**, *122*, 12035–12036.

75 Q. Liu, N. W. Luedtke, Y. Tor. A simple conversion of amines into monosubstituted ureas in organic and aqueous solvents. *Tetrahedron Lett.* **2001**, *42*, 1445–1447.

76 S. R. Kirk, N. W. Luedtke, Y. Tor. 2-Aminopurine as a real-time probe of enzymatic cleavage and inhibition of hammerhead ribozymes. *Bioorg. Med. Chem.* **2001**, *9*, 2295–2301.

77 T. J. Baker, N. W. Luedtke, Y. Tor, M. Goodman. Synthesis and anti-HIV activity of guanidinoglycosides. *J. Org. Chem.* **2000**, *65*, 9054–9058.

78 A. Litovchick, A. G. Evdokimov, A. Lapidot. Aminoglycoside-arginine conjugates that bind TAR RNA: synthesis, characterization, and antiviral activity. *Biochemistry* **2000**, *39*, 2838–2852.

3

RNA Targeting by Bleomycin

Sidney M. Hecht

Although the bleomycin (BLM) group antitumor antibiotics are best known for their properties as sequence-selective DNA-cleaving agents [1–8], several possible biochemical targets have actually been reported for these compounds. These include DNA and RNA nucleases, DNA ligase, both DNA and RNA polymerases [9–12], as well as lipid peroxidation [13–16]. Of particular note was the report by Magliozzo *et al.* in 1989 that bleomycin could release free bases from transfer RNAs [17]. The study of RNA cleavage by bleomycin has been studied by a few different laboratories; this chapter summarizes the status of those studies.

3.1
Activation of Bleomycin for Polynucleotide Degradation

Bleomycin-mediated DNA degradation is an oxidative process that requires oxygen and a redox-active metal ion [1–8]. The structural domain believed to be involved in metal binding and oxygen activation is indicated in Fig. 3.1 for bleomycin A_2. While any of several metal ions can support DNA cleavage by bleomycin in an experimental setting, under physiological conditions the relevant metal ion is likely to be Fe [18] or possibly Cu [19]. Bleomycin cleaves double-stranded DNA predominantly at 5'-GT-3' and 5'-GC-3' sequences; a subset of these sites is cleaved with high efficiency. Two sets of products are formed as a consequence of DNA degradation by activated metallobleomycins; both are believed to result from the abstraction of an H atom from C4' H of deoxyribose [1–8]. It may be noted that C4' H resides in the minor groove of DNA, the dimensions of which may be smaller than the size of the metal-binding domain of metallobleomycins. In fact, it has been suggested [20] that the 5'-GC-3' and 5'-GT-3' selectivity of cleavage by metallobleomycins reflects the fact that these sequences correspond to the widest, shallowest part of the minor groove in B-form DNA [21] (i.e. that structural element best able to accommodate the metal-binding domain). Consistent with this suggestion, it has been found that structurally altered DNAs predicted to have widened minor grooves exhibited enhanced cleavage at those sites [22–24].

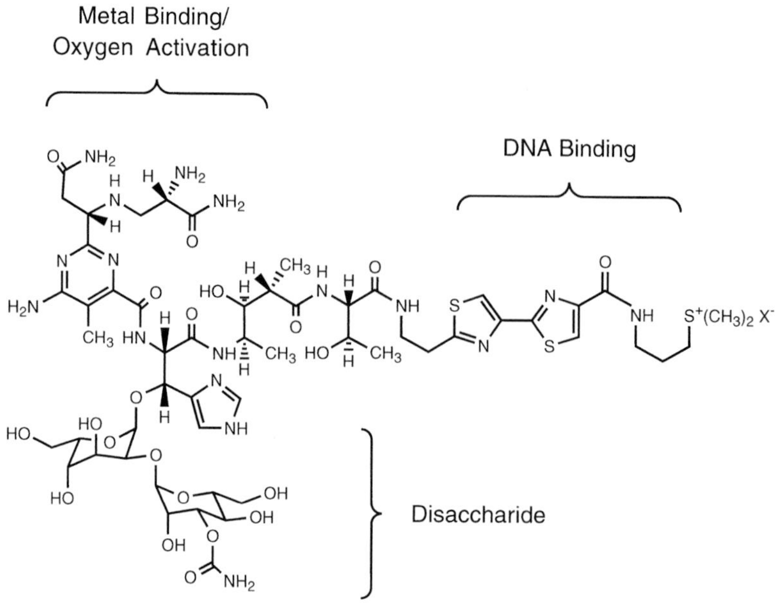

Fig. 3.1. Structure of bleomycin A_2. The individual domains are labeled.

3.2
Bleomycin-mediated Cleavage of Transfer RNAs and tRNA Precursor Transcripts

As noted above, Magliozzo *et al.* reported the first study of the cleavage of tRNAs by activated bleomycin [17]. The experiments involved relatively high concentrations of (non-radiolabeled) tRNAs and blenoxane, the clinically used mixture consisting primarily of BLM A_2 and BLM B_2. Treatment of unfractionated yeast tRNAs, or *Escherichia coli* tRNA^{Tyr} and tRNA^{Phe}, with activated Fe·BLM afforded tRNA breakage in a dose-dependent fashion. Significant cleavage was observed when 3 mM tRNA^{Phe} was treated with 0.3 mM activated Fe·BLM; also formed were products that co-migrated with adenine and uracil.

A survey of the susceptibility of a number of individual tRNAs and tRNA precursor transcripts was carried out by the Hecht laboratory using ^{32}P-end labeled substrates. Most of the substrates studied were not substrates for cleavage by Fe·BLM at any reasonable concentration. However, *Bacillus subtilis* tRNA^{His} precursor transcript was found to be cleaved with reasonable efficiency by 3 μM Fe(II)·BLM A_2 (Fig. 3.2) [25]. An ostensibly similar tRNA precursor, *E. coli* tRNA^{Tyr} precursor, was not a substrate for cleavage by Fe·BLM even when much higher concentrations were employed. It is also interesting that the tRNA^{His} precursor was cleaved at a single major site, unlike the cleavage of any DNA restriction fragment studied [1–8], all of which are cleaved at numerous sites.

The survey of tRNA structures revealed Fe·bleomycin-mediated cleavage of a *Schizosaccharomyces pombe* amber suppressor tRNA^{Ser} construct at two major sites,

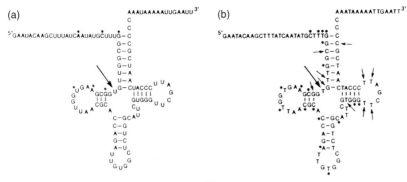

Fig. 3.2. (a) Structure of *Bacillus subtilis* tRNA^{His} precursor, indicating the major (arrow) and minor (asterisks) sites of cleavage by Fe(II)·bleomycin. (b) Cleavage of a DNA transcript having the same primary sequence as tRNA^{His} precursor.

one of which was in the D-loop while the other was within the 5'-leader sequence [26]. In comparison, no cleavage was noted for *E. coli* tRNA^{Cys} or yeast mitochondrial tRNA^{Asp} or tRNA$_f^{Met}$ precursor constructs [26]. Of particular interest were the results of treatment of *E. coli* tRNA$_1^{His}$ with Fe·BLM. Cleavage was observed at three major sites, but none of these was the same as the major or minor cleavage sites noted for *B. subtilis* tRNA^{His} precursor transcript [25, 26].

A more detailed study of the effect of tRNA conformation on susceptibility to cleavage by Fe·BLM was carried out using the crystallographically defined yeast cytoplasmic tRNA^{Phe} and tRNA^{Asp}, as well as *in vitro* tRNA transcripts related to these two mature tRNAs [27]. The two sites of cleavage of mature tRNA^{Asp} and four sites of cleavage of mature tRNA^{Phe} are shown in Fig. 3.3. In Fig. 3.4, four tRNA transcripts structurally related to RNA^{Asp} and tRNA^{Phe} are shown. Transcript A has the same sequence as tRNA^{Asp} with the exception of a G_1–C_{72} base pair introduced to facilitate *in vitro* transcription. This transcript has been shown to be activated efficiently by yeast aspartyl-tRNA synthetase [29]. As noted in Fig. 3.4, this transcript was cleaved at four sites, none of which was the same as the two sites cleaved in mature tRNA^{Asp}. A second tRNA^{Asp} substrate utilized was transcript F, which has an altered D-loop and a fifth nucleotide in the variable loop; both of these are characteristic of yeast tRNA^{Phe} [30].

As indicated in Fig. 3.4, this transcript was cleaved at five sites, namely U_{17}, G_{19}, A_{21}, U_{25}, and G_{50}. The cleavage at G_{19} was at the same site as one of the major cleavage sites of yeast tRNA^{Phe} and the cleavage of the transcript at G_{50} was close to the U_{52} cleavage site in mature tRNA^{Phe}. However, the remaining cleavage sites in transcript F resembled neither the cleavage pattern of tRNA^{Phe} nor tRNA^{Asp}.

Also studied were transcripts B and D, each of which contains only one of the two alterations by which transcripts A and F differ. Transcript B, containing an additional uridine between A_{46} and U_{47}, was cleaved at eight sites, indicating that the addition of a single uridine rendered the substrate significantly more susceptible to cleavage by Fe·BLM. The eight cleavage sites included U_{16}, C_{29}, and C_{38},

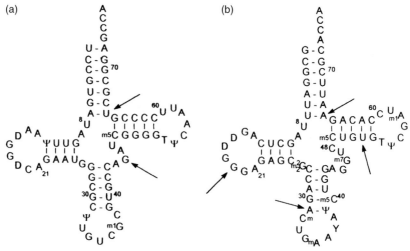

Fig. 3.3. Structures of (a) yeast tRNA^{Asp} and (b) yeast tRNA^{Phe} (right). The sites of Fe(II)·BLM A$_2$-mediated cleavage are indicated by arrows.

all of which were also cleaved in transcript A. In comparison, transcript D, having a D-loop identical with transcript F, was cleaved only at a single position (A$_{46}$) which was the single cleavage site in transcript A not cleaved in transcript B. In spite of the similarity of transcript D to transcript F, they showed no cleavage site in common.

Thus the overall picture that emerges from this survey is one of great sensitivity to changes in tRNA tertiary structure. While 5'-G-pyr-3' sequences were probably the most common sequence cleaved by Fe·BLM, many other sequences were also cleaved. Further, at least at the level of secondary structure analysis, cleavages occurred in nominally single- and double-stranded regions. More careful analysis at the level of tertiary structure suggests that many of the sites susceptible to cleavage reside at the junction between single- and double-stranded structures (i.e. adjacent to minor groove-like structures exhibiting local widening).

3.3
Other RNA Targets for Bleomycin

Early studies of RNA cleavage by bleomycin failed to detect any effect [31–35]. The earliest report of attempted cleavage of an RNA–DNA heteroduplex by Fe·BLM was described by Haidle and Bearden in 1975; they found that Fe·BLM degraded only the poly(dT) strand of a poly(rA)·poly(dT) heteroduplex [34]. This finding was confirmed and extended by Krishnamoorthy *et al.* using both poly(rA)·poly(dT) and poly(dA)·poly(rU) as substrates [36, 37]. Nonetheless, a subsequent study by Morgan and Hecht, using an RNA–DNA heteroduplex substrate formed by reverse

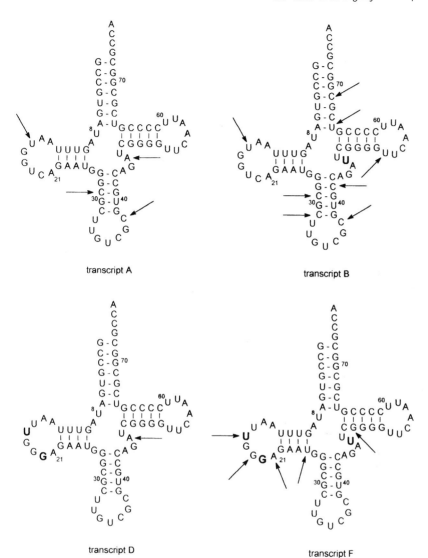

Fig. 3.4. Structure of four tRNA transcripts illustrating (arrows) the sites of cleavage by Fe(II)·BLM A_2. Nomenclature for the transcripts was based on Giegé *et al.* [28]. Transcript A has the same sequence as mature yeast tRNA^{Asp}, but contained a G_1-C_{72} base pair to facilitate transcription in lieu of the normal U_1-A_{72} base pair. Relative to transcript A, transcript B contained an additional nucleotide (U_{47}) in the variable loop, while transcript D has an altered dihydrouridine loop. Transcript F contained both the additional nucleotide in the variable loop and an altered dihydrouridine loop.

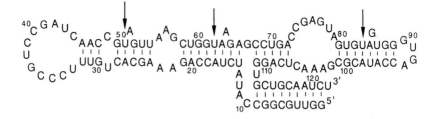

yeast 5S RNA

Fig. 3.5. Three sites of cleavage (arrows) of yeast 5S ribosomal RNA by Fe(II)·bleomycin.

transcription of *E. coli* 5S ribosomal RNA, clearly demonstrated cleavage at a limited number of sites on both the DNA and RNA strands [38]. It is interesting that the sites of cleavage on the DNA strand of the heteroduplex were found to be a subset of those cleaved on a DNA duplex having the same sequence.

The BLM-mediated cleavage of a ribosomal RNA, namely yeast 5S ribosomal RNA, has also been studied [26]. The three sites of cleavage are summarized in Fig. 3.5. All of these involved the uridine residue in a 5′-GUA-3′ sequence. Interestingly each of the cleaved sequences was flanked by a one-base bulge one or two nucleotides to the 3′-side of the cleavage site, a feature that is reminiscent of DNA cleavage in proximity to bulges [22].

As regards cleavage of messenger RNAs, two reports document that these can be substrates for cleavage by Fe·BLM. The first involved the treatment of a 347-nucleotide substrate corresponding to the 5′-end of HIV-1 reverse transcriptase mRNA. This mRNA was found to be cleaved at no less than four sites by Fe·BLM [25].

Subsequently, Dix *et al.* [39] studied the cleavage of the iron regulatory element of ferritin mRNA using Fe·BLM as a probe. The wild-type sequence was cleaved solely at 5′-GU_{17}-3′, which is believed to be at the junction of a single- and double-stranded region within the stem–loop structure. A functional mutant of the iron regulatory element in which the flanking region was altered by disruption of phylogenetically conserved base pairs, was cleaved at A_{10} and A_{11}, i.e. within the double-stranded region on the opposite side of the stem–loop structure.

3.4
Characteristics of RNA Cleavage by Fe·BLM

In comparison with DNA cleavage, Fe·bleomycin-mediated RNA cleavage has a number of unique characteristics. While the cleavage of both substrates requires Fe^{2+} and oxygen for bleomycin activation, RNA cleavage is much more selective than DNA cleavage [20]. Further, while the RNA cleavage sites involve a disproportionate number of 5′-G·pyr-3′ sequences, tertiary structure appears to be quite important in determining the sites of cleavage of RNA by Fe·BLM.

It should also be noted that the various RNA substrates described above are not cleaved with equal facility by Fe(II)·BLM. The best two substrates are the *B. subtilis* tRNA^{His} precursor [25] and the yeast 5S ribosomal RNA [26]. In addition to being susceptible to Fe·BLM-mediated cleavage at low micromolar concentrations, cleavage of these species was not eliminated in the presence of physiological concentrations of Mg^{2+} [26]. This contrasts with the cleavage of species such as yeast tRNA^{Phe}, which require greater concentrations of Fe·BLM and are not substrates in the presence of even 0.5 mM Mg^{2+} [40].

It seemed possible that some of the differences in effects of Fe·BLM on RNA as compared with DNA might be explained by differences in efficiency of binding of activated Fe·BLM. Accordingly, an experiment was designed in which the cleavage of *B. subtilis* tRNA^{His} precursor at 5'-GU$_{35}$-3' was monitored in the presence of increasing concentrations of the self-complementary dodecanucleotide d(GCGT$_3$A$_3$GCG), which also has a single preferred cleavage site (at 5'-GC$_{11}$-3') and binds BLM quite efficiently [41]. As shown in Fig. 3.6, when the tRNA^{His} precursor was present at a concentration of 2.3 μM, cleavage at 5'-GU$_{35}$-3' was readily

Fig. 3.6. Cleavage of tRNA^{His} precursor by Fe(II)·BLM in the presence of a DNA oligonucleotide. *B. subtilis* tRNA^{His} precursor was treated with Fe(II)·BLM A$_2$ in the presence of 5'-d(CGCT$_3$A$_3$GCG)-3'. Lane 1, tRNA^{His} precursor alone (270 μM final nucleotide concentration); lane 2, 10 μM Fe^{2+}; lane 3, 10 μM BLM A$_2$; lanes 4–13, 10 μM Fe(II)·BLM A$_2$ + 400, 300, 200, 150, 100, 50, 10, 5, 0.5, and 0 μM dodecanucleotide, respectively.

apparent in the presence of 10 µM d(CGCT$_3$A$_3$GCG). Thus activated Fe·BLM clearly binds to this good RNA substrate no less efficiently than it binds to a good DNA duplex substrate [26].

While the most obvious difference between DNA and RNA structures is at the level of the mononucleotide constituents, it is also true that the secondary and tertiary structures assumed by RNAs generally have no DNA counterparts. Therefore, we prepared "tDNA^{His}" (i.e. a DNA molecule having the same primary sequence as *B. subtilis* tRNA^{His} precursor). While there was no assurance that the tDNA would actually assume secondary and tertiary structures analogous to those of a tRNA, earlier studies with a tDNA analog of *E. coli* tRNA$_f^{Met}$ having a riboadenosine at the 3′-end revealed that this species was activated by methionyl-tRNA synthetase [42]. Further, the tDNA analogs of *E. coli* tRNA^{Phe} and tRNA^{Lys} were also bound and activated by the cognate aminoacyl-tRNA synthetases [43].

As shown in Fig. 3.7, treatment of tDNA^{His} with 0.5 µM Fe(II)·BLM A$_2$ afforded a major cleavage band at 5′-GT$_{35}$-3′ (i.e. exactly analogous to the major site of cleavage of tRNA^{His} precursor) [44]. Interestingly, higher concentrations of Fe(II)·BLM A$_2$ produced additional cleavage products and complete consumption of the intact tDNA. In comparison, no significant cleavage of tRNA^{His} precursor was apparent at Fe(II)·BLM A$_2$ concentrations below 2.5 µM, and concentrations as high as 250 µM produced no greater degradation of the tRNA^{His} precursor. The actual sites of tDNA^{His} and tRNA^{His} cleavage are summarized in Fig. 3.2. In addition to sharing the same major site of cleavage at U(T)$_{35}$, it is interesting that cleavage of the tDNA at less efficient cleavage sites involved a number of sequences other than the 5′-G·pyr-3′ sequence that dominates the cleavage of duplex DNA by bleomycin. These argue that both DNA and RNA undergo cleavage by activated Fe·BLM that is based primarily on the overall shape of the oligonucleotide target.

Competition experiments analogous to that shown in Fig. 3.6 were also carried out. Treatment of radiolabeled tRNA^{His} (7–8 µM nucleotide concentration) with 25 µM Fe(II)·BLM A$_2$ in the presence of unlabeled tDNA^{His} at concentrations up to 80 µM had no significant effect on cleavage of the radiolabeled tRNA. When the experiment was repeated with unlabeled tRNA^{His}, no diminution of cleavage was observed in the presence of 16 µM unlabeled tRNA^{His}, but cleavage was virtually absent in the presence of 80 µM unlabeled tRNA^{His}. In contrast, while cleavage of 7–8 µM tDNA^{His} at T$_{35}$ by 1.25 µM Fe(II)·BLM A$_2$ was diminished only by 80 µM unlabeled tDNA^{His}, it was significantly diminished by the presence of even 8 µM tRNA^{His} and virtually absent at higher tRNA^{His} concentrations. The clear conclusion is that Fe(II)·BLM A$_2$ binds more tightly to tRNA^{His} than to tDNA^{His} [44].

Another interesting facet of tDNA^{His} cleavage by Fe(II)·BLM A$_2$ is shown in Tab. 3.1, which records the effect of Mg^{2+} on tRNA^{His} and tDNA^{His} cleavage. The diminution of cleavage of the two polynucleotides caused by Mg^{2+} is not dramatically different for the two substrates. Thus the effects of Mg^{2+} on Fe·BLM-mediated RNA cleavage are related to tertiary structure rather than any fundamental property of RNA (versus DNA) structure.

In view of the finding that the affinity of Fe(II)·BLM A$_2$ for tRNA^{His} was actually

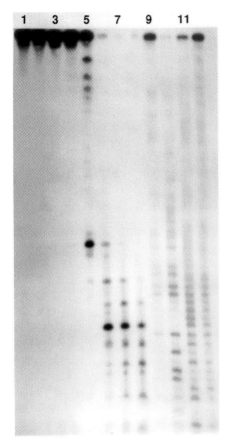

Fig. 3.7. Fe(II)·bleomycin-mediated cleavage of tDNA^{His} substrate. Lane 1, DNA alone (approx. 2.1 μM nucleotide concentration); lane 2, 2.5 μM BLM; lane 3, 2.5 μM Fe^{2+}; lanes 4–8, 0.25, 0.5, 1.25, 2.5, and 5.0 μM Fe(II)·BLM, respectively; lane 9, G lane; lane 10, G + A lane; lane 11, C + T lane, lane 12, C lane.

Tab. 3.1. Effect of Mg^{2+} on Fe·BLM-mediated cleavage of tRNA^{His} and tDNA^{His} precursors.

[Mg^{2+}] (mM)	Percentage cleavage	
	tDNA^{His}	tRNA^{His}
0	10.4	13.5
0.25	8.4	7.2
0.5	4.4	4.3
2.0	3.5	1.8
5.0	1.9	1.8

greater than that for tDNA^{His}, but the DNA substrate nonetheless underwent cleavage at lower concentrations of added Fe·BLM and to a greater extent, it seems reasonable to consider the factors that limit RNA cleavage. Since RNA cleavage is limited neither by binding affinity nor by the effects of divalent cation, two explanations seem possible. One of these is that Fe·BLM binds to RNA in a fashion that is relatively unproductive from the perspective of RNA cleavage. The other possibility is that reactive intermediates are formed from the RNA substrate but do not result in strand scission. Crich and Mo have suggested that the intermediate radicals produced from RNA sugars may have greater stability than those derived from DNA [45]; conceivably they could be "repaired" by H atom abstraction from some other species. Alternatively, it is possible that the degradation of RNA affords lesions analogus to the alkali-labile lesions in DNA [1–8, 46–48] that do not lead directly to strand scission. The latter possibility has been explored in the context of the actual chemistry of RNA degradation.

3.5
Chemistry of Bleomycin-mediated RNA Cleavage

The chemistry of BLM-mediated RNA cleavage is significantly more complex than that of DNA. Perhaps the most surprising finding was that of Keck and Hecht [49] which demonstrated a metal ion- and oxygen-independent pathway for RNA strand scission. The transformations occurred at all phosphodiester bonds within 5′-pyr·pur-3′ sequences not involving modified nucleobases. The products of the transformation were shown to be 5′-oligonucleotides terminating in nucleoside 2,3′-cyclic phosphates and 3′-oligonucleotides having free 5′-OH groups (i.e. the products expected to result from phosphoryl transfer initiated by the 2′-OH group of the pyrimidine nucleotide). The facility of this transformation was comparable to that of the oxidative transformation of RNAs by activated Fe·BLM. In common with oxidative cleavage of RNA by metalloBLMs, the "nuclease-like" cleavage was also found to be diminished in the presence of Mg^{2+}; for yeast tRNA^{Phe} no strand scission was observed when [Mg^{2+}] was greater than 200 μM [49].

This transformation required the presence of a free primary amino group within the β-aminoalanineamide side chain. It was quite similar to the pattern of strand scission obtained with imidazole–intercalator conjugates prepared as RNase mimics [50], as well as a number of other reagents such as polyvinylpyrrolidone [51, 52]. The mechanism is not entirely clear, but the transformation occurs at sites susceptible to spontaneous hydrolysis [53], and presumably reflects the ability of the 2′-OH group on the pyrimidine nucleoside to assume a conformation conducive to phosphoryl transfer.

As regards the oxidative chemistry of RNA, the earliest evidence was provided by Magliozzo *et al.*, who reported the isolation of adenine and uracil nucleobases, identified by co-migration with authentic samples on TLC [17]. Also reported were base propenals, identified on the basis of their colorimetric response upon treatment with thiobarbituric acid. Given the presence of 2′-OH groups in the ribose moieties of the substrate nucleotides, the actual structures of the base propenals would presumably have to be hydroxylated analogs of the base propenals formed from DNA [1–8].

The release of uracil concomitant with cleavage of U_{35} in the *B. subtilis* tRNA^{His} precursor was established by the use of a tRNA^{His} transcript incorporating both [^{32}P]CMP and [5,6-^{3}H]uridine radiolabels. Cleavage of the tRNA precursor substrate was monitored on a polyacrylamide gel, while release of [^{3}H]uracil was determined in parallel by HPLC analysis [54]. [^{3}H]Uracil equivalent to 83% of the number of tRNA strand breaks was detected. By the use of both 5′- and 3′-^{32}P end-labeled yeast 5S ribosomal RNAs, it was determined that the 5′-oligonucleotide cleavage products migrated on polyacrylamide gels as though they had 3′-phosphoroglycolate moieties [1–8, 41], while the 3′-oligonucleotide products co-migrated with oligonucleotides having 5′-phosphate termini [26]. These products, of course, are the same as those formed from DNA by the action of activated metalloBLMs and would be fully consistent with oxidation initiated by abstraction of an H atom from the C4′ position of ribose.

A more detailed analysis was carried out by the use of self-complementary deoxyoligonucleotide containing a single *ribo*-cytidine or *ara*-cytidine (i.e. CGrCTAGCG) (Fig. 3.8). As shown in the figure, treatment with Fe(II)·BLM under aerobic conditions afforded a product having the same chromatographic properties as CpGpCH$_2$COOH [25, 26] (i.e. the product anticipated from initial abstraction of C4′ H).

From a chemical perspective, the radical formed by abstraction of C1′ H of (deoxy)ribose would be expected to be of reasonable stability and it was suggested a number of years ago that this pathway might also be utilized by activated BLM in effecting DNA degradation [55]. Although no products of DNA degradation resulting from C1′ H abstraction have yet been reported, it has been noted that it might be logical to anticipate the formation of such products from an A-form duplex in which C1′ H resides prominently within the minor groove [20]. In fact, Absalon *et al.* [37] attempted to identify such products from the DNA strand of a homo-polymeric DNA–RNA heteroduplex, but concluded that they were not formed. In comparison, activated Fe·BLM treatment of either the chimeric octanucleotide shown in Fig. 3.9, or else the isomeric C$_3$-*ara* octanucleotide, afforded products in addition to those formed by abstraction of C4′ H [55]. As shown in Fig. 3.9, one of the formed products could be trapped as a dinucleotide quinoxaline derivative following treatment with 1,2-diaminobenzene. The isolation of this product constitutes strong supporting evidence for the abstraction of C1′ H by activated Fe·BLM. The CpG-quinoxaline derivative constituted about 10% of the products derived from degradation of C$_3$-*ribo* CGCTAGCG, and 58% of those formed from C$_3$-*ara* CGCTAGCG.

As is clear from Fig. 3.9, the intermediate resulting from H atom abstraction from the C1′ position would afford free nucleobase, but not lead directly to strand scission. An analogous product, the alkali-labile lesion [46–48] is a well-characterized product of Fe·BLM-mediated DNA degradation following C4′ H abstration, and could conceivably be formed from RNA as well. Unfortunately, due to

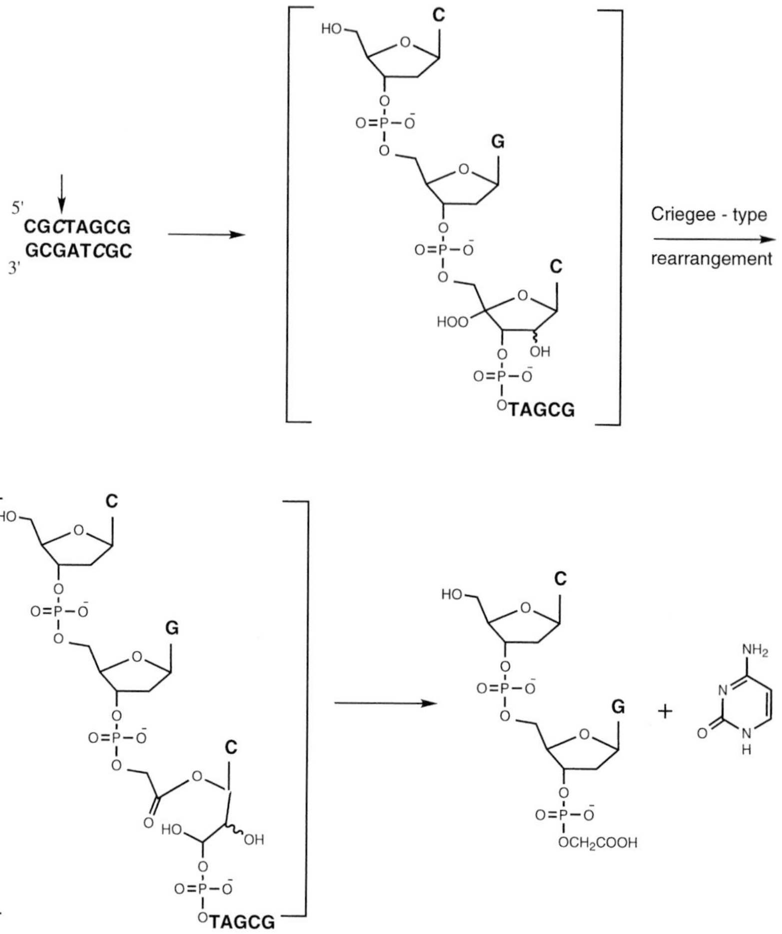

Fig. 3.8. Oxidative degradation of chimeric octanucleotides initiated by abstraction of C4' H.

the ability of all RNA phosphodiester linkages to alkali and other nucleophilic reagents, the selective cleavage of BLM-induced RNA lesions is not straightforward. However, evidence for the formation of a lesion of this type has been obtained by successive treatments of the RNA substrate with Fe(II)·BLM and then with hydrazine [54].

3.6
Significance of RNA as a Target for Bleomycin

While DNA has long been thought to constitute the critical therapeutic locus of action for bleomycin, there are observations which argue that this may not be the sole locus of action for the drug. Unresolved issues include the relatively poor cor-

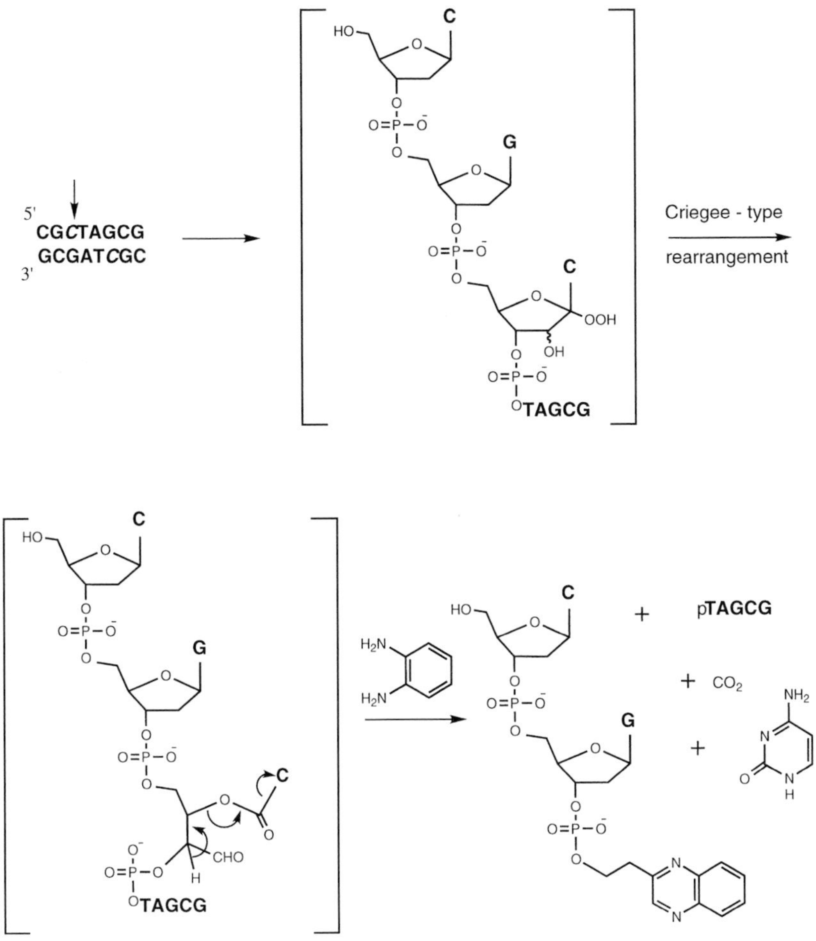

Fig. 3.9. Oxidative degradation of chimeric octanucleotides initiated by abstraction of C1′ H.

relation between the ability of individual BLM congeners to mediate DNA strand scission and their ability to inhibit the growth of cultured KB cells, as well as the remarkably facile repair of BLM-mediated damage to chromatin [57].

That the lipid membrane may constitute an additional locus of action for BLM is suggested by the ability of the drug to effect lipid peroxidation [13–16] and by the finding that dibucaine, a local anesthetic known to increase membrane flexibility, rendered cultured KB cells susceptible to inhibition by a BLM analog that was incapable of DNA cleavage [58]. To the extent that the uptake of BLM by a cell is limiting to the expression of BLM cytotoxicity [59], lipid peroxidation could obviously alter uptake and thereby contribute importantly to potency of action.

There are reasons to consider RNA as a therapeutic target for BLM as well. If drug uptake is limiting for the expression of cytotoxicity. The presence of RNA targets in the cytoplasm may facilitate the action of the drug since the nuclear membrane is likely to constitute a hindrance to nuclear entry as well. Additional advantages to an RNA target might be thought to include the apparent lack of intensive packaging of cytoplasmic RNAs, and the limited mechanisms for RNA repair. At least for a bacterial cell, it has been established that BLM is actually capable of intracellular RNA strand scission [60].

A key issue is the ability to kill cancer cells by destruction of one or more cellular RNAs. In fact, nature has provided good examples of cell killing by means of RNA targeting. These include the cytotoxic protein ricin, which functions by depurination of 28S ribosomal RNA [61]. Another good example is onconase, a cytotoxic member of the RNase A superfamily that appears to function at the level of tRNA degradation [62, 63].

One interesting strategy to resolve the issue of the actual locus of action of bleomycin would be to identify BLM analogs capable of functioning only at a single locus (i.e. either RNA or DNA degradation). The behavior of such species as antitumor agents could potentially result in the identification of the primary locus at which BLM expresses its antitumor effects. In this context it is worth noting that regardless of the actual locus of action of BLM itself, the elaboration of analogs of BLM capable of targeting critical RNAs with good selectivity could afford antitumor agents that function at novel cellular loci.

Acknowledgments

The work from the Hecht laboratory discussed in this chapter was supported by NIH Research Grants CA53913, CA76297 and CA77284, awarded by the National Cancer Institute.

References

1 HECHT, S. M. The chemistry of activated bleomycin. *Acc. Chem. Res.* **1986**, *19*, 383–391.

2 STUBBE, J., KOZARICH, J. W. Mechanisms of bleomycin-induced DNA degradation. *Chem. Rev.* **1987**, *87*, 1107–1136.

3 NATRAJAN, A., HECHT, S. M. Bleomycin: mechanism of polynucleotide recognition and oxidative degradation, in *Molecular Aspects of Anticancer Drug–DNA Interactions*, NEIDLE, S., WARING, M. (eds), Macmillan, London, 1993, 197–242.

4 KANE, S. A., HECHT, S. M. Polynucleotide recognition and degradation by bleomycin. *Prog. Nucleic Acid Res. Mol. Biol.* **1994**, *49*, 313–352.

5 HECHT, S. M. Bleomycin group antitumor agents, in *Cancer Chemotherapeutic Agents*, FOYE, W. O. (ed.), American Chemical Society, Washington, DC, 1995, 369–388.

6 HECHT, S. M. Bleomycin: new perspectives on the mechanism of action. *J. Nat. Prod.* **2000**, *63*, 158–168.

7 BURGER, R. M. Cleavage of nucleic acids by bleomycin. *Chem. Rev.* **1998**, *98*, 1153–1170.

8 CLAUSSEN, C. A., LONG, E. C. Nucleic acid recognition by metal complexes of bleomycin. *Chem. Rev.* **1999**, *99*, 2797–2816.

9 TANAKA, N., YAMAGUCHI, H., UMEZAWA, H. Mechanism of action of phleomycin. I. Selective inhibition of the DNA synthesis in *E. coli* and HeLa cells. *J. Antibiot.* **1963**, *16A*, 86–91.

10 FALASCHI, A., KORNBERG, A. Phleomycin, an inhibitor of DNA polymerase. *Fedn. Proc.* **1964**, *23*, 940–945.

11 MUELLER, W. E., ZAHN, R. K. Effect of bleomycin on DNA, RNA, protein, chromatin and on cell transformation by oncogenic RNA viruses. *Prog. Biochem. Pharmacol.* **1976**, *11*, 28–47.

12 OHNO, T., MIYAKI, M., TAGUCHI, T., OHASHI, M. Actions of bleomycin on DNA ligase and polymerases. *Prog. Biochem. Pharmacol.* **1976**, *11*, 48–58.

13 GUTTERIDGE, J. M. C., FU, X.-C. Enhancement of bleomycin-iron free radical damage to DNA by antioxidants and their inhibition of lipid peroxidation. *FEBS Lett.* **1981**, *123*, 71–74.

14 EKIMOTO, H. TAKAHASHI, K., MATSUDA, A., TAKITA, T., UMEZAWA, H. Lipid peroxidation by bleomycin-iron complexes *in vitro*. *J. Antibiot.* **1985**, *38*, 1077–1082.

15 NAGATA, R., MORIMOTO, S., SAITO, I. Iron-peplomycin catalyzed oxygenation of linoleic acid. *Tetrahedron Lett.* **1990**, *31*, 4485–4488.

16 KIKUCHI, H., TETSUKA, T. On the mechanism of lipoxygenase-like action of bleomycin-iron complexes. *J. Antibiot.* **1992**, *45*, 548–555.

17 MAGLIOZZO, R. A., PEISACH, J., CIRIOLO, M. R. Transfer RNA is cleaved by activated bleomycin. *Mol. Pharmacol.* **1989**, *35*, 428–432.

18 SAUSVILLE, E. A., STEIN, R. W., PEISACH, J., HORWITZ, S. B. Properties and products of the degradation of DNA by bleomycin and iron (II). *Biochemistry* **1978**, *17*, 2746–2754.

19 EHRENFELD, G. M., SHIPLEY, J. B., HEIMBROOK, D. C. et al. Copper-dependent cleavage of DNA by bleomycin. *Biochemistry* **1987**, *26*, 931–942.

20 HECHT, S. M. RNA degradation by bleomycin, a naturally occurring bioconjugate. *Bioconjugate Chem.* **1994**, *5*, 513–526.

21 DICKERSON, R. E. What do we really know about B-DNA? in *Structure and Methods*, Vol. 3, *DNA and RNA. Proceedings of the Sixth Conversation in Biomolecular Stereodynamics*, Sarma, R. H., Sarma, M. H. (eds) Adenine Press, Schenectady, New York, 1990, 1–38.

22 WILLIAMS, L. D., GOLDBERG, I. H. Selective strand scission by intercalating drugs at DNA bulges. *Biochemistry* **1988**, *27*, 3004–3011.

23 GOLD, B., DANGE, V., MOORE, M. A., EASTMAN, A., VAN DER MAREL, G. A., VAN BOOM, J. H., HECHT, S. M. Alteration of bleomycin cleavage specificity in a platinated DNA oligomer of defined structure. *J. Am. Chem. Soc.* **1988**, *110*, 2347–2349.

24 KANE, S. A., HECHT, S. M., SUN, J.-S., GARESTIER, T., HÉLÈNE, C. Specific cleavage of a DNA triplex helix by Fe^{II}·bleomycin. *Biochemistry* **1995**, *34*, 16715–16724.

25 CARTER, B. J., DE VROOM, E., LONG, E. C., VAN DER MAREL, G. A., VAN BOOM, J. H, HECHT. S. M. Site-specific cleavage of RNA by Fe(II)·bleomycin. *Proc. Natl Acad. Sci. USA* **1990**, *87*, 9373–9377.

26 HOLMES, C. E., CARTER, B. J., HECHT, S. M. Characterization of iron(II)·bleomycin-mediated RNA strand scission. *Biochemistry* **1993**, *32*, 4293–4307.

27 HOLMES, C. E., ABRAHAM, A. T., HECHT, S. M., FLORENTZ, C., GIEGÉ, R. Fe·bleomycin as a probe of RNA conformation. *Nucleic Acids Res.* **1996**, *24*, 3399–3406.

28 GIEGÉ, R., FLORENTZ, C., GARCIA, A. et al. Exploring the aminoacylation function of transfer-RNA by macromolecular engineering approaches-involvement of conformational features in the charging process of yeast transfer RNA^{Asp}. *Biochimie* **1990**, *72*, 453–461.

29 PERRET, V., GARCIA, A., GROSJEAN, H., EBEL, J.-P., FLORENTZ, C., GIEGÉ, R. Relaxation of a transfer-RNA

specificity by removal of modified nucleotides. *Nature* **1990**, *344*, 787–789.

30 PERRET, V., FLORENTZ, C., PUGLISI, J. D., GIEGÉ, R. Effect of conformational features on the aminoacylation of transfer-RNAs and consequences on the permutation of transfer-RNA specificities. *J. Mol. Biol.* **1992**, *226*, 323–333.

31 SUZUKI, H., NAGAI, K., AKUTSU, E., YAMAKI, T., TANAKA, N., UMEZAWA, H. On the mechanism of action of bleomycin: strand scission of DNA caused by bleomycin and its binding to DNA *in vitro. J. Antibiot.* **1970**, *23*, 473–480.

32 MÜLLER, W. E. G., YAMAZAKI, Z., BRETER, H.-J., ZAHN, R. K. Action of bleomycin on DNA and RNA. *Eur. J. Biochem.* **1972**, *31*, 518–525.

33 HAIDLE, C. W., KUO, M. T., WEISS, K. K. Nucleic acid-specificity of bleomycin. *Biochem. Pharmacol.* **1972**, *21*, 3308–3312.

34 HAIDLE, C. W., BEARDEN, J., JR. Effect of bleomycin on an RNA–DNA hybrid. *Biochem. Biophys. Res. Commun.* **1975**, *65*, 815–821.

35 HORI, M. Interaction of bleomycin with DNA. *Bleomycin: Chemical, Biochemical and Biological Aspects*, HECHT, S. M. (ed.), Springer-Verlag, New York, 1979, 195–206.

36 KRISHNAMOORTHY, C. R., VANDERWALL, D. E., KOZARICH, J. W., STUBBE, J. Degradation of DNA-RNA hybrids by bleomycin: evidence for DNA strand specificity and for possible structural modification of chemical mechanism. *J. Am. Chem. Soc.* **1988**, *110*, 2008–2009.

37 ABSALON, M. J., KRISHNAMOORTHY, C. R., McGALL,G., KOZARICH, J. W., STUBBE, J. Bleomycin mediated degradation of DNA-RNA hybrids does not involve C-1' chemistry. *Nucleic Acids Res.* **1992**, *20*, 4179–4185.

38 MORGAN, M., HECHT, S. M. Iron(II)bleomycin-mediated degradation of a DNA-RNA heteroduplex. *Biochemistry* **1994**, *33*, 10286–10293.

39 DIX, D. J., LIN, P.-N., McKENZIE, A. R., WALDEN, W. E., THEIL, E. C. The

influence of the base-paired flanking region on structure and function of the ferritin mRNA iron regulatory element. *J. Mol. Biol.* **1993**, *231*, 230–240.

40 HÜTTENHOFER, A., HUDSON, S., NOLLER, H. F., MASCHARAK, P. K. Cleavage of tRNA by Fe(II)-bleomycin. *J. Biol. Chem.* **1992**, *267*, 24471–24475.

41 SUGIYAMA, H., KILKUSKIE, R. E., CHANG, L.-H., MA, L.-T., HECHT, S. M., VAN DER MAREL, G. A., VAN BOOM, J. H. DNA strand scission by bleomycin: catalytic cleavage and strand selectivity. *J. Am. Chem. Soc.* **1986**, *108*, 3852–3854.

42 PERREAULT, J. P., PON, R. T., JIANG, M., USMAV, N., PIKA, J., OGILIVIE, K. K., CEDERGREN, R. The synthesis and functional-evaluation of RNA and DNA polymers having the sequence of *Escherichia coli* tRNA^fMet. *Eur. J. Biochem.* **1989**, *186*, 87–93.

43 KHAN, A. S., ROE, B.A. Aminoacylation of synthetic DNAs corresponding to *Escherichia coli* phenylalanine and lysine tRNAs. *Science* **1988**, *241*, 74–79.

44 HOLMES, C. E., HECHT, S. M. Fe·bleomycin cleaves a transfer RNA precursor and its "transfer DNA" analog at the same major site. *J. Biol. Chem.* **1993**, *268*, 25909–25913.

45 CRICH, D., MO, X.-S. Nucleotide C3',4'-radical cations and the effect of a 2'-oxygen substituent. The DNA/RNA paradox. *J. Am. Chem. Soc.* **1997**, *119*, 249–250.

46 SUGIYAMA, H., XU, C., MURUGESAN, N., HECHT, S. M. Structure of the alkali-labile product formed during Fe(II)·bleomycin-mediated DNA strand scission. *J. Am. Chem. Soc.* **1985**, *107*, 4104–4105.

47 RABOW, L. E., STUBBE, J., KOZARICH, J. W., GERLT, J. A. Identification of the alkali-labile product accompanying cytosine release during bleomycin-mediated degradation of d(CGCGCG). *J. Am. Chem. Soc.* **1986**, *108*, 7130–7131.

48 SUGIYAMA, H., XU, C., MURUGESAN, N., HECHT, S. M., VAN DER MAREL, G. A., VAN BOOM, J. H. Chemistry of the alkali-labile lesion formed from iron(II)-bleomycin and d(CGCTTTAAAGCG). *Biochemistry* **1988**, *27*, 58–67.

49 KECK, M. V., HECHT, S. M. Sequence-specific hydrolysis of yeast tRNA^{Phe} mediated by metal free bleomycin. *Biochemistry* **1995**, *34*, 12029–12037.

50 PODYMINOGIN, M. A., VLASSOV, V. V., GIEGÉ, R. Synthetic RNA-cleaving molecules mimicking ribonuclease A active center. Design and cleavage of tRNA transcripts. *Nucleic Acids Res.* **1993**, *21*, 5950–5956.

51 KIERZEK, R. Hydrolysis of oligoribonucleotides-influence of sequence and length. *Nucleic Acids Res.* **1992**, *20*, 5073–5077.

52 KIERZEK R. Nonenzymatic hydrolysis of oligoribonucleotides. *Nucleic Acids Res.* **1992**, *20*, 5079–5084.

53 DOCK-BREGEON, A. C., MORAS, D. Conformational changes and dynamics of tRNAs: evidence from hydrolysis patterns. *Cold Spring Harbor Symp. Quant. Biol.* **1987**, *52*, 113–121.

54 HOLMES, C. E., DUFF, R. J., VAN DER MAREL, G. A., VAN BOOM, J. H., HECHT, S. M. On the chemistry of RNA degradation by Fe(II)·BLM. *Bioorg. Med. Chem.* **1997**, *5*, 1235–1248.

55 HECHT, S. M. Symposium summary, in *Bleomycin: Chemical, Biochemical and Biological Aspects*, HECHT, S. M. (ed.), Springer-Verlag, New York, 1979, 1–23.

56 DUFF, R. J., DE VROOM, E., GELUK, A., HECHT, S. M., VAN DER MAREL, G. A., VAN BOOM, J. H. Evidence for C-1' hydrogen abstraction from modified oligonucleotides by Fe·bleomycin. *J. Am. Chem. Soc.* **1993**, *115*, 3350–3351.

57 BERRY, D. E., CHANG, L.-H., HECHT, S. M. DNA damage and growth inhibition in cultured human cells by bleomycin congeners. *Biochemistry* **1985**, *24*, 3207–3214.

58 BERRY, D. E., KILKUSKIE, R. E., HECHT, S. M. Damage induced by bleomycin in the presence of dibucaine is not predictive of cell growth inhibition. *Biochemistry* **1985**, *24*, 3214–3219.

59 PODDEVIN, B., ORLOWSKI, S., BELEHRADEK, J., JR., MIR, L. M. Very high cytotoxicity of bleomycin introduced into the cytosol of cells in culture. *Biochem. Pharmacol.* **1991**, *42*, S-67–S-75.

60 HECHT, S. M. RNA as a therapeutic target for bleomycin, in *The Many Faces of RNA*, EGGLESTON, D. S., PRESCOTT, C. D., PEARSON, N. D. (eds), Academic Press, London, 1998, 3–17.

61 ENDO, Y., MITSUI, K., MOTIZUKI, M., TSURUGI, K. The mechanism of action of ricin and related toxic lectins on eukaryotic ribosomes. *J. Biol.. Chem.* **1987**, *262*, 5908–5912.

62 LIN, J.-J., NEWTON, D. L., MIKULSKI, S. M., KUNG, H.-F., YOULE, R. J., RYBAK, S. M. Characterization of the mechanism of cellular and cell free protein synthesis inhibition by an anti-tumor ribonuclease. *Biochem. Biophys. Res. Commun.* **1994**, *204*, 156–162.

63 IORDANOV, M. S., RYABININA, O. P., WONG, J. *et al.* Molecular determinants of apoptosis induced by the cytotoxic ribonuclease onconase: evidence for cytotoxic mechanisms different from inhibition of protein synthesis. *Cancer Res.* **2000**, *60*, 1983–1994.

4
Inhibitors of the Tat–TAR Interactions

Chimmanamada U. Dinesh and Tariq M. Rana

4.1
Introduction

Acquired immune deficiency syndrome (AIDS) is one of the major causes of death in the recent history of humanity and the AIDS pandemic is not going away. As of the end of 2001, an estimated 40 million people worldwide – 37.2 million adults and 2.7 million children younger than 15 years – were living with AIDS. In spite of remarkable medical advances, HIV-1 infections are still increasing. Antiretroviral therapies have brought down the mortality rate significantly in developed countries. Unfortunately, the highest burden of AIDS is in nations that have the most limited medical resources. Worldwide, 95% of HIV-1 infections are in the developing world.

Enormous progress has been made in the development of antiretroviral agents. An understanding of the HIV replication mechanism first led to the development of "reverse transcriptase" (RT) inhibitors and later to the "protease" inhibitors. The nucleoside analog zidovudine (AZT) was one of the first RT inhibitors to be used for the treatment of AIDS. Several nucleoside, non-nucleoside, nucleotide RT inhibitors and protease inhibitors are currently available for AIDS therapy. However, no chemotherapeutic regimen is curative as yet.

Current AIDS therapies face three major problems: (1) first-line drugs are not effective in some patients, (2) newer medications have major side effects, (3) new drug-resistant strains of HIV are emerging. Therefore, there is a great need to find new drugs and treatment strategies. Given the pathogenesis of HIV mutants capable of resisting triple drug therapies, the identification of drugs that target HIV proteins other than reverse transcriptase and protease is a high priority for the development of new drugs.

HIV-1 is a complex retrovirus that encodes six regulatory proteins, including Tat and Rev, essential for viral replication. Inhibition of Tat and Rev function is an attractive target for new antiviral therapies. Both Tat and Rev are RNA-binding proteins and require specific interactions with RNA structures called TAR and RRE, respectively, for their function. In line with the focus of this chapter, Tat and its target RNA (TAR RNA) will be discussed in the following section.

4.2
Mechanism of Transcriptional Activation by Tat

The HIV replication cycle can be divided into two distinct phases (Fig. 4.1, see p. 60). During an early pre-integration stage, virus infects the cell and reverse transcriptase converts viral RNA into double-stranded DNA and the proviral genome is transported to the nucleus and integrated into the host genome. In the second post-integration stage, viral gene expression takes place from integrated proviral genome which is followed by viral assembly and maturation.

After the integration step of the HIV life cycle, the HIV proviral genome is transcribed by the human cellular machinery, including RNA polymerase II and other transcription accessory factors. Transcription from the viral long terminal repeat (LTR) is a complex process and is elegantly regulated at the elongation stage of transcription. The HIV-1 encodes a transcriptional activator protein, Tat, which is expressed early in the viral life cycle and is essential for viral gene expression, replication, and pathogenesis (for reviews, see Refs. [1–6]). Tat enhances the processivity of RNA Pol II elongation complexes that initiate in the HIV long terminal repeat (LTR) region. Mutational analysis of HIV-1 Tat protein has identified two important functional domains: an arginine-rich region that is required for binding to TAR RNA, and an activation domain that mediates the interactions with cellular machinery [7, 8]. Recent studies showed that Tat *trans*-activation function is mediated by a nuclear Tat-associated kinase, TAK [1, 2, 6]. The *trans*-activation domain of Tat interacts with TAK [9, 10], which was recently shown to be identical to the kinase subunit of P-TEFb [11, 12]. Tat interacts with the cyclinT1 (CycT1) subunit of P-TEFb and recruits the kinase complex to TAR RNA. Recruitment of P-TEFb to TAR has been proposed to be both necessary and sufficient for activation of transcription elongation from the HIV-1 long terminal repeat promoter [13]. The mechanism of Tat activation is summarized in Fig. 4.2 (see p. 60).

Neither CycT1 nor the P-TEFb complex bind TAR RNA in the absence of Tat, and thus the binding is highly cooperative for both Tat and P-TEFb [14, 15]. Tat appears to contact residues in the C-terminal boundary of the CycT1 cyclin domain that are not critical for binding of cyclinT1 to CDK9 [13, 16–23]. Mutagenesis studies showed that the CycT1 sequence containing amino acids 1–303 was sufficient to form complexes with Tat–TAR and CDK9 [13, 16–23]. Thus, the assembly of this complex appears to involve a series of adaptive interactions between the *trans*-activation and arginine-rich motif (RNA-binding) domains of Tat and their respective protein (CycT1) and nucleic acid (TAR) partners during transcription. Recent fluorescence resonance energy transfer studies using fluorescein-labeled TAR RNA and a rhodamine-labeled Tat protein showed that CycT1 remodels the structure of Tat to enhance its affinity for TAR RNA, and that TAR RNA further enhances interaction between Tat and CycT1 [24].

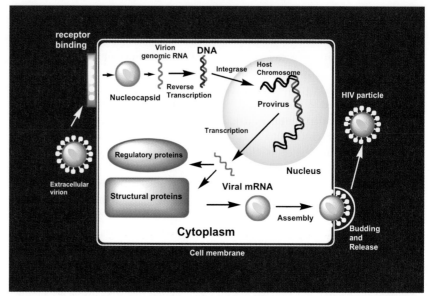

Fig. 4.1. Replication cycle of HIV. Virus initiates contact with receptors that is followed by entry and uncoating. Viral RNA is converted into DNA by the viral reverse transcriptase and the proviral genome is transported into the nucleus and integrated into the host genome. RNA polymerase II transcribes viral mRNA and the transcription of the proviral genome requires the HIV regulatory protein Tat, which is expressed at low levels during early life cycle and is imported to the nucleus to activate transcription of viral genes. At later stages of the life cycle, structural proteins are expressed and progeny virus particles are assembled and released by a process of budding and subsequent maturation into infectious virus.

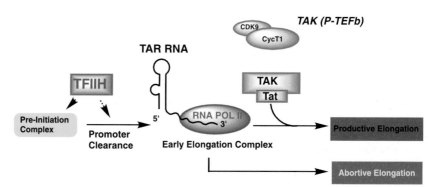

Fig. 4.2. Mechanism of transcriptional activation by Tat. Human RNA polymerase II initiates transcription from HIV promoter and TFIIH kinase assists in promoter clearance steps. TFIIH leaves the elongation complex when 30–36 nucleotides mRNA are transcribed [128]. Elongation is inefficient in the absence of Tat protein. Tat recruits P-TEFb kinase complex (TAK – <u>T</u>at-<u>a</u>ssociated <u>k</u>inase) to TAR RNA via contacting the RNA and cyclinT1 component of the complex. Then, CDK9 kinase phosphorylates RNA polymerase II and other substrates in the elongation complex, which leads to processive transcription elongation [129].

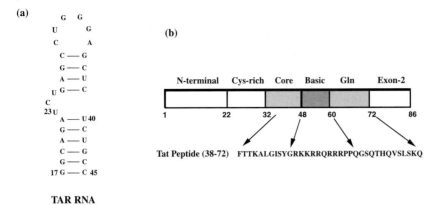

TAR RNA

trans-activation response element

Fig. 4.3. (a) Sequence and secondary structure of TAR RNA used in structural studies. TAR RNA spans the minimal sequences that are required for Tat responsive-ness *in vivo* [29] and for *in vitro* binding of Tat-derived peptides [31]. Wild-type TAR contains two non-wild-type base-pairs to increase transcription by T7 RNA polymerase [130]. Nucleotides critical for Tat binding are outlined. (b) Regions of the HIV-1 Tat protein and sequence of the Tat (38–72) peptide that recognizes Tat with high affinity and specificity.

4.3
Tat–TAR Interactions

As described above, the Tat protein is a potent transcriptional activator of the HIV-1 long terminal repeat promoter element. A regulatory element between +1 and +60 in the HIV-1 long terminal repeat which is capable of forming a stable stem–loop structure designated TAR is critical for Tat function.

Tat proteins are small arginine-rich RNA-binding proteins. HIV-1 Tat is encoded by two exons containing 86–101 amino acids in different HIV-1 isolates. Amino acids encoded by the first exon are both necessary and functional for TAR RNA binding and *trans*-activation *in vivo*. Tat protein is composed of several functional regions (Fig. 4.3). A cysteine-rich region (amino acids 22–37) contains seven cysteine residues; a "core" sequence (37–48) contains hydrophobic amino acids; a basic RNA-binding region (48–59) contains six arginines and two lysines and is a characteristic of a family of sequence-specific RNA-binding proteins; a glutamine-rich region at the C-terminus of the first exon contains several regularly spaced glutamines. In lentiviral proteins, only the basic and core regions are conserved. Although the integrity of the Cys-rich region is essential for *trans*-activation, this region does not appear to be directly involved in TAR RNA recognition. Based on mutational analysis, Tat can be divided into two functional domains. The first domain is the activation domain (amino acids 1–47) or cofactor-binding domain, which is functionally autonomous and is active when recruited to the HIV-1 LTR via a heterologous RNA-binding protein [25]. The second functional domain contains the basic region and is required for both RNA binding and nuclear localization activities of Tat [26].

HIV-1 Tat protein acts by binding to the TAR (*trans*-activation responsive) RNA element, a 59-base stem–loop structure located at the 5′ ends of all nascent HIV-1 transcripts [27]. TAR RNA was originally localized to nucleotides +1 to +80 within the viral LTR [28]. Subsequent deletion studies have established that the region from +19 to +42 incorporates the minimal domain that is both necessary and sufficient for Tat responsiveness *in vivo* (Fig. 4.3) [29]. TAR RNA contains a six-nucleotide loop and a three-nucleotide pyrimidine bulge which separates two helical stem regions [30].

Tat protein recognizes the trinucleotide bulge in TAR RNA. Key elements required for TAR recognition by Tat have been defined by extensive mutagenesis, chemical probing, and peptide-binding studies [31–35]. Tat interacts with U23 and two other bulge residues, C24 and U25, which act as spacers because they can be replaced by other nucleotides or linkers [34, 36]. In addition to the trinucleotide bulge region, two base pairs above and below the bulge also contribute significantly to Tat binding [32, 34]. Phosphate contacts below the bulge at positions 22, 23, and 40 are critical for Tat interactions [33, 34, 37]. Chemical crosslinking studies showed that a TAR duplex containing a trisubstituted pyrophosphate replacing the phosphate at 38–39 reacted specifically with Lys51 in the basic region of Tat(37–72) peptide [38]. Site-specific photo-crosslinking experiments on Tat–TAR complex using 4-thiouracil as a photoactive nucleoside showed that Tat interacts with U23, U38, and U40 in the major groove of TAR RNA [39]. In a recent study, 6-thio-G was incorporated at specific sites in TAR RNA seqence and Tat–TAR photo-crosslinking experiments were performed [40]. Results of these experiments provide direct evidence that, during RNA–protein recognition, Tat is in close proximity to O^6 of G21 and G26 in the major groove of TAR RNA. Taken together, these studies establish that Tat binds TAR RNA at the trinucleotide bulge region and interacts with two base pairs above and below the bulge in the major groove of RNA.

The basic regions of Tat proteins are directly involved in RNA binding. Short basic peptide mimics of this region bind TAR RNA in the bulge region [31–34, 41]. A number of studies determined the sequence requirements within the basic region of Tat for RNA binding. *In vivo* experiments using full-length Tat and *in vitro* studies with synthetic peptides indicate that conservation of overall positive charge, including several arginines, is essential for TAR RNA recognition. For example, Tat–TAR interactions are not affected by interchanging the basic region sequences of Tat and Rev [42]. Amino acid substitutions that recreate the consensus sequence of the RNA-binding region (R/KXXRRXRR, where R is arginine, K is lysine, and X is any amino acid) reconstitute a functional Tat protein. These results are inconsistent with *in vitro* peptide studies showing that high-affinity TAR RNA-binding requires three out of four arginines within the RRXRR stretch, although any one arginine can be substituted suggesting some redundancy [41].

4.4
RNA as a Small Molecule Drug Target

Protein–nucleic acid interactions are involved in many cellular functions, including transcription, RNA splicing, and translation. Readily accessible synthetic molecules that can bind with high affinity to specific sequences of single- or double-stranded nucleic acids have the potential to interfere with these interactions in a controllable way, making them attractive tools for molecular biology and medicine. Successful approaches used thus far include duplex-forming (antisense) [43] and triplex-forming (anti-gene) oligonucleotides [44–46], peptide nucleic acids (PNA) [47], and pyrrole-imidazole polyamide oligomers [48, 49]. Each class of compounds employs a readout system based on simple rules for recognizing the primary or secondary structure of a linear nucleic acid sequence. Another approach employs carbohydrate-based ligands, calicheamicin oligosaccharides, which interfere with the sequence-specific binding of transcription factors to DNA and inhibit transcription *in vivo* [50, 51]. While antisense oligonucleotides and PNA employ the familiar Watson–Crick base-pairing rules, two others, the triplex-forming oligonucleotides and the pyrrole-imidazole polyamides, take advantage of straightforward rules to read the major and minor grooves, respectively, of the double helix itself.

In addition to its primary structure, RNA has the ability to fold into complex tertiary structures consisting of such local motifs as loops, bulges, pseudoknots, and turns [52, 53]. It is not surprising that, when they occur in RNAs that interact with proteins, these local structures are found to play important roles in protein–RNA interactions (Fig. 4.4). This diversity of local and tertiary structure, however, makes it impossible to design synthetic agents with general, simple-to-use recognition rules analogous to those for the formation of double- and triple-helical nucleic acids. Since RNA–RNA and protein–RNA interactions can be important in viral and microbial disease progression, it would be advantageous to have a general method for rapidly identifying synthetic compounds for targeting specific RNA structures. A particular protein-binding RNA structure can be considered as a molecular receptor not only for the protein with which it interacts but also for synthetic compounds, which may prove to be antagonists of the protein–RNA interaction.

Two examples of such interactions are the Tat–TAR and Rev–RRE recognition, which are essential elements in the mechanism of HIV-1 gene expression. The following sections will describe recent developments in the identification of ligands for inhibition of Tat–TAR interactions.

4.5
Ligands for TAR RNA

4.5.1
TAR RNA Bulge Binders

As discussed above, Tat protein binds a trinucleotide bulge sequence in TAR RNA. Therefore, it is obvious that a Tat antagonist should be able to recognize the bulge

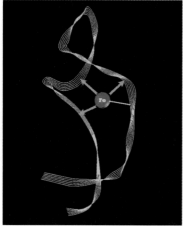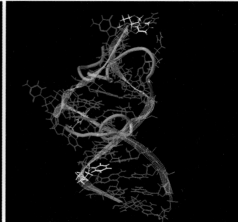

Fig. 4.4. TAR RNA folding and Tat–TAR interactions. (a) Tertiary folding of TAR RNA probed by RNA self-cleavage method using a tethered iron chelate [131]. Iron(II) is covalently attached at a specific site in the RNA sequence and arrows show the regions of RNA cleaved by the production of localized hydroxyl radicals. (b) Proposed model for Tat peptide–TAR interactions showing protein orientation and the proximity of various nucleotides to the peptide in the RNA–protein complex. Ribbon structure of TAR RNA is shown in yellow lines and nucleotides in red [55]. Ribbon structure of the Tat peptide (ribbon/tube) and the N-terminal Phe38 are shown in cyan color. Ribbon structure of the Tat peptide is drawn from Tat protein structure [132]. Orientation of the Tat peptide is based on previous photo-crosslinking results indicating that Lys41 and Arg57 are close to U42 and U31, respectively [66,130]. Lys41 and Arg57 side chains are shown in green. U42 is shown in yellow and U31 in atom-by-type colors. Structures of RNA and protein were visualized using Insight II software on an IRIS workstation.

structure and, ideally, should be able to bind TAR bulge with affinities higher than that of Tat protein. Hamy and co-workers identified a peptidic compound, CGP64222, that was able to bind TAR RNA with high affinities [54]. NMR studies suggested that CGP64222 binds the bulge region of TAR and induces conformational change in TAR resulting in a structure very similar to that of a Tat–peptide-bound TAR RNA [54, 55]. This nine-residue oligomer, a hybrid peptide/peptoid, was screened and identified by a deconvolution combinatorial library method. The Tat activity in a cellular Tat-dependent *trans*-activation assay was inhibited with 10–30 μM CGP64222. The structure of CGP64222 is shown in Fig. 4.5.

In another study, Hamy *et al.* [56] reported the identification of low-molecular weight Tat antagonists, their affinities for TAR RNA, and biological activities. A series of compounds on the basis of published structural data of the molecular interactions between TAR and Tat-derived peptide was synthesized. This new class of Tat antagonists contains two different functional motifs, a polyaromatic motif for stacking interactions with TAR RNA and a polycationic anchor for contacts with the phosphate backbone of RNA. A varying linker to connect the stacking motif with the RNA-binding motif was used. The most active compound competed with Tat–TAR complex formation with a competition dose CD_{50} of 22 nM *in vitro* and

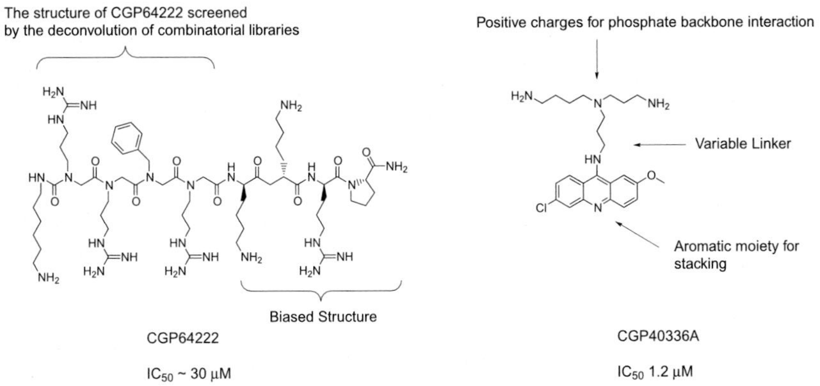

The structure of CGP64222 screened
by the deconvolution of combinatorial libraries

Positive charges for phosphate backbone interaction

Biased Structure

CGP64222

IC$_{50}$ ~ 30 µM

Variable Linker

Aromatic moiety for
stacking

CGP40336A

IC$_{50}$ 1.2 µM

Fig. 4.5. Structures of TAR RNA bulge-binding molecules that
inhibit Tat–TAR interactions [54, 56].

blocked Tat activity in a cellular system with an IC$_{50}$ of 1.2 µM. Figure 4.5 shows
the structure of the active compound. From structure–activity relationship studies,
two new features of Tat–TAR inhibitors became clear: (1) Modification of the linker
length has a mild effect on activity and the structure of polyamine moiety is critical
for Tat–TAR inhibition. (2) The position of the polyaromatic ring for a substitution
of the linker is important and the type of chemical bond between the linker and
the polyaromatic motif is also crucial for activity.

4.5.2
Targeting Multiple Sites in TAR RNA

To identify small organic molecules that inhibit HIV-1 replication by blocking Tat–
TAR interactions, Mei *et al.* [57] screened their research compound libraries and
reported three inhibitors of Tat–TAR interactions that target TAR RNA and not the
protein. These three Tat–TAR inhibitors include neomycin, quinoxaline, and ami-
noquinozaline. Chemical structures of these compounds, IC$_{50}$, and their binding
sites on TAR RNA are outlined in Fig. 4.6. Each of these inhibitors recognizes a
different structural region in TAR RNA such as the bulge, lower stem, and the loop
sequence.

In another attempt to discover Tat–TAR inhibitors, Mei *et al.* [58] screened their
corporate compound library containing approximately 150,000 compounds. Selective Tat–TAR inhibitors were screened by *in vitro* high-throughput screening assays
and inhibitory activities were determined by gel mobility shift assays, scintillation
proximity assays, filtration assays, and electrospray ionization mass spectrometry
(ESI-MS). After *in vitro* assays, Tat-activated reporter gene analyses were employed
to investigate the cellular activities of the primary Tat–TAR inhibitors. Approximately 500 Tat–TAR inhibitors were selected from *in vitro* assays and 50 compounds exhibited dose-dependent cellular activities with IC$_{50}$ values ≤50 µM.

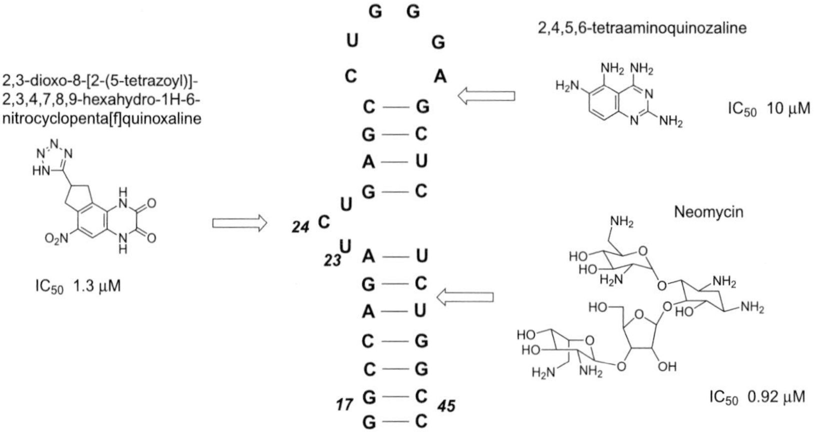

Fig. 4.6. Structures and IC_{50} values of three TAR RNA ligands. Putative RNA-binding sites are indicated [57].

Among them, approximately 20 compounds were relatively non-toxic (therapeutic index, TC_{50}/IC_{50}, ≥ 5) and considered selective for Tat-dependent transcription.

4.5.3
Targeting RNA with Peptidomimetic Oligomers

4.5.3.1 Backbone modification

We have recently begun to examine TAR RNA recognition by unnatural biopolymers [59, 60]. We synthesized oligocarbamates and oligourea containing the basic arginine-rich region of Tat by solid-phase synthesis methods, and tested for TAR RNA binding. The oligocarbamate backbone consists of a chiral ethylene backbone linked through relatively rigid carbamate groups [61]. Oligoureas have backbones with hydrogen-bonding groups, chiral centers, and a significant degree of conformational restriction. Introducing additional side chains at the backbone NH sites can further modify biological and physical properties of these oligomers (Fig. 4.7).

A Tat-derived oligourea binds specifically to TAR RNA with affinities significantly higher than the wild-type Tat peptide. To synthesize Tat-derived oligourea on solid support, we used activated *p*-nitrophenyl carbamates and protected amines in the form of azides, which were reduced with $SnCl_2$-thiophenol-triethylamine on solid support [62, 63]. After HPLC purification and characterization by mass spectrometry, the oligourea was tested for TAR RNA binding. The Tat-derived oligourea was able to bind TAR RNA and failed to bind a mutant TAR RNA without the bulge residues [60].

Equilibrium dissociation constants of the oligourea–TAR RNA complexes were measured using direct and competition electrophoretic mobility assays. Dissociation constants were calculated from multiple sets of experiments which showed

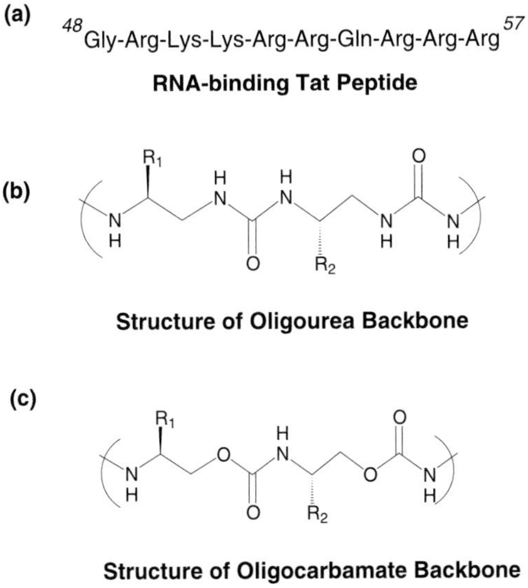

(a)

^{48}Gly-Arg-Lys-Lys-Arg-Arg-Gln-Arg-Arg-Arg^{57}

RNA-binding Tat Peptide

(b)

Structure of Oligourea Backbone

(c)

Structure of Oligocarbamate Backbone

Fig. 4.7. (a) The Tat-derived peptide, amino acids 48–57, contains the RNA-binding region of Tat protein. Structure of the oligourea (b) and oligocarbamate (c) backbone. Sequence of the oligourea and oligocarbamate correspond to the Tat peptide shown in (a), except the addition of an L-Tyr amino acid at the C-terminus of oligourea [59, 60].

that the oligourea binds TAR RNA with a K_D of 0.11 ± 0.07 μM. To compare the RNA-binding affinities of the oligourea to natural peptide, we synthesized a Tat-derived peptide (Tyr47 to Arg57) containing the RNA-binding domain of Tat protein. Dissociation constants of the Tat peptide–RNA complexes were determined from multiple sets of experiments under the same conditions used for oligourea–TAR RNA complexes. These experiments showed that the Tat peptide (47–57) binds TAR RNA with a K_D of 0.78 ± 0.05 μM. A relative dissociation constant (K_{REL}) can be determined by measuring the ratios of wild-type Tat peptide to the oligourea dissociation constants (K_D) for TAR RNA. Our results demonstrate that the calculated value for K_{REL} was 7.09, indicating that the urea backbone structure significantly enhanced the TAR-binding affinities of the unnatural biopolymer, and this difference in K_D values could be more dramatic because gel mobility shift methods severely underestimate absolute peptide–RNA-binding affinities [41].

Site-specific photo-crosslinking and competition experiments showed that a small Tat-derived oligourea binds TAR RNA specifically with high affinity and interacts in the major groove of TAR RNA similar to Tat peptides. Due to the difference in backbone structure, oligoureas may differ from peptides in hydrogen-bonding properties, lipophilicity, stability, and conformational flexibility. Moreover, oligoureas are resistant to proteinase K degradation. These characteristics of oligoureas may be useful in improving pharmacokinetic properties relative to peptides. RNA recognition by an oligourea provides a new approach for the design of drugs, which will modulate RNA–protein interactions.

4.5.3.2 D-Peptides

Due to the difference in chirality, D-peptides are resistant to proteolytic degradation and cannot be efficiently processed for major histocompatibility complex class II-restricted presentation to T helper cells (T$_H$ cells). Consequently, D-peptides would not induce a vigorous humoral immune response that impairs the activity of L-peptide drugs [64]. The D-peptide ligands may provide useful starting points for the design or selection of novel drugs.

Can D-peptides recognize naturally occurring nucleic acid structures? To test this hypothesis, we synthesized a D-Tat peptide, Tat(37–72), containing the basic arginine-rich region of Tat by solid-phase peptide synthesis methods. After HPLC purification and characterization by mass spectrometry, the D-Tat peptide was tested for TAR RNA binding. Similar to L-Tat, the D-Tat peptide was able to bind TAR RNA and failed to bind a mutant TAR RNA without the bulge residues. Equilibrium dissociation constants of the D-Tat–TAR RNA complexes were measured using direct electrophoretic mobility shift assays [65, 66]. Dissociation constants were calculated from eight sets of experiments which showed that the D-Tat peptide binds TAR RNA with a K_D of 0.22 µM. Under similar experimental conditions, L-Tat(37–72) binds TAR RNA with a K_D of 0.13 µM [67].

To test the effects of D-Tat peptide on HIV-1 transcription in a cell-free system, *in vitro* transcription reactions were performed using HeLa cell nuclear extract and linearized HIV-1 DNA template [68]. Tat produced a large increase in the synthesis of correctly initiated 530 nucleotide runoff RNA transcripts {69]. Tat stimulated transcription at concentrations ranging from 50 ng to 300 ng per 10 µL reactions. Quantitation revealed that 100 ng of Tat per reaction produced a 10–12-fold stimulation of HIV-1 transcription. Control experiments showed that Tat did not significantly increase transcription from an HIV-1 promoter with a mutated TAR element either in the stem region (G26 to C26) or in the loop sequence (U31 to G31). Increasing amounts of D-Tat resulted in a significant decrease in Tat-mediated transcriptional activation. In the presence of 1 µg D-Tat (approximately equal to 3 times the wild-type Tat), more than 80% Tat *trans*-activation was inhibited [69]. The amount of recovered transcripts and the efficiency of transcription were normalized by including a labeled RNA not originating from HIV-1 LTR. To determine the specificity of *trans*-activation inhibition by D-Tat, a mutant D-Tat peptide, Gly44–Gln72, where all Arg residues in the RNA-binding region were substituted with Ala, was synthesized. The mutant D-Tat was unable to bind TAR RNA in electrophoretic mobility shift experiments and did not inhibit Tat *trans*-activation *in vitro*. These results indicate that D-Tat is able to specifically inhibit Tat *trans*-activation *in vitro*.

It has been previously established that Tat peptides containing the basic domain are taken up by cells within less than 5 minutes and accumulate in the cell nucleus [70]. Since the D-Tat peptide also contains the basic domain of Tat, we reasoned that this peptide would be rapidly taken up by HeLa cells and accumulate in the nucleus. Once D-Tat peptide reaches the nucleus, it would compete with Tat for TAR binding and lead to inhibition of Tat function. To test this hypothesis, we added D-Tat during transfection of pSV2-Tat [71] and pAL [72] plasmids into HeLa cells containing an integrated LTR-CAT reporter [73]. Plasmids pSV2Tat and pAL express the first exon of Tat protein and luciferase enzyme, respectively. Transfection of pSV2Tat enhanced transcription as determined by CAT activity. Increasing amounts of the D-Tat resulted in a decrease of CAT activity while luciferase activity was not affected [69]. Tat *trans*-activation was inhibited more than 60% by 5 μg (approx. 0.5 μM) D-Tat peptide. Further addition of D-Tat did not further inhibit Tat *trans*-activation, probably because maximum peptide uptake efficiency is reached at 5 μg of D-Tat. To rule out the possibility that the observed inhibition of *trans*-activation is due to some non-specific toxicity of the D-peptide or reduction of the pSV2Tat plasmid uptake, transcription of luciferase gene was monitored. Transcription of luciferase gene was not affected by D-Tat peptide as measured by luciferase enzymatic activity assays. Cell viability assays showed that cells were not killed by D-Tat treatment. Specificity of the inhibition was tested by adding a mutant D-Tat peptide, Gly44–Gln72, where all Arg residues in the RNA-binding region were substituted with Ala during transfection of plasmids and analyzing the CAT and luciferase activities as described above for D-Tat. This mutant D-Tat peptide did not inhibit Tat *trans*-activation. Thus, these results indicate that the D-Tat peptide specifically inhibits *trans*-activation by Tat protein *in vivo* [69].

These findings show that a small Tat-derived D-peptide binds TAR RNA and selectively inhibits Tat *trans*-activation. It remains to be determined whether a broad range of RNA–protein interactions can be selectively targeted. These results present an example of the application of D-peptides as artificial regulators of cellular processes involving RNA–protein interactions *in vivo*.

4.6
Combinatorial Library Approach in the Discovery of Small Molecule Drugs Targeting RNA

4.6.1
Combinatorial Chemistry

The demand for a variety of chemical compounds for identifying and optimizing new drug candidates has increased dramatically. In the past, traditional mass screening of natural products from plants, marine organisms, and synthetic compounds has been successful in identifying a lead chemical structure. In order to support this demand, chemists have developed new methodologies that are accelerating the drug discovery process. Combinatorial synthesis is one of the most promising approaches to the synthesis of a large collection of diverse molecules because vast libraries of molecules having different chemical identities are synthesized in a short period of time [74–79]. There is considerable evidence to show that combinatorial chemistry plays an important role in the lead discovery process. For example, a variety of biological targets such as proteases [63, 80, 81], protein kinases [82–84], cathepsin D [85], and SH3 domain [86–88] have been screened with this new technology.

Combinatorial synthesis has been primarily facilitated by the application of solid-phase synthesis [89, 90]. Each substrate is linked to a solid support (a polymer bead), and it is possible to synthesize a variety of products that are spatially separated, and thus reagents and by-products not bound to the beads may be removed simply by filtration. In short, the combinatorial synthesis is faster, and thus more efficient and much cheaper than classical organic synthesis, and can give up to thousand or even million of products simultaneously.

The combinatorial drug discovery process has three major parts: (1) The generation of a large collection of diverse molecules, known as combinatorial libraries, by systematic synthesis of a variety of building blocks. (2) Screening of such libraries with biological targets to identify novel lead compounds. (3) Determining the chemical structures of active compounds. Therefore, the combinatorial discovery process requires not only the rational design and synthesis of combinatorial libraries of molecular diversity, but also the development of screening methodologies for library evaluation.

A variety of libraries have been prepared by synthetic and biological methods such as phage display, polysomes, and plasmids [91, 92]. Two major synthetic methods are the iterative deconvolution approach [93, 94] and the one-bead one-compound method [95]. First, iterative deconvolution is a chemical method in which the chemical structure of the active compound in a combinatorial library is characterized in an iterative manner. It involves screening of combinatorial library pools, identification of the active sublibrary pool, resynthesis of sublibraries, and rescreening of the resynthesized sublibraries. This method has the advantage that it affords fully characterizable, non-modified structure, solution-phase libraries which afford more realistic interaction results than solid-support bound libraries. However, the time-consuming nature of having to resynthesize and reassay sublibraries and the potential inconsistencies during resynthesis have led to a search for alternative combinatorial methods. Recently, Hamy *et al.* [54, 56] have used this method to identify inhibitors of HIV-1 replication. The one-bead–one-compound method is based on the fact that the combinatorial bead library contains single beads displaying only one type of compound although there may be up to approx. 100 pmol (approx. 10^{13} molecules) of the same molecules on a single 90-μm-diameter polymer bead.

This approach has several unique features: (1) A large combinatorial library is synthesized by a split synthesis method. (2) Each library member (compound) is spatially separated in the solution and all the library compounds can be screened independently at the same time. (3) Once active beads are screened, the chemical structure of the active beads may be determined directly by using NMR, mass spectrometry, and HPLC [96, 97] or by an encoding method [98, 99].

4.6.2
Split Synthesis

The beauty of split synthesis is that it is a simple and very efficient synthetic method (Fig. 4.8) [100]. A sample of support material (bead) is divided into a

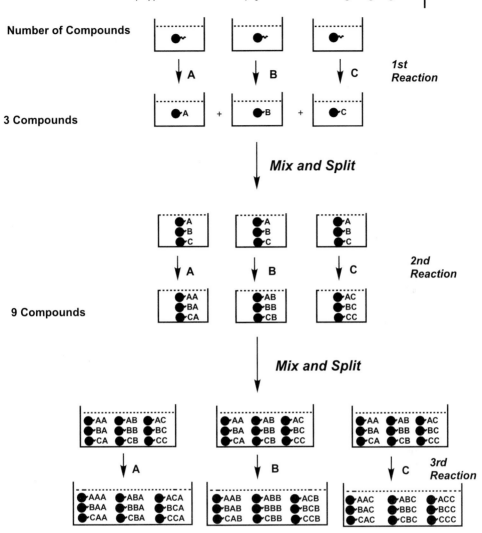

27 Compounds

= (3 Vessels) ³ Reactions

Fig. 4.8. Split synthesis using three monomers would produce as many as 27 different timers.

number of equal portions (n) and each reaction vessel is individually reacted with a specific substrate. After completion of the substrate reaction, it is subsequently washed to remove by-products and excess reagent. The individual reaction products are recombined, the whole is throughly mixed, and divided again into portions. Further reaction with a set of reagents gives complete sets of possible dimer

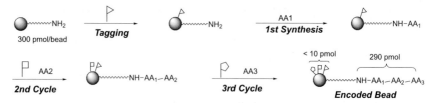

Fig. 4.9. Schematic representation of the encoding process.

combinatorial libraries, and this whole process may then be repeated as necessary (total x times). After this split synthesis, the number of library compounds obtained arises from the exponential increase in molecular diversity, in this case n to the power of x (n^x). For example, split synthesis using 20 different D and L amino acids at each site of a pentapeptide would produce as many as $20^5 = 3,200,000$ different compounds.

4.6.3
Encoding

The idea of encoding synthetic information with a chemical tag was first proposed by Brenner and Lerner [101]. Encoding involves attaching unique arrays of readily analyzable chemical tags to each bead that designate the particular set of reagents used in the split synthesis of that specific bead [99, 102]. Thus analyzing any bead for its tag content yields the history for the synthesis of that specific bead (Fig. 4.9). Although one could use a different tag for each reagent, it is much simpler to use a mixture of tags because tag mixtures of N different tags can encode 2^N different reactions. For example, only six tags are needed to encode as many as 64 different reactions. While almost any kind of chemical can be used for encoding, there are practical problems because tags need to be chemically inert to library synthesis and reliably analyzed in picogram quantity from a single positive bead. Numerous tagging methods have been developed such as oligonucleotide, peptide, and halophenyl tag approaches [74, 102, 103]. Among these methods, halophenyl tagging was chosen because it fulfills the following requirements:

1. An assurance of fast structural determination by GC/ECD (gas chromatrography using electron capture detector). In early stages of encoding strategies, biopolymer tags were used [104]. The problems of molecular structure determination using biopolymers are the stability of the tag to the vigorous organic reaction conditions and the amount of compound on one bead may not allow for characterization by Edman degradation or PCR. This problem can be solved using photocleavable halophenyl tags that can afford the structural determination in 10 minutes using femtomole quantity of tags on the bead.
2. Affordability of a variety of tagging compounds. Using different spacers and halophenyl compounds, 30–40 tagging compounds are easily obtained with commercially available starting materials.

3. Simplicity of the tag cleavage reactions. The photocleavable halophenyl tag is cleaved by UV irradiation and it does not require any chemical reactions.
4. Sensitivity of halophenyl derivatives. A few femtomole quantity of halophenyl functionality has been detected with GC/ECD analysis.

4.6.4
On-bead Screening and Identification of Structure-specific TAR-Binding Ligands

Previous studies using combinatorial chemistry to identify new ligands to block the TAT–TAR interaction have relied on a variety of complex methods that are labor intensive or require expensive robotics equipment [58]. For the most part, these methods originated in the study of individual protein–nucleic acid interaction experiments. Moreover, in some cases time-consuming deconvolution strategies are also needed to identify the individual compounds responsible for the properties found in a mixture of compounds tested together [54, 56]. We decided to investigate on-bead screening methods that have been previously used with success on small organic receptors. This entailed covalently attaching the dye Disperse Red to the TAR RNA (Fig. 4.10) and incubating it in a suspension of library beads made from the split synthesis method. Diffusion of low-molecular-weight receptors into a bead of TentaGel resin is known to be rapid, whereas one might expect that a macromolecule such as a protein or large nucleic acid might be excluded from the bead interior where the bulk of the peptide is displayed. Nevertheless, we have found that the dye–TAR conjugate was able to enter the beads and bind in a structure-dependent manner. Peptides specific for portions of TAR other than the bulge region were blocked by using a relatively large concentration of an unlabeled TAR analog lacking the natural 3-nucleotide bulge. A small amount of detergent and using a low RNA concentration (250 nM) also minimizes non-specific binding. Another advantage to our method is the use of chemically encoded beads [105].

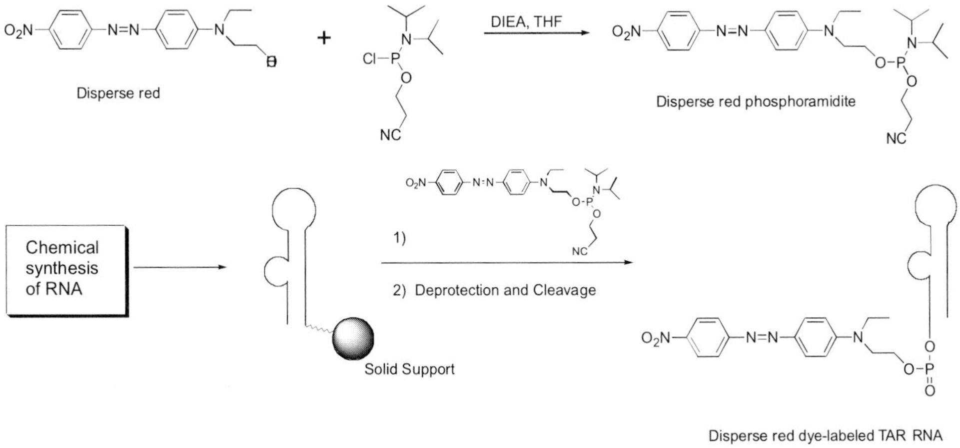

Fig. 4.10. Covalent attachment of the dye Disperse Red to the TAR RNA.

Tab. 4.1. RNA-binding ligands.

ID no.	Structures	Frequency	Color	K_D (nM)[a]	K_{REL}[b]
1	NH$_2$-(L)Lys-(D)Lys-(L)Asn-OH	2	Red	420 ± 44	1.73
2	NH$_2$-(L)Lys-(D)Lys-(D)Asn-OH	1	Pink	4173 ± 208	0.17
3	NH$_2$-(L)Lys-(L)Lys-(L)Asn-OH	2	Pink	3224 ± 183	0.23
4	NH$_2$-(L)Arg-(D)Lys-(L)Ala-OH	1	Pink	2640 ± 219	0.28
5	NH$_2$-(L)Arg-(D)Lys-(D)Val-OH	1	Pink	10,434 ± 594	0.07
6	NH$_2$-(L)Arg-(D)Lys-(L)Arg-OH	1	Pink	878 ± 80	0.83
7	NH$_2$-(D)Thr-(D)Lys-(L)Asn-OH	1	Pink	564 ± 80	1.29
8	NH$_2$-(D)Thr-(D)Lys-(L)Phe-OH	1	Pink	2087 ± 244	0.35

[a] K_D values were determined from four independent experiments.
[b] $K_{REL} = K_D$ of a basic Tat peptide (727 ± 74 nM)/K_D of inhibitor.
The bead library peptides were all N-acetylated. Peptides used for *in vitro* and *in vivo* experiments were not N-acetylated and contained free N-termini.
Adapted from Ref. [106].

Once a dye-stained bead is selected, the identification of the RNA-binding small molecule is rapid and straightforward (Fig. 4.11). Although many binding experiments are conducted simultaneously, the compounds remain discrete, each in its own assay vessel, the bead produced by the split-synthesis method [106].

4.6.5
Ligand Sequence Analysis

Upon incubating the dye–TAR conjugate with the library we picked a set of beads and decoded the structure of RNA-binding ligands (Tab. 4.1). To verify that our assay reflected RNA–trimer interaction and to determine the affinity of these trimer ligands for TAR RNA, we resynthesized trimer peptides **1–8** (Tab. 4.1) and measured their dissociation constants with wild-type TAR RNA. The results in Tab. 4.1 confirm that the on-bead assay mimics RNA binding because ligand **1** has the highest affinity for TAR RNA (K_D = 441 nM). To compare the RNA-binding affinities of eight ligands to natural Tat peptide, we synthesized a Tat-derived peptide (Gly48 to Arg57) containing the RNA-binding region of Tat protein. Dissociation constants of the Tat peptide–RNA complexes were determined under the

Fig. 4.11. (a) Schematic presentation of screening and decoding of the combinatorial library. TAR RNA was labeled with a red dye, Disperse Red, at its 5′ end by chemical synthesis and incubated with the trimer library. (b) A portion of the beads in the library after incubating with red-dye-labeled TAR RNA. The dark bead in the center was identified as ligand **1**. (c) The equilibrium interaction between dye-labeled TAR RNA and a tripeptide tethered to beads. A suspension of beads containing ligand **1** in Tris-HCl buffer (400 μL) was incubated with dye-labeled TAR RNA (1 μM) at 4°C for 5 h. Beads were stained red upon TAR RNA binding (left). Red beads became colorless when excess of unlabeled TAR RNA (middle) or ligand **1** (right) was added, indicating the displacement of red–TAR RNA from the beads. (Reprinted from Hwang *et al.* [106].)

same conditions used for trimer ligand–TAR RNA complexes. These experiments showed that the Tat peptide (48–57) binds TAR RNA with a K_D of 698 nM. A relative dissociation constant (K_{REL}) can be determined by measuring the ratios of the Tat peptide to trimer ligand dissociation constants (K_D) for TAR RNA. These re-

(a)

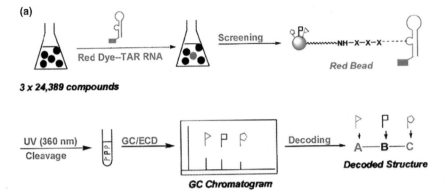

(b)

(c)

sults are shown in Tab. 4.1. Ligand **1** binds TAR RNA with affinities higher than that of the wild-type Tat peptide. These results indicate that selection frequency reflects ligand activity and, if a large enough library sample is used, could be used as an indicator of ligand affinity for RNA [106].

It was interesting that ligands **1** and **7** were the tightest-binding tripeptides found, suggesting a consensus sequence of X-(D)Lys-(L)Asn. Furthermore, two diastereomers of ligand **1**, peptides **2** and **3**, were found in the assay, **3** being the only homochiral sequence. RNA–peptide binding measurements revealed that the dissociation constants for these two diastereomeric sequences were approximately 7 times higher than for the strongest sequence, **1**. This sharp loss of binding energy among diastereomers indicates that the binding interaction is highly stereospecific and not merely the result of a non-specific lysine-phosphate backbone attraction. Another interesting feature of these results is that all but one of the TAR-binding sequences found were heterochiral and would have been missed by other techniques such as phage display which only use the proteinogenic amino acids. Use of D- and L-amino acids together yields a richer stereochemical variety of ligands, in addition to the diversity imparted by using the alpha-amino acids.

4.6.6
Heterochiral Small Molecules Target TAR RNA Bulge

The interaction sites of our trimer ligand **1** on TAR RNA were determined by NMR experiments [106]. NMR spectra of free TAR and TAR complex with ligand **1** were recorded. Due to the spectral overlap, it was impossible to follow all but a few well-isolated resonances by conventional one-dimensional experiments. Therefore, we carried out two-dimensional NOESY and TOCSY experiments. All spectra were recorded on a Bruker AMX-500 NMR spectrometer operating at 500 MHz for ^{1}H equipped with triple resonance probe and H-broadband inverse detection probes. The 1D and 2D ^{1}H spectra were recorded at approximately 1–2 mM RNA and ligand concentrations in 5 mM phosphate buffer (pH 5.5), with up to 100 mM NaCl. All other conditions were the same as described earlier by Aboul-ela *et al.* [55]. We obtained a set of complete TAR RNA assignments. Increasing amounts of ligand **1** were added to TAR RNA and the spectral changes were monitored by two-dimensional TOCSY experiments. The TOCSY spectrum contains a region where only pyrimidine H5–H6 resonances are found and this region has a well-dispersed 2D spectrum. Resonances in the free RNA and RNA–ligand complexes were assigned by NOESY experiments. Results of our TOCSY experiments are shown in Fig. 4.12. Resonances in the bulge region, U23 and C24, were shifted. All other resonances were not affected significantly by the addition of the ligand.

To address the question whether spectral changes at U23 and C24 were due to specific ligand **1** binding or the result of perturbation by a non-specific exogeneous ligand, we performed NMR experiments in the presence of a basic tripeptide containing L-Lys amino acids. Our results showed that the Lys-peptide did not cause shift of resonances in the bulge region including U23 and C24 (data not shown), indicating that ligand **1** specifically interacts with TAR RNA at the bulge region.

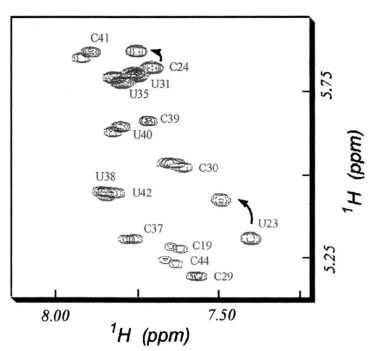

1

Fig. 4.12. Titration of wild-type TAR RNA by increasing amounts of the ligand 1, (L) Lys-(D) Lys-(L) Asn (Upper). (Lower) Superposition of TOCSY spectra at increasing concentration of 1 (free RNA, pink and 1:1 ligand to RNA ratio, blue) shows that only resonances at bulge U23 and C24 are affected by the ligand 1 binding. Adapted from Hwang et al. [106].

Interestingly, ligand **1**–TAR RNA interactions are different from TAR RNA–peptide or TAR–Arg complexes reported in previous NMR studies [55, 107] because there were no detectable interactions with G26 and A27 regions as observed in previous studies. Another TAR-binding ligand, CGP64222, which contains Arg side chains in its sequence also causes conformational change in TAR and creates an RNA structure that is similar to TAR–peptide structure [54]. These results indicate that the ligand **1** is the first ligand that binds specifically to the bulge of TAR RNA in a manner different from previously reported TAR ligands and Tat peptides. These findings suggest an intriguing possibility that small molecules that interact with TAR RNA and induce a conformational change in TAR resulting in a structure different from that of Tat–TAR complex could be used to lock RNA in a nonfunctional structure.

4.6.7
Inhibition of Tat *trans*-Activation *in vivo*

To test whether this small molecule–RNA interaction could be used to control HIV-1 gene expression *in vivo*, we used HL3T1 cells, a HeLa cell line derivative containing an integrated HIV-1 LTR promoter and CAT reporter gene [73]. We added different amounts of ligand **1** during transfection of pSV2–Tat [71] and pAL [72] plasmids into HL3T1 cells. Plasmids pSV2Tat and pAL express the first exon of Tat protein and luciferase enzyme, respectively. Luciferase reporter gene provides an internal control. Transfection of HeLa cells with pSV2Tat enhanced transcription as determined by CAT activity. Increasing amounts of the ligand **1** resulted in a decrease of CAT activity while luciferase activity was not affected. In the presence of 700 nM concentrations of ligand **1**, more than 90% of Tat *trans*-activation was inhibited [106]. To rule out the possibility that the observed inhibition of *trans*-activation could be due to some non-specific toxicity of the ligand **1** or reduction of the pSV2Tat plasmid uptake, transcription of luciferase gene was monitored. Transcription of luciferase gene was not affected by the ligand **1**. Cell viability assays showed that ligand **1** treatment was not toxic to the cells. Further control experiments showed that weaker TAR RNA-binding ligands such as L-argininamide and a scrambled Tat peptide containing D-amino acids had no inhibitory effect on Tat *trans*-activation [69].

4.7
Cyclic Structures as RNA-targeting Drugs

Efforts to develop cyclic peptide-based drugs increased manifold after the antibiotic gramicidin S was found to be a cyclic decapeptide. Many antibiotics and toxins are also known to be cyclic amino acid sequences. Cyclization of amino acid sequences results in increased metabolic stability, potency, receptor selectivity, and bioavailability [108–110]. Cyclic peptides have been used as synthetic immunogens [111], transmembrane ion channels [112], potent vaccine for diabetes [113], antigens for Herpes Simplex Virus [114], inhibitor against alpha-amylase [115], pancreatic trip-

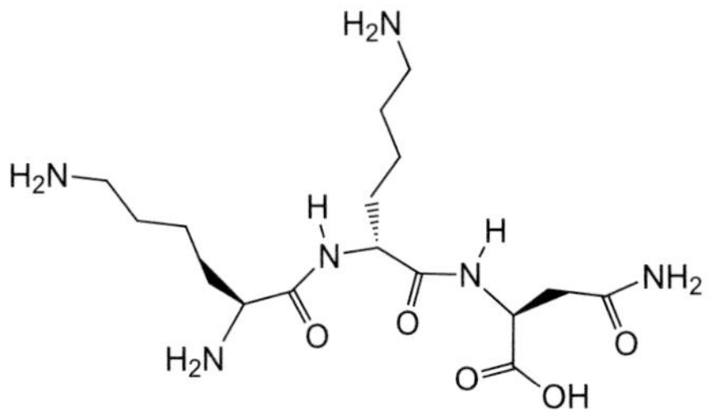

Fig. 4.13. The structure of tripeptide **(1)** and cyclic peptide **(2)**. Reprinted from Tamilarasu *et al.* [127].

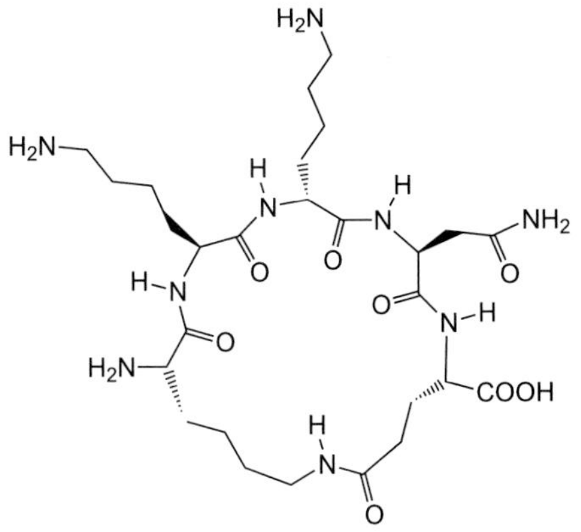

sin [116], integrin $\alpha_v\beta_3$ [117], and as protein stabilizer [118]. Side chain to side chain lactamization has been utilized to improve receptor selectivity in enkephalins [119], cholecystokinin [120], melanotropin [121], tachykinin [122], RGD-dependent adhesion proteins [123], and many other biological systems [124, 125]. If designed carefully without causing drastic changes in the conformation of active peptides, the rigid geometry of the cyclic peptides enhances the binding affinity towards a selected target molecule compared to their linear counterparts.

To improve pharmacokinetic properties of the ligand **1** described above, we planned to synthesize a cyclic peptide derivative (Fig. 4.13) based on the ligand **1** structure. In general, homodetic cyclic peptides are made by head-to-tail, N-terminal to side chain or side chain to side chain coupling methods on solid supports [126]. Our modeling studies suggested that the cyclic peptide derived from side chain to side chain coupling method would least affect the conformation of our linear ligand **1**, therefore, we designed a synthetic method based on this strategy [127]. This cyclic peptide inhibited transcriptional activation by Tat protein in human cells with an IC$_{50}$ of approximately 40 nM. Cyclic peptides that can target specific RNA structures provide a new class of small molecules that can be used to control cellular processes involving RNA–protein interactions *in vivo*.

4.8
Summary and Perspective

Tat and Rev proteins and other auxiliary factors intricately control regulation of HIV-1 gene expression. The relevance of RNA structure and RNA–protein interactions to the regulation of gene expression suggests the design of drugs that specifically target regulatory RNA sequences. There has been a considerable effort to discover specific inhibitors of Tat–TAR and Rev–RRE interactions over the past few years and a number of high-affinity RNA ligands have been identified. Combinatorial chemical synthesis is one of the most promising approaches to the synthesis of large collections of diverse molecules. The use of encoded combinatorial chemistry to screen ligands for specific RNA targets would allow the discovery of novel molecules containing diverse structures. To avoid the problem of rapid hydrolysis by host enzymes, peptides and drug molecules containing unnatural linkages can be designed. Structural and activity analysis of the lead molecules will provide new insights into designing drugs with improved properties. The discovery of highly selective and cell permeable inhibitors of RNA–protein interactions would greatly assist us in understanding the functional significance of RNA–protein interactions *in vivo*.

An important consideration in identification of inhibitors of RNA–protein interactions, which is often overlooked, is that the inhibitor does not always have to bind RNA at the binding site of the protein. Although it is true that flexibility of the RNA structure makes it difficult to design molecules for specific RNA sequences, however, this flexibility of RNA folding can be exploited to lock an RNA into a non-functional structure. For example, a TAR ligand can bind the sequences below or above the bulge, or loop region and cause conformational change in TAR RNA that is not recognized by Tat and cellular proteins involved in transcriptional activation. Similarly, TAR bulge-binding ligands can be identified that interact with the RNA in a manner different from that of Tat. Locking RNA into a non-functional structure could also be kinetically more favorable than a ligand–RNA interaction that competes with protein for the same site.

With the increasing wealth of knowledge being generated in the field of RNA–protein recognition and identification of new RNA targets, it would be exciting to see how chemists and biologists use this information to discover drugs that target specific RNA structures and manipulate RNA–protein interactions to control biological processes.

Acknowledgments

This work was supported by grants from the National Institutes of Health (AI 41404 and AI 45466).

References

1 JONES, K. A. Taking a new TAK on Tat transactivation. *Genes Dev.* **1997**, *11*, 2593–2599.

2 CULLEN, B. R. HIV-1 auxiliary proteins: making connections in a dying cell. *Cell* **1998**, *93*, 685–692.

3 EMERMAN, M., MALIM, M. HIV-1 regulatory/accessory genes: keys to unraveling vial and host cell biology. *Science* **1998**, *280*, 1880–1884.

4 JEANG, K.-T., XIAO, H., RICH, E. A. Multifaceted activities of the HIV-1 transactivator of transcription, Tat. *J. Biol. Chem.* **1999**, *274*, 28837–28840.

5 KARN, J. TACKLING TAT. *J. Mol. Biol.* **1999**, *293*, 235–254.

6 TAUBE, R., FUJINAGA, K., WIMMER, J., BARBORIC, M., PETERLIN, B. M. Tat transactivation: a model for the regulation of eukaryotic transcriptional elongation. *Virology* **1999**, *264*, 245–253.

7 KUPPUSWAMY, M., SUBRAMANIAN, T., SRINIVASAN, A., CHINNADURAI, G. Multiple functional domains of Tat, the *trans*-activator of HIV-1, defined by mutational analysis. *Nucleic Acids Res.* **1989**, *17*, 3551–3561.

8 MADORE, S. J., CULLEN, B. R. Genetic analysis of the cofactor requirement for human immunodeficiency virus type 1 Tat function. *J. Virol.* **1993**, *67*, 3703–3711.

9 HERRMANN, C., RICE, A. Specific interaction of the human immunodeficiency virus Tat protein with a cellular protein kinase. *Virology* **1993**, *197*, 601–608.

10 HERRMANN, C. H., RICE, A. P. Lentivirus Tat proteins specifically associate with a cellular protein kinase, TAK, that hyperphosphorylates the carboxyl-terminal domain of the large subunit of RNA polymerase II: candidate for a Tat cofactor. *J. Virol.* **1995**, *69*, 1612–1620.

11 ZHU, Y., PE'ERY, T., PENG, J. *et al.* Transcription elongation factor P-TEFb is required for HIV-1 Tat transactivation in vitro. *Genes Dev.* **1997**, *11*, 2622–2632.

12 MANCEBO, H. S. Y., LEE, G., FLYGARE, J. *et al.* P-TEFb kinase is required for HIV Tat transcriptional activation in vivo and in vitro. *Genes Dev.* **1997**, *11*, 2633–2644.

13 BIENIASZ, P. D., GRDINA, T. A., BOGERD, H. P., CULLEN, B. R. Recruitment of cyclin T1/P-TEFb to an HIV type 1 long terminal repeat promoter proximal RNA target is both necessary and sufficient for full activation of transcription. *Proc. Natl Acad. Sci. USA* **1999**, *96*, 7791–7796.

14 WEI, P., GARBER, M. E., FANG, S.-M., FISCHER, W. H., JONES, K. A. A novel CDK9-associated C-type cyclin interacts directly with HIV-1 Tat and mediates its high-affinity, loop specific binding to TAR RNA. *Cell* **1998**, *92*, 451–462.

15 GARBER, M. E., WEI, P., JONES, K. A. HIV-1 Tat interacts with cyclin T1 to direct the P-TEFb CTD kinase complex to TAR RNA. *Cold Spring Harbor Symp. Quant. Biol.* **1998**, *63*, 371–380.

16 GARBER, M. E., WEI, P., KEWALRAMANI, V. N. *et al.* The interaction between HIV-1 Tat and human cyclin T1 requires zinc and a critical cysteine residue that is not conserved in the murine CycT1 protein. *Genes Dev.* **1998**, *12*, 3512–3527.

17 FUJINAGA, K., CUJEC, T., PENG, J. *et al.* The ability of positive transcription elongation factor b to transactivate human immunodeficiency virus transcription depends on a functional kinase domain, cyclin T1 and Tat. *J. Virol.* **1998**, *72*, 7154–7159.

18 BIENIASZ, P. D., GRDINA, T. A., BOGERD, H. P., CULLEN, B. R. Recruitment of a protein complex containing Tat and cyclin T1 to TAR governs the species specificity of HIV-1 Tat. *EMBO J.* **1998**, *17*, 7056–7065.

19 IVANOV, D., KWAK, Y. T., NEE, E., GUO, J., GARCIA-MARTINEZ, L. F., GAYNOR, R. B. Cyclin T1 domains involved in complex formation with Tat and TAR RNA are critical for tat-activation. *J. Mol. Biol.* **1999**, *288*, 41–56.

20 BIENIASZ, P. D., GRDINA, T. A., BOGERD, H. P., CULLEN, B. R. Analysis of the effect of natural sequence variation in Tat and in cyclin T on the formation and RNA binding properties of Tat-cyclin T complexes. *J. Virol.* **1999**, *73*, 5777–5786.

21 WIMMER, J., FUJINAGA, K., TAUBE, R. *et al.* Interactions between Tat and TAR and human immunodeficiency virus replication are facilitated by human cyclin T1 but not cyclin T2a or T2b. *Virology* **1999**, *255*, 182–189.

22 FUJINAGA, K., TAUBE, R., WIMMER, J., CUJEC, T., PETERLIN, B. Interactions between human cyclin T, Tat, and the transactivation response element (TAR) are disrupted by a cysteine to tyrosine substitution found in mouse cyclin T. *Proc. Natl Acad. Sci. USA* **1999**, *96*, 1285–1290.

23 CHEN, D., FONG, Y., ZHOU, Q. Specific interaction of Tat with the human but not rodent P-TEFb complex mediates the species-specific Tat activation of HIV-1 transcription. *Proc. Natl Acad. Sci. USA* **1999**, *96*, 2728–2733.

24 ZHANG, J., TAMILARASU, N., HWANG, S. *et al.* HIV-1 TAR RNA enhances the interaction between tat and cyclin T1. *J. Biol. Chem.* **2000**, *275*, 34314–34319.

25 SELBY, M. J., PETERLIN, B. M. Trans-activation by HIV-1 Tat via a heterologous RNA binding protein. *Cell* **1990**, *62*, 769–776.

26 DINGWALL, C., ERNBERG, I., GAIT, M. J. *et al.* HIV-1 Tat protein stimulates transcription by binding to the stem of the TAR RNA Structure. *EMBO J.* **1990**, *9*, 4145–4153.

27 BERKHOUT, B., SILVERMAN, R. H., JEANG, K. T. Tat trans-activates the Human Immunodeficiency Virus through a nascent RNA target. *Cell* **1989**, *59*, 273–282.

28 ROSEN, C. A., SODROSKI, J. G., HASELTINE, W. A. Location of Cis-acting regulatory sequences in the Human T cell Lymphotropic Virus type III (HTLV-III/LAV) long terminal repeat. *Cell* **1985**, *41*, 813–823.

29 JAKOBOVITS, A., SMITH, D. H., JAKOBOVITS, E. B., CAPON, D. J. A

Discrete element 3 of Human Immunodeficiency Virus 1 (HIV-1) and HIV-2 mRNA initiation sites mediates transcriptional activation by an HIV trans activator. *Mol. Cell. Biol.* **1988**, *8*, 2555–2561.

30 BERKHOUT, B., JEANG, K.-T. Trans activation of human immuno-deficiency virus type 1 is sequence specific for both the single-stranded bulge and loop of the trans-acting-responsive hairpin: a quantitative analysis. *J. Virol.* **1989**, *63*, 5501–5504.

31 CORDINGLEY, M. G., LA FEMINA, R. L., CALLAHAN, P. L. *et al.* Sequence-specific interaction of Tat protein and Tat peptides with the transactivation-responsive sequence element of Human Immunodeficiency Virus type 1 in vitro. *Proc. Natl Acad. Sci. USA* **1990**, *87*, 8985–8989.

32 WEEKS, K. M., CROTHERS, D. M. RNA recognition by Tat-derived peptides: interaction in the major groove? *Cell* **1991**, *66*, 577–588.

33 CALNAN, B. J., BIANCALANA, S., HUDSON, D., FRANKEL, A. D. Analysis of arginine-rich peptides from the HIV Tat protein reveals unusual features of RNA protein recognition. *Genes Dev.* **1991**, *5*, 201–210.

34 CHURCHER, M. J., LAMONT, C., HAMY, F. *et al.* High affinity binding of TAR RNA by the Human Immuno-deficiency Virus Type-1 *tat* protein requires base-pairs in the RNA stem and amino acid residues flanking the base region. *J. Mol. Biol.* **1993**, *230*, 90–110.

35 BERKHOUT, B., JEANG, K.-T. Detailed mutational analysis of TAR RNA: critical spacing between the bulge and loop recognition domains. *Nucleic Acids Res.* **1991**, *19*, 6169–6176.

36 SUMNER-SMITH, M., ROY, S., BARNETT, R. *et al.* Critical chemical features in *trans*-acting-responsive RNA are required for interaction with human immunodeficiency virus tye 1 tat protein. *J. Virol.* **1991**, *65*, 5196–5202.

37 HAMY, F., ASSELINE, U., GRASBY, J. *et al.* Hydrogen-bonding contacts in the major groove are required for Human Immunodeficiency Virus Type-1 *tat*

protein recognition of TAR RNA. *J. Mol. Biol.* **1993**, *230*, 111–123.

38 NARYSHKIN, N. A., FARROW, M. A., IVANOVSKAYA, M. G., ORETSKAYA, T. S., SHABAROVA, Z. A., GAIT, M. J. Chemical cross-linking of the human immunodeficiency virus type 1 Tat protein to synthetic models of the RNA recognition sequence TAR containing site-specific trisubstituted pyrophosphate analogues. *Biochemistry* **1997**, *36*, 3496–3505.

39 WANG, Z., RANA, T. M. RNA conformation in the Tat-TAR complex determined by site-specific photo-cross-linking. *Biochemistry* **1996**, *35*, 6491–6499.

40 WANG, Z., RANA, T. M. RNA-protein interactions in the Tat-*trans*-activation response element complex determined by site-specific photo-cross-linking. *Biochemistry* **1998**, *37*, 4235–4243.

41 LONG, K. S., CROTHERS, D. M. Interaction of human immunodeficiency virus type 1 Tat-derived peptides with TAR RNA. *Biochemistry* **1995**, *34*, 8885–8895.

42 SUBRAMANIAN, T., GOVINDARAJAN, R., CHINNADURAI, G. Heterologous basic domain substitutions in the HIV-1 Tat protein reveal an arginine-rich motif required for transactivation. *EMBO J.* **1991**, *10*, 2311–2318.

43 MILLER, P. S. Development of antisense and antigene oligonucleotide analogs. *Prog Nucleic Acid Res. Mol. Biol.* **1996**, *52*, 261–291.

44 BEAL, P. A., DERVAN, P. B. Second structural motif for recognition of DNA by oligonucleotide-directed triplhelix formation. *Science* **1991**, *251*, 1360–1363.

45 MAHER, L. J. D., WOLD, B., DERVAN, P. B. Oligonucleotide-directed DNA triple-helix formation: an approach to artificial repressors? *Antisense Res. Dev.* **1991**, *1*, 277–281.

46 HELENE, C., THUONG, N. T., HAREL-BELLAN, A. Control of gene expression by triple helix-forming oligonucleotides. The antigene strategy. *Ann. NY Acad. Sci.* **1992**, *660*, 27–36.

47 NIELSEN, P. E. Applications of peptide nucleic acids. *Curr. Opin. Biotechnol.* **1999**, *10*, 71–75.

48 GOTTESFELD, J. M., NEELY, L., TRAUGER, J. W., BAIRD, E. E., DERVAN, P. B. Regulation of gene expression by small molecules. *Nature* **1997**, *387*, 202–205.

49 WHITE, S., SZEWCZYK, J. W., TURNER, J. M., BAIRD, E. E., DERVAN, P. B. Recognition of the four Watson-Crick base pairs in the DNA minor groove by synthetic ligands. *Nature* **1998**, *391*, 468–471.

50 HO, S. N., BOYER, S. H., SCHREIBER, S. L., DANISHEFSKY, S. J., CRABTREE, G. R. Specific inhibition of formation of transcription complexes by a calicheamicin oligosaccharide: a paradigm for the development of transcriptional antagonists. *Proc. Natl Acad. Sci. USA* **1994**, *91*, 9203–9207.

51 LIU, C., SMITH, B. M., AJITO, K. *et al.* Sequence-selective carbohydrate-DNA interaction: dimeric and monomeric forms of the calicheamicin oligosaccharide interfere with transcription factor function. *Proc. Natl Acad. Sci. USA* **1996**, *93*, 940–944.

52 CHASTAIN, M., TINOCO, I., JR. Structural elements in RNA. *Prog. Nucleic Acid Res. Mol. Biol.* **1991**, *41*, 131–177.

53 CHOW, C. S., BOGDAN, F. M. A structural basis for RNA-ligand interactions. *Chem. Rev.* **1997**, *97*, 1489–1514.

54 HAMY, F., FELDER, E., HEIZMANN, G. *et al.* An inhibitor of the TAT/TAR RNA interaction that effectively suppresses HIV-1 replication. *Proc. Natl Acad. Sci. USA* **1997**, *94*, 3548–3553.

55 ABOUL-ELA, F., KARN, J., VARANI, G. The structure of the human immunodeficiency virus type-1 TAR RNA reveals principles of RNA recognition by Tat protein. *J. Mol. Biol.* **1995**, *253*, 313–332.

56 HAMY, F., BRONDANI, V., FLORSHEIMER, A., STARK, W., BLOMMERS, M. J. J., KLIMKAIT, T. A new class of HIV-1 Tat antogonist acting through Tat-TAR inhibition. *Biochemistry* **1998**, *37*, 5086–5095.

57 MEI, H.-Y., CUI, M., HELDSINGER, A. *et al.* Inhibitors of protein-RNA complexation that target the RNA: specific recognition of human immunodeficiency virus type I TAR TNA by small organic molecules. *Biochemistry* **1998**, *37*, 14204–14212.

58 MEI, H., MACK, D., GALAN, A. *et al.* Discovery of selective, small-molecule inhibitors of RNA complexes – I. The Tat protein/TAR RNA complexes required for HIV-1 transcription. *Bioorganic Med. Chem.* **1997**, *5*, 1173–1184.

59 WANG, X., HUQ, I., RANA, T. M. HIV-1 TAR RNA recognition by an unnatural biopolymer. *J. Am. Chem. Soc.* **1997**, *119*, 6444–6445.

60 TAMILARASU, N., HUQ, I., RANA, T. M. High affinity and specific binding of HIV-1 TAR RNA by a Tat-derived oligourea. *J. Am. Chem. Soc.* **1999**, *121*, 1597–1598.

61 CHO, C. Y., MORAN, E. J., CHERRY, S. R. *et al.* An unnatural biopolymer. *Science* **1993**, *261*, 1303–1305.

62 KIM, J. M., BI, Y. Z., PAIKOFF, S. J., SCHULTZ, P. G. The solid phase synthesis of oligoureas. *Tetrahedron Lett.* **1996**, *37*, 5305–5308.

63 KICK, E., ELLMAN, J. Expedient method for the solid-phase synthesis of aspartic acid protease inhibitors directed toward the generation of libraries. *J. Med. Chem.* **1995**, *38*, 1427–1430.

64 DINTZIS, H. M., SYMER, D. E., DINTZIS, R. Z., ZAWADZKE, L. E., BERG, J. M. A comparison of the immunogenicity of a pair of enantiomeric proteins. *Proteins* **1993**, *16*, 306–308.

65 FRIED, M., CROTHERS, D. M. Equilibria and kinetics of lac repressor-operator interactions by polyacrylamide gel electrophoresis. *Nucleic Acids Res.* **1981**, *9*, 6505–6525.

66 WANG, Z., RANA, T. M. Chemical Conversion of a TAR RNA-binding fragment of HIV-1 Tat protein into a site-specific crosslinking agent. *J. Am. Chem. Soc.* **1995**, *117*, 5438–5444.

67 HUQ, I., WANG, X., RANA, T. M. Specific recognition of HIV-1 TAR RNA by a D-Tat peptide. *Nature Struct. Biol.* **1997**, *4*, 881–882.

68 WANG, Z., RANA, T. M. DNA damage-dependent transcriptional arrest and termination of RNA polymerase II elongation complexes in DNA template containing HIV-1 promoter. *Proc. Natl Acad. Sci. USA* **1997**, *94*, 6688–6693.

69 HUQ, I., PING, Y.-H., TAMILARASU, N., RANA, T. M. Controlling human immunodeficiency virus type 1 gene expression by unnatural peptides. *Biochemistry* **1999**, *38*, 5172–5177.

70 VIVES, E., BRODEN, P., LEBLEU, B. A truncated HIV-1 Tat protein basic domain rapidly translocates through the plasma membrane and accumulates in the cell. *J. Biol. Chem.* **1997**, *272*, 16010–16017.

71 FRANKEL, A. D., PABO, C. O. Cellular uptake of the tat protein from human immunodeficiency virus. *Cell* **1988**, *55*, 1189–1194.

72 NORDEEN, S. K. Luciferase reporter gene vectors for analysis of promoters and enhancers. *Biotechniques* **1988**, *6*, 454–457.

73 FELBER, B. K., PAVLAKIS, G. N. A quantitative bioassay for HIV-1 based *trans*-activation. *Science* **1988**, *239*, 184–187.

74 STILL, W. C. Discovery of sequence-selective peptide binding by synthetic receptors using encoded combinatorial libraries. *Acc. Chem. Res.* **1996**, *29*, 155–163.

75 ARMSTRONG, R., COMBS, A., TEMPEST, P., BROWN, S., KEATING, T. Multiple-component condensation strategies for combinatorial library synthesis. *Acc. Chem. Res.* **1996**, *29*, 123–131.

76 BROWN, D. Future pathway for combinatorial chemistry. *Mol. Diversity* **1996**, *2*, 217–222.

77 DEWITT, S., CZARNIK, A. Combinatorial organic synthesis using Parke-Davis's DIVERSOMER method. *Acc. Chem. Res.* **1996**, 114–122.

78 ELLMAN, J. Design, synthesis, and evaluation of small-molecule libraries. *Acc. Chem. Res.* **1996**, *29*, 132–143.

79 WENTWORTH JR, P., JANDA, K. Generating and analyzing combinatorial chemistry libraries. *Curr. Opin. Biotechnol.* **1998**, *9*, 109–115.

80 LAM, P., JADHAV, P., EYERMANN, C. *et al.* Rational design of potent, bioavailable, nonpeptide cyclic ureas as HIV protease inhibitors. *Science* **1994**, *263*, 380–384.

81 LI, J., MURRAY, C., WASZKOWYCZ, B., YOUNG, S. Targeted molecular deversity in drug discovery: integration of structure-based design and combinatorial chemistry. *Drug Discovery Today* **1998**, *3*, 105–112.

82 NORMAN, T., GRAY, N., KOH, J., SCHULTZ, P. A structure-based library approach to kinase inhibitors. *J. Am. Chem. Soc.* **1996**, *118*, 7430–7431.

83 GRAY, N. S., WODICKA, L., THUNNISSEN, A.-M. W. H. *et al.* Exploiting chemical libraries, structure, and genomics in the search of kinase inhibitors. *Science* **1998**, *281*, 533–538.

84 LAM, K., SROKA, T., CHEN, M., ZHAO, Y., LOU, Q., WU, J., ZHAO, Z. Application of "one-bead one-compound" combinatorial library methods in signal trasduction research. *Life Sci.* **1998**, *62*, 1577–1583.

85 KICK, E., ROE, D., SKILLMAN, A. *et al.* Structure-based design and combinatorial chemistry yield low nanomolar inhibitors of cathepsin D. *Chem. Biol.* **1997**, *4*, 297–307.

86 FENG, S., CHEN, J., YU, H., SIMON, J., SCHREIBER, S. Two binding orientations for peptides to the Src SH3 domain: development of a general model for SH3-ligand interactions. *Science* **1994**, *266*, 1241–1247.

87 KAPOOR, T., ANDREOTTI, A., SCHREIBER, S. Exploring the specificity pockets of two homologous SH3 domains using structure-based, split-pool synthesis and affinity-based selection. *J. Am. Chem. Soc.* **1998**, *120*, 23–29.

88 MORKEN, J., KAPOOR, T., FENG, S., SHIRAI, F., SCHREIBER, S. Exploring the leucine-proline binding pocket of the src SH3 domain using structure-based, split-pool synthesis and affinity-based selection. *J. Am. Chem. Soc.* **1998**, *120*, 30–36.

89 GRAVERT, D., JANDA, K. Organic synthesis on soluble polymer supports: liquid-phase methodologies. *Chem. Rev.* **1997**, *97*, 489–509.

90 HERMKENS, P., OTTENHEIJM, H., REES, D. Solid-phase organic reactions II: a review of the literature Nov 95–Nov 96. *Tetahedron* **1997**, *53*, 5643–5678.

91 CORTESE, R. *Combinatorial Libraries: Synthesis, Screening and Application*, Walter de Gruyter, New York, 1996.

92 OSBORNE, S., ELLINGTON, A. Nucleic acid selection and the challenge of combinatorial chemistry. *Chem. Rev.* **1997**, *97*, 349–370.

93 PURAS LUTZKE, R., EPPENS, N., WEBER, P., HOUGHTEN, R., PLASTERK, R. Identification of a hexapeptide inhibitor of the human immuno-deficiency virus integrase protein by using a combinatorial chemical library. *Proc. Natl Acad. Sci. USA* **1995**, *92*, 11456–11460.

94 WILSON-LINGARDO, L., DAVIS, P., ECKER, D. *et al.* Deconvolution of combinatorial libraries for drug discovery: experimental comparison of pooling strategies. *J. Med. Chem.* **1996**, *39*, 2720–2726.

95 LAM, K., SALMON, S., HERSH, E., HRUBY, V., KAZMIERSKI, W., KNAPP, R. A new type of synthetic peptide library for identifying ligand-binding activity. *Nature* **1991**, *354*, 82–84.

96 YOUNGQUIST, R., FUENTES, G., LACEY, M., KEOUGH, T. Generation and screening of combinatorial peptide libraries designed for rapid sequencing by mass spectrometry. *J. Am. Chem. Soc.* **1995**, *117*, 3900–3906.

97 NI, Z., MACLEAN, D., HOLMES, C., MURPHY, M., RUHLAND, B., JACOBS, J., GORDON, E., GALLOP, M. Versatile approach to encoding combinatorial organic syntheses using chemically robust secondary amine tags. *J. Med. Chem.* **1996**, *1996*, 1601–1608.

98 O'DONNELL, M. J., ZHOU, C., SCOTT, W. L. Solid-phase unnatural peptide synthesis (UPS). *J. Am. Chem. Soc.* **1996**, *118*, 6070–6071.

99 CZARNIK, A. Encoding strategies in combinatorial chemistry. *Proc. Natl Acad. Sci. USA* **1997**, *94*, 12738–12739.

100 FURKA, A., SEBESTYEN, F., ASGEDOM, M., DIBO, G. General method for rapid synthesis of multicomponent peptide mixtures. *Int. J. Pept. Prot. Res.* **1991**, *37*, 487–493.

101 BRENNER, S., LERNER, R. A. Encoded combinatorial chemistry. *Proc. Natl Acad. Sci. USA* **1992**, *89*, 5381–5183.

102 LAM, K., LEBL, M., KRCHNAK, V. The one-bead-one compound combinatorial library method. *Chem. Rev.* **1997**, *97*, 411–448.

103 MACLEAN, D., SCHULLEK, J., MURPHY, M., NI, Z., GORDON, E., GALLOP, M. Encoded combinatorial chemistry: synthesis and screening of a library of highly functionalized pyrrolidines. *Proc. Natl Acad. Sci. USA* **1997**, *94*, 2805–2810.

104 JANDA, K. D. Tagged versus untagged libraries: Methods for the generation and screening of combinatorial chemical libraries. *Proc. Natl Acad. Sci. USA* **1994**, *91*, 10779–10785.

105 OHLMEYER, M. H. J., SWANSON, R. N., DILLARD, L. W. *et al.* Complex synthetic chemical libraries indexed with molecular tags. *Proc. Natl Acad. Sci. USA* **1993**, *90*, 10922–10926.

106 HWANG, S., TAMILARASU, N., RYAN, K. *et al.* Inhibition of gene expression in human cells through small molecule-RNA interactions. *Proc. Natl Acad. Sci. USA* **1999**, *96*, 12997–13002.

107 PUGLISI, J. D., TAN, R., CALNAN, B. J., FRANKEL, A. D., WILLIAMSON, J. R. Conformation of the TAR RNA-arginine complex by NMR spectroscopy. *Science* **1992**, *257*, 76–80.

108 SCOTT, C. P., ABEL-SANTOS, E., WALL, M., WAHNON, D. C., BENKOVIC, S. J. Production of cyclic peptides and proteins in vivo. *Proc. Natl Acad. Sci. USA* **1999**, *96*, 13638–13643.

109 OLIGINO, L., LUNG, F.-D. T., SASTRY, L. *et al.* Nonphosphorylated peptide ligands for the Grb2 Src homology 2 domain. *J. Biol. Chem.* **1997**, *272*, 29046–29052.

110 GUDMUNDSSON, O. S., NIMKAR, K., GANGWAR, S., SIAHAAN, T., BORCHARDT, R. T. Phenylpropionic acid-based cyclic prodrugs of opioid peptides that exhibit metabolic stability to peptidases and excellent cellular permeation. *Pharm. Res.* **1999**, *16*, 16–23.

111 BRUGGHE, H. F., TIMMERMANS, H. A. M., VAN UNEN, L. M. A. *et al.* Simultaneous multiple synthesis and selective conjugation of cyclized peptides derived from a surface loop of a meningococcal class 1 outer membrane protein. *Int. J. Pept. Protein Res.* **1994**, *43*, 166–172.

112 CHALOIN, L., MERY, J., VAN MAU, N., DIVITA, G., HEITZ, F. Synthesis of a template-associated peptide designed as a transmembrane ion channel former. *J. Pept. Sci.* **1999**, *5*, 381–391.

113 BEREZHKOVSKIY, L., PHAM, S., REICH, E.-P., DESHPANDE, S. Synthesis and kinetics of cyclization of MHC class II-derived cyclic peptide vaccine for diabetes. *J. Pept. Res.* **1999**, *54*, 112–119.

114 MEZO, G., MAJER, Z., VALERO, M. L., ANDREU, D., HUDECZ, F. Synthesis of cyclic Herpes simplex virus peptides containing 281–284 epitope of glycoprotein D-1 in endo- or exo-position. *J. Pept. Sci.* **1999**, *5*, 272–282.

115 ONO, S., HIRANO, T., YASUTAKE, H. *et al.* Biological and structural properties of cyclic peptides derived from the alpha-amylase inhibitor tendamistat. *Biosci. Biotechnol. Biochem.* **1998**, *62*, 1621–1623.

116 KASHER, R., OREN, D. A., BARDA, Y., GILON, C. Miniaturized proteins: the backbone cyclic proteinomimetic approach. *J. Mol. Biol.* **1999**, *292*, 421–429.

117 HAUBNER, R., FINSINGER, D., KESSLER, H. Stereoisomeric peptide libraries and peptidomimetics for designing selective inhibitors of the alpha(v)beta3 integrin for a new cancer therapy. *Angew. Chem. Int. Ed. Engl.* **1997**, *36*, 1374–1389.

118 IWAI, H., PLUCKTHUN, A. Circular-beta-lactamase: stability enhancement

by cyclizing the backbone. *FEBS Lett.* **1999**, *459*, 166–172.

119 SCHILLER, P. W., NGUYEN, T. M.-D., LEMIEUX, C., MAZIAK, L. A. Synthesis and activity profiles of novel cyclic opioid peptide monomers and dimers. *J. Med. Chem.* **1985**, *28.*

120 CHARPENTIER, B., DOR, A., ROY, P. *et al.* Synthesis and binding affinities of cyclic and related linear analogues of CCK8 selective for central receptors. *J. Med. Chem.* **1989**, *32*, 1184–1190.

121 AL-OBEIDI, F., CASTRUCCI, A. M. D. L., HADLEY, M. E., HRUBY, V. J. Potent and prolonged acting cyclic lactam analogues of alpha-melanotropin: design based on molecular dynamics. *J. Med. Chem.* **1989**, *32.*

122 HOLZEMANN, G., JONCZYK, A., EIERMANN, V., PACHLER, K. G. R., BARNICKEL, G., REGOLI, D. Conformation-based design of two cyclic physalaemin analogues. *Biopolymers* **1991**, *31*, 691–697.

123 PEISHOFF, C. E., ALI, F. E., BEAN, J. W. *et al.* Investigation of conformational specificity at GPIIB/IIIA: evaluation of conformationally constrained RGD peptides. *J. Med. Chem.* **1992**, *35*, 3962–3969.

124 LIN, M., CHAN, M. F., BALAJI, V. N., CASTILLO, R. S., LARIVE, C. K. Synthesis and conformational analysis of cyclic pentapeptide endothelin antagonists. *Int. J. Pept. Protein Res.* **1996**, *48*, 229–239.

125 MORIKIS, D., ASSA-MUNT, N., SAHU, A., LAMBRIS, J. D. Solution structure of Compstatin, a potent complement inhibitor. *Protein Sci.* **1998**, *7*, 619–627.

126 BLACKBURN, C., KATES, S. A. Solid-phase synthesis of cyclic peptides. *Methods Enzymol.* **1997**, *289*, 175–198.

127 TAMILARASU, N., HUQ, I., RANA, T. M. Design, synthesis, and biological activity of a cyclic peptide: an inhibitor of HIV-1 Tat-TAR interactions in human cells. *Bioorganic Med. Chem. Lett.* **2000**, *10*, 971–974.

128 PING, Y.-H., RANA, T. M. Tat-associated kinase (P-TEFb): a component of transcription preinitiation and elongation complexes. *J. Biol. Chem.* **1999**, *274*, 7399–7404.

129 PING, Y.-H., RANA, T. M. DSIF and NELF interact with RNA Pol II elongation complex and HIV-1 Tat stimulates P-TEFb-mediated phosphorylation of RNA Polymerase II and DSIF during transcription elongation. *J. Biol. Chem.* **2001**, *276*, 12951–12958.

130 WANG, Z., WANG, X., RANA, T. M. Protein orientation in the Tat-TAR complex determined by psoralen photocross-linking. *J. Biol. Chem.* **1996**, *271*, 16995–16998.

131 HUQ, I., TAMILARASU, N., RANA, T. M. Visualizing tertiary folding of RNA and RNA-protein interactions by a tethered iron chelate: analysis of HIV-1 Tat-TAR complex. *Nucleic Acids Res.* **1999**, *27*, 1084–1093.

132 BAYER, P., KRAFT, M., EJCHART, A., WESTENDORP, M., FRANK, R., RÖSCH, P. Structural studies of HIV-1 Tat protein. *J. Mol. Biol.* **1995**, *247*, 529–535.

5

DNA and RNA Recognition and Modification by Gly-Gly-His-Derived Metallopeptides

Eric C. Long and Craig A. Claussen

5.1
Introduction

5.1.1
General Considerations

Despite their central position in the evolution of DNA- and RNA-binding proteins and certain natural products, peptides containing the 20 ribosomally translated amino acids represent a relatively untapped resource with respect to the design of low-molecular-weight nucleic acid-binding molecules. This status is due, in part, to the general lack of defined three-dimensional structure within most short, 3–5 amino acid peptides in aqueous solution [1]. The above consequence is unfortunate given the molecular-recognition capabilities of amino acids. Peptides contain guanidinium, amide, amine, planar hydrophobic moieties, and a host of other side chain functional groups that can be harnessed to interact with a nucleic acid structure through hydrogen bonding, electrostatics, hydrophobic, and van der Waals forces [2]. Indeed, this potential was recognized in previous studies of amino acid– and peptide–DNA association phenomena [3–7].

Through the above interactions, and in a fashion complementary to the DNA-binding properties of distamycin, netropsin [8–10], and linear or hairpin poly-amides [11–14], designed peptides of the appropriate three-dimensional architecture theoretically have the wherewithal to recognize DNA or RNA via not only their nucleotide sequences, but also their groove structures, stacked Watson–Crick base pairs, phosphodiester backbone or some combination of these features. Importantly, the potential for an appropriately structured peptide to target a nucleic acid may lead to strategies to augment the activity of existing DNA- or RNA-binding agents or to develop lead compounds resulting in wholly organic structures with biological properties. Peptides have a distinct advantage in the development of nucleic acid-binding compounds through their relative ease of individual or combinatorial synthesis [15–17], the ability to alter readily the chirality of select α-carbon stereo-centers, and their potential transformation into peptidomimetic [18] or biosynthetic agents. Overall, the key to exploiting amino acids and peptides in the design of low-molecular-weight nucleic acid-binding molecules appears to be the inclusion of strategies which enable a peptide to be structured appropriately in solution and to also contain a judicious choice of amino acid side chain functional groups.

5.1.2
Metallopeptides in the Study of Nucleic Acid Recognition

With the goal of exploring the nucleic acid recognition capabilities of peptides containing naturally occurring amino acids, ongoing efforts from this laboratory employ transition metal ions to generate low-molecular-weight peptides with well-defined three-dimensional architectures [19, 20]. In a fashion that is analogous to other metal complexes that often display selective DNA [21–26] and RNA [25, 27] binding and, at times, antitumor properties [28–34], the use of transition metal ions in the generation of structured, nucleic acid-binding peptides provides several distinct advantages including: (1) the ability to construct complexes with well-defined geometries and (2) the inclusion of metal-centered redox properties enabling the final complex to mark a location of binding through an act of DNA or RNA strand scission [21, 22]. Thus, similar to the zinc-finger strategy that has evolved naturally in transcription factors [35], metallopeptides create well-defined peptides that present functional groups to a nucleic acid target leading to selective or perhaps specific binding interactions.

As described in the subsequent sections of this chapter, the metal binding properties of simple tripeptides of the form NH_2-Xaa-Xaa-His-$CONH_2$ (where Xaa is an α-amino acid) create a core scaffold upon which to present amino acids and their pendant side chains to a nucleic acid target [20]. Systems of this type permit metal ligation throughout the entire peptide sequence defining rigidly in space the position of all function groups and minimizing structurally undefined regions. The long-term goal of this experimental strategy is to understand, at the level of a low-molecular-weight complex, details concerning preferred amino acid– and peptide–DNA/RNA contacts, groove preferences, and how particular spatial arrangements of amino acid side chains can influence the efficiency and selectivity of nucleic acid binding. In this way, knowledge pertaining to protein– *and* low-molecular-weight agent–nucleic acid recognition principles can be gained and perhaps exploited in future generations of designed DNA- or RNA-targeted agents [36–40].

5.2
Interactions of Gly-Gly-His-Derived Metallopeptides with DNA

5.2.1
Natural Occurrence and Metal-binding Properties

Peptides of the general form NH_2-Xaa-Xaa-His-$CONH_2$, or in their least substituted form, Gly-Gly-His, represent the consensus sequence of the N-terminal

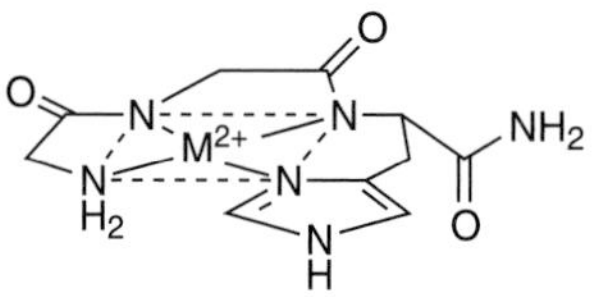

Fig. 5.1. The structure of M(II)·Xaa-Xaa-His metallopeptides.

Cu(II) or Ni(II) chelating domain of the serum albumins [41]. In addition to the albumins, this sequence also appears in neuromedins C and K, human sperm protamine P2a, and the histatins [41]. At or above physiological pH, Gly-Gly-His and other Xaa-Xaa-His tripeptides bind avidly to Cu(II) or Ni(II) with dissociation constants of 10^{-16}–10^{-17} M [41, 42]. Metal ion chelation by the tripeptide occurs (Fig. 5.1) through the terminal amine, two intervening deprotonated amide nitrogens and the histidine imidazole nitrogen, resulting in distorted square planar or square planar complexes with Cu(II) [42] and Ni(II) [43], respectively. While this mode of binding indicates that Cu(II)/Ni(II)·Xaa-Xaa-His complex formation should not be overly influenced by the amino acid side chains found at either the first or second peptide positions, as noted previously [41], some substitutions actually assist in increasing the metal-binding affinity of the core peptide [44].

Importantly, for purposes of nucleic acid recognition, using the peptide bond as a basis for metal ion ligation, as described above, allows the side chain functional groups contained within a chosen peptide to be presented at the periphery of the total complex and available for interaction with a nucleic acid target. Inversion of chirality at individual α-carbon stereocenters also enables key side chain functional groups to be projected from different locations of the core metallopeptide. Thus, like the "metallopeptide" portion of the antitumor agent bleomycin, which is responsible for the 5′-GC and 5′-GT selectivity exhibited by the drug [29, 45], M(II)·Xaa-Xaa-His metallopeptides have the ability to present a well-defined core structure that permits interactions to occur between the amino acid side chains presented by the complex and a nucleic acid target. In addition, these metallopeptides do not form isomers at the metal center, eliminating the need to resolve individual species prior to DNA recognition studies.

5.2.2
Development as a DNA-Cleavage Agent

The study of Xaa-Xaa-His metallopeptides with nucleic acids began with a report of the activity of Gly-Gly-His in the presence of Cu(II) + ascorbate against Ehrlich ascites tumor cells [46] and a further study [47] that demonstrated DNA strand scission by Cu(II)·Gly-Gly-His + ascorbate (Fig. 5.2). DNA strand scission in these cases was attributed to the generation of hydroxyl radicals by the Cu(II) center in a reaction that actually is attenuated by the peptide–ligand [48, 49]. Following these initial reports, the DNA-cleavage activity of Cu(II)·Gly-Gly-His was harnessed in the generation of an N-terminal affinity cleavage appendage to the DNA-binding

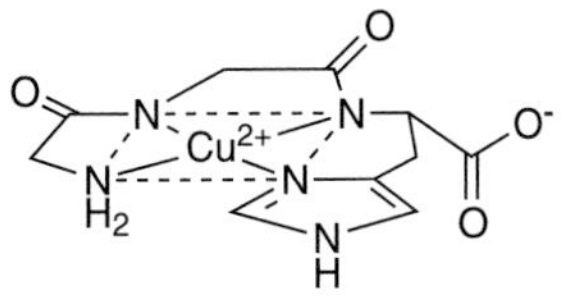

Fig. 5.2. The structure of Cu(II)·Gly-Gly-His containing a carboxylated-terminus.

domain of Hin recombinase [Gly-Gly-His(Hin 139–190), synthesized via modified *t*Boc protocols] (Fig. 5.3). Cu(II)·Gly-Gly-His(Hin 139–190) + sodium ascorbate/hydrogen peroxide [50] and Ni(II)·Gly-Gly-His(Hin 139–190) + magnesium monoperoxyphthalic acid (MMPP), iodosobenzene, or hydrogen peroxide [51, 52] were observed to induce DNA strand scission at the site of Hin recombinase binding, as directed by the Hin 139–190 portion of this conjugate. Observations made with Cu(II)/Ni(II)·Gly-Gly-His(Hin 139–190) indicated that the metallopeptide portion likely resided in the DNA minor groove and generated a non-diffusible oxidant resulting in deoxyribose modification.

Importantly, Ni(II)·Gly-Gly-His(Hin 139–190) was found to be considerably more efficient than the corresponding Cu(II) complex in producing a highly focused oxidation of the sugar–phosphate backbone. Freely diffusing radical intermediates akin to those generated by Fe(II)·EDTA systems [21, 22] were not evident, suggesting the formation of a ligand- or metallopeptide-based radical that permitted the chemistry to be directed to the DNA backbone. In addition, the use of Ni(II) as the activating metal in the presence of exogenous chemical oxidants avoided complicating background DNA damage produced by the radical generating ability of free Cu(II) + ascorbate [47] in the absence of a chelating peptide.

With the report of the above activity, further examples of Cu(II) and Ni(II) complexes of Xaa-Xaa-His-conjugated proteins have been generated with the aim of using the metallotripeptide portion as an affinity cleavage appendage. These systems include N-terminal Gly-Gly-His-modified Sp1 [53] and Gly-Lys-His-modified Fos [54], both of which were generated through biosynthetic protocols. In each of these examples, and Gly-Gly-His(Hin 139–190), the metallo-Gly-Gly-His appendage is passive with respect to DNA recognition and association by the much larger protein portion. Activation of the metallopeptide moiety of the conjugate resulted in DNA cleavage at the protein-binding site. Along with these examples of engineered protein conjugates, studies with a truncated peptide model of human

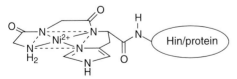

Fig. 5.3. The structural arrangement within Ni(II)·Gly-Gly-His(Hin 139–190) and other N-terminal Gly-Gly-His-modified proteins for DNA affinity cleavage.

protamine P2, a protein that contains an N-terminal Xaa-Xaa-His motif in its native sequence [55], found that Cu(II) or Ni(II) binding actually promoted the association of this peptide mimic with DNA and in the presence of hydrogen peroxide resulted in oxidative DNA damage [56, 57]. These studies suggested a plausible mechanism for the carcinogenicity and toxicity of Ni(II).

In addition to the study of DNA-binding proteins, the Cu(II)/Ni(II)·Xaa-Xaa-His motif has also found utility in the development and examination of other nucleic acid binding and cleaving molecules. Examples include Cu(II)·Gly-Gly-His-modified netropsin [58], low-molecular-weight peptides [59], anthraquinone [60], peptide nucleic acids (PNAs) [61], oligonucleotides [62, 63], and naphthalene diimides [64]. Along with their ability to induce nucleic acid strand scission, Ni(II)·Gly-Gly-His and its conjugates have the wherewithal to carry out directed protein cleavage [65] and to modify and cross-link proteins [66–71].

In light of the utility of Cu(II)/Ni(II)·Gly-Gly-His-containing systems and their ability to generate a non-diffusible oxidant, this laboratory sought early on to extend the capabilities of this motif through the development of a means to place it at any location within a synthetic polypeptide strand [72, 73]. In its native form, the positioning of a Gly-Gly-His motif is limited to the N-termini of natural polypeptide structures due to the required terminal amine functional group as a fourth nitrogen donor. To circumvent this requirement (Fig. 5.4): (1) a synthetic Gly-His-(Xaa)$_n$-resin-bound peptide can be coupled to an orthogonally protected Orn residue (N^δ-Boc-N^α-Fmoc-protected Orn in the case of *t*Boc peptide synthesis protocols) followed by (2) selective deprotection of the Orn side chain and continued solid phase peptide synthesis via the δ-amine of Orn to generate (Xaa)$_n$-(δ)-Orn-Gly-His-(Xaa)$_n$ peptides. After final deprotection, this connectivity permits the α-amino group of Orn and the adjacent Gly and His residues to bind metal ions in a fashion identical to an N-terminal Gly-Gly-His regardless of its location in a synthetic peptide. It was found also that the DNA-cleavage properties of this modified motif were essentially identical to an N-terminal Gly-Gly-His model peptide [72].

Resulting from studies of model peptides containing the δ-Orn-Gly-His sequence, this laboratory noticed their ability to mediate *selective* DNA strand scission in the presence of Ni(II). These observations suggested that the core M(II)·Xaa-Xaa-His structure, even in the absence of a defined DNA-binding domain, exhibited DNA site-selectivity thus presenting an opportunity to explore peptide-DNA interactions in a systematic fashion.

5.2.3
DNA Binding and Modification by Ni(II)·Xaa-Xaa-His Metallopeptides

5.2.3.1 Selective minor groove recognition and binding
Given the structural and chemical properties of the Ni(II)·Xaa-Xaa-His motif, this laboratory chose to examine the DNA recognition potential of these metallotripeptides as independent, low-molecular-weight complexes [74]. In these and subsequent studies, Ni(II) complexes were chosen over those containing Cu(II) due to their reported ability to produce an efficient, non-diffusible oxidizing equiv-

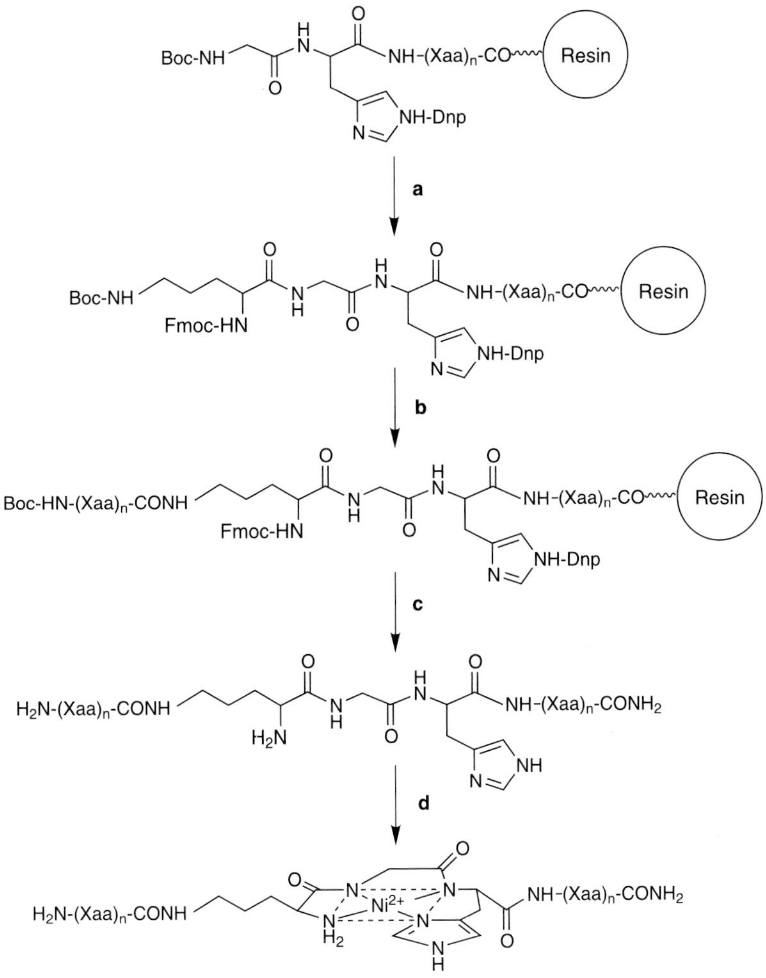

Fig. 5.4. Stepwise solid-phase peptide synthesis of Orn-containing metal-binding oligopeptides. Synthesis involves: (a) coupling of Boc-Orn(Fmoc) to resin-bound peptides terminating in Gly-His; (b) additional cycles of Boc-benzyl peptide synthesis via the side chain of Orn; (c) side chain deprotection and cleavage from the resin; (d) HPLC purification and addition of Ni(OAc)$_2$.

alent [51, 52] and also due to the lack of reactivity of Ni(II) versus Cu(II) in the absence of a coordinated tripeptide–ligand. Additionally, in light of one previous examination of Ni(II)·Gly-Gly-His containing the carboxylate analog of the peptide –ligand (i.e. NH$_2$-Gly-Gly-His-COOH), which did not reveal significant amounts of DNA modification [75], we chose to examine carboxamide peptide–ligands in these and all subsequent experiments (i.e. NH$_2$-Xaa-Xaa-His-CONH$_2$). Carboxamide peptide–ligands create overall charge-neutral complexes in the case of Ni(II)·Gly-Gly-His (due to the coordination of two deprotonated amides) or posi-

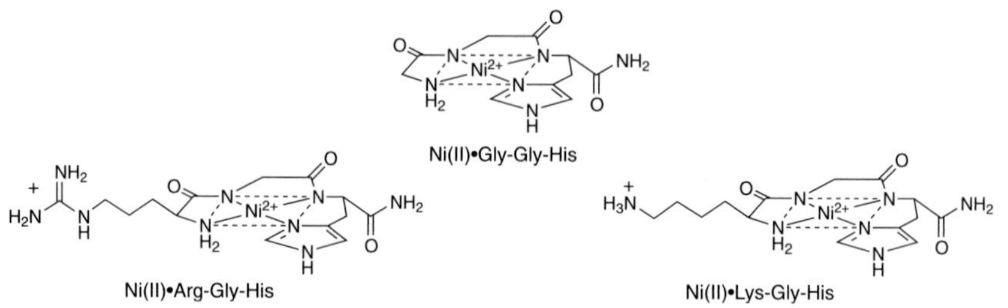

Fig. 5.5. The structures of Ni(II)·Gly-Gly-His, Ni(II)·Arg-Gly-His, and Ni(II)·Lys-Gly-His-bearing carboxamide termini.

tively charged complexes upon inclusion of Arg or Lys within NH_2-Xaa-Xaa-His-$CONH_2$. As a further factor in this choice, carboxamide-containing peptide ligands do not decarboxylate [43] in the presence of metal ions, thus creating a more stable ligand to support the redox activity of the metal center.

Initially, we chose to examine the DNA site-selectivity of three Ni(II)·Xaa-Xaa-His metallotripeptides (Fig. 5.5) containing carboxamide termini, Ni(II)·Gly-Gly-His, Ni(II)·Lys-Gly-His, and Ni(II)·Arg-Gly-His [74]. This choice of amino acid substitutions stemmed from the desire to compare the activity of an unsubstituted, charge-neutral species with two overall positively charged metallopeptides containing amino acid substitutions found commonly in DNA-binding proteins [2]. Using ^{32}P end-labeled restriction fragments and polyacrylamide gel electrophoresis (PAGE), the selectivities of the above complexes were compared in 10 mM sodium cacodylate buffer, pH 7.5, containing 50 μM calf thymus carrier DNA. Upon activation in the presence of equimolar amounts of $KHSO_5$ (with respect to [metallopeptide]) for one minute without a post-reaction chemical work-up, direct and selective DNA strand scission was observed (Fig. 5.6). In the case of Ni(II)·Gly-Gly-His, the DNA-cleavage patterns revealed a preference for AT-containing regions of the DNA duplex, while with Lys or Arg, the overall cleavage intensity was increased due to the positively charged nature of these complexes and highly directed to AT-rich regions. The DNA-binding affinities of the positively charged complexes, determined through distamycin competition experiments and ethidium displacement assays, were estimated later [76] to be in the $\sim 10^6 - 10^7$ M^{-1} range with Ni(II)·Gly-Gly-His being approximately 2- to 3-fold lower in affinity. While the observation of an AT-selective interaction parallels other known DNA-binding molecules targeted to the minor groove [77–79], the level of discrimination exhibited by these metallopeptides nonetheless was surprising given their relatively small size. Overall, these studies revealed for the first time that the M(II)·Xaa-Xaa-His motif alone was capable of exhibiting site-selective DNA strand scission.

In conjunction with the initial study described above, the influence of amino acid stereochemistry was also probed through substitutions of two D-amino acids into the Xaa-Xaa-His framework [74]; D-His was substituted for L-His in the C-terminal peptide position of Ni(II)·Lys-Gly-His and L/D-Asn residues were substituted for the Gly residues in Ni(II)·Gly-Gly-L/D-His. These alterations were

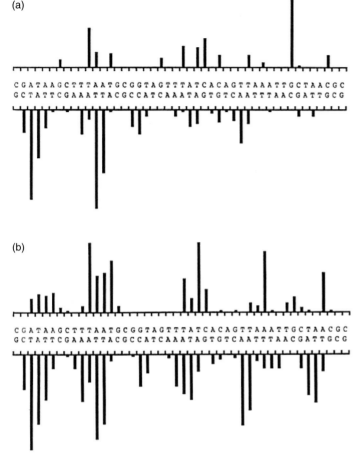

Fig. 5.6. Histogram illustrating the typical site-selectivity of DNA cleavage induced by Ni(II)·Gly-Gly-His (a) and Ni(II)·Arg-Gly-His (b) in a DNA restriction fragment substrate. Bar lengths are proportional to the intensity of direct DNA strand scission observed. Essentially identical results are obtained for Ni(II)·Arg-Gly-His and Ni(II)·Lys-Gly-His.

chosen to place an amide functionality, also found in protein–DNA interactions [2], at various locations around the metallopeptide periphery. Upon ʟ → ᴅ-His substitution within Ni(II)·Lys-Gly-His, a noticeable decrease in the focused AT-selectivity was observed in experiments that employed restriction fragment substrates. DNA cleavage was now observed to occur at mixed GC- and AT-rich regions in addition to the stronger AT-sites observed with the corresponding ʟ-His diastereomer. A potential basis for this decrease in selectivity was suggested through molecular modeling: ᴅ-His substitutions re-position the C-terminal amide functionality to be co-planar with the bulk of the metallopeptide creating a more flattened structure (Fig. 5.7). Relative to metallopeptides containing ʟ-His, this structural rearrangement might be accommodated more easily by additional minor groove

Fig. 5.7. Molecular models illustrating the structural changes that occur upon L → D-His substitution within an Ni(II)·Xaa-Xaa-His metallopeptide; the metallopeptides are oriented edge-on with the C-terminal amide to the left of each structure (darkened). The metallopeptide on the left contains an L-His residue, the metallopeptide on the right contains a D-His residue.

sequences. Interestingly, substitution of D-His to create Ni(II)·Gly-Gly-D-His(Hin 139–190) also resulted in an alternative pattern of DNA cleavage at the binding site of Hin recombinase, suggesting some form of diasterofacial selectivity upon metallopeptide activation or DNA association [51, 52].

In the case of Asn substitutions within Ni(II)·Gly-Gly-His, two of the resulting charge-neutral metallopeptides, Ni(II)·Asn-Gly-His and Ni(II)·Gly-Asn-His, resulted in DNA-cleavage patterns similar to the selectivity and intensity produced by Ni(II)·Gly-Gly-His [74]. Also, consistent with earlier observations with Ni(II)·Lys-Gly-D-His, Ni(II)·Gly-Asn-D-His appeared somewhat less selective. In contrast, however, to the other Asn-substituted metallopeptides, Ni(II)·Gly-D-Asn-His exhibited a noticable selectivity for several 5′-CCT sites and a less pronounced selectivity for a 5′-CCA and 5′-CCG site within the restriction fragment substrate used. Interestingly, these sites were avoided by the other metallopeptides examined in these early studies and occurred at junctures between two AT-rich regions. Again, molecular modeling suggested that in contrast to the other metallopeptides in this series, Ni(II)·Gly-D-Asn-His was unique in its projection of two amide functionalities toward the same face of the planar metallopeptide. As suggested at the time [74], this arrangement may lead to an overall metallopeptide shape or pattern of hydrogen bond donor/acceptors that permitted increased discrimination amongst available binding sites presented by the DNA helix.

Along with demonstrating the selective DNA recognition of Ni(II)·Xaa-Xaa-His metallopeptides, the results obtained through these initial experiments also revealed several important aspects of their DNA-binding and -modification properties under our stated conditions: (1) Direct DNA strand scission was observed at preferred regions *and* isolated examples of all four available nucleobase sites (Fig. 5.6) without a post-reaction chemical work-up, supporting the previous suggestion that Ni(II)·metallopeptide activation results in a non-diffusible oxidant of the deoxyribose moiety without a particular nucleobase reactivity [51, 52]. (2) Although not necessary to reveal a pattern of DNA modification, the intensity of DNA cleavage at a site of metallopeptide interaction can be slightly enhanced by post-reaction *n*-butylamine or piperidine treatment pointing to the generation of direct strand breaks *and* the formation of alkaline-labile lesions reminiscent of those generated by bleomycin through deoxyribose C4′-hydroperoxidation and C4′-hydroxylation, respectively [28–32]. (3) While alternating or mixed AT-rich regions are preferred, homopolymeric AT sites (i.e. poly(dA)·poly(dT) sequences) are avoided, suggesting that the narrowing of the minor groove widths at such sites inhibits metallopeptide binding [80]. (4) Cleavage of both complementary strands of a duplex DNA substrate revealed a distinct 3′-asymmetric cleavage pattern (Fig. 5.6) indicative of minor groove residence in other systems [21, 22]. (5) Metallopeptide-induced DNA cleavage can be inhibited in the presence of the well-documented minor groove binding agent distamycin [8–10], implying a competition for similar binding sites.

Overall, the above studies revealed that Ni(II)·Xaa-Xaa-His metallopeptides, as independent low-molecular-weight complexes, are capable of selectively recognizing DNA in a fashion that is influenced by the nature of the peptide ligand, the side chain functional groups present, and their chirality and placement around the periphery of the total metal complex. In addition, the recognition properties and characteristics of the DNA damage observed supported the notion that Ni(II)·Xaa-Xaa-His metallopeptides associate with the minor groove of the DNA helix, exhibit sensitivity to minor groove structural features, and produce DNA strand scission through an interaction with the deoxyribose positions accessible from this groove location.

Recently, studies with Cu(II)·Gly-Gly-His, Cu(II)·Lys-Gly-His, and Cu(II)·Arg-Gly-His have also been performed in an attempt to probe directly the binding and orientation of these metallopeptides when they associate with DNA [81]. In this study, the metallopeptides were examined by DNA fiber EPR spectroscopy and molecular modeling. DNA fiber EPR spectroscopy can determine the positioning and orientation of a DNA-bound, EPR-active Cu(II) center through the angular dependence of their EPR line shapes with respect to an oriented DNA fiber.

Initially, it was confirmed that all the above-named metallopeptides exhibited EPR values in frozen solution or in frozen DNA pellets that were consistent with identical, planar tetradentate structures as also found with the corresponding Ni(II)-coordinated metallopeptides. Subsequently, EPR spectra of the Cu(II)·Xaa-Gly-His metallopeptides and the Cu(II) complex of NH_2-Gly-Gly-His-COOH bound to B-form DNA-fibers at room temperature showed a conspicuous and consistent angular dependence to their line shapes, indicating that the nature of the Xaa amino acid does not influence the orientation of the DNA-bound species. Examination of actual and simulated spectra indicated that the angular dependence of the EPR spectra fit well assuming that the $g_{//}$ axes of the Cu(II) complexes were tilted about 50° from the DNA fiber axis. This finding indicates that the mean coordination planes of these metallopeptides were oriented approximately 40° relative to the DNA helical axis, suggesting that the metallopeptides were stereo-specifically oriented with respect to a DNA groove.

In contrast to B-form DNA fibers, the EPR spectra of the metallopeptides bound to A-form DNA fibers suggested that the complexes were randomly oriented with respect to the DNA helix. This lack of orientation with A-form DNA is not surprising due to the transformation of the distinct narrow and deep minor groove found in B-form DNA to a wide and shallow feature as found in the A-form [82].

In light of the results described above, models were constructed using B-form DNA oligonucleotides and the crystal structure of Cu(II)·Gly-Gly-His [42] with appropriate amino acid substitutions. As illustrated for Cu(II)·Arg-Gly-His (Fig. 5.8)

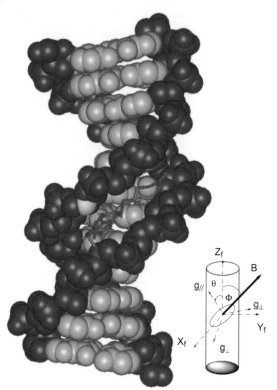

Fig. 5.8. A molecular model of Cu(II)·Arg-Gly-His (red) docked in the minor groove of B-form DNA with the $g_\parallel$ axis of the complex oriented 50° relative to the DNA helix axis. With this orientation, the mean plane of the metallopeptide is oriented 40° relative to the helix axis. The metallopeptide is easily accommodated by the minor groove and in this orientation the metal center (yellow) is positioned close to a C4′ H (yellow).

with the His imidazole and N-terminal amine inserted into the minor groove of B-form DNA, based on evidence provided by the study of the corresponding Ni(II)·Xaa-Xaa-His metallopeptides [74, 83], the width of the minor groove can accommodate well and actually direct a bound metallopeptide to orient its coordination plane approximately 40° relative to the DNA helical axis. With such a positioning, the metallopeptide is well constrained between the walls of the minor groove and permits the His imidazole functional group to interact with the floor of the minor groove via hydrogen bonding (e.g., with the O2 of thymine or the N3 of adenine [14]). Evidence for this type of His interaction is derived from the following: (1) Ni(II)·Xaa-Xaa-His metallopeptides containing a His N3 methyl substituent do not cleave DNA [83]; (2) the N3 pyrrole nitrogen of His is a very good hydrogen bond donor upon metal complexation [84]; and (3) earlier DNA fiber EPR investigations have indicated that His residues were a key factor in DNA binding [85]. The orientation described above also permits the N-terminal amino acid side chain to interact with the floor and walls of the minor groove and the second amino acid in the peptide–ligand to interact with features of the exterior surface of the DNA such as the phosphate backbone. Importantly, with the metallopeptides docked as described within the B-form minor groove, the metal center is positioned to interact with the C4′ H of a deoxyribose ring of a proximal nucleotide residue.

5.2.3.2 Minor groove-directed deoxyribose oxidation

To further confirm the location of the Ni(II)·Xaa-Xaa-His metallopeptide–DNA association and the mechanism of direct DNA strand scission observed under the conditions used to examine their site-selective binding, this laboratory carried out an analysis of the products formed upon direct DNA strand scission [86]. Previously, mechanistic investigations carried out with Ni(II)·Gly-Gly-His(Hin 139–190) activated with MMPP, iodosobenzene, or hydrogen peroxide suggested that the metallopeptide appendage resided in the minor groove at the site of Hin recombinase binding, generating, upon chemical activation, a non-diffusible oxidant that produced deoxyribose-centered damage [51, 52]. Evidence for deoxyribose damage obtained at that time included: (1) the formation of cleaved DNA fragments bearing 3′- and 5′-phosphorylated termini; (2) a kinetic isotope effect ($k_H/k_D = 1.6$) upon C4′-deuteration of a target nucleotide; and (3) the observation that DNA strand scission was enhanced upon post-reaction chemical work-up with *n*-butylamine reminiscent of the documented behavior of the C4′-hydroxylated deoxyribose lesion generated by Fe(II)·bleomycin.

With two representative metallotripeptides, Ni(II)·Gly-Gly-His and Ni(II)·Lys-Gly-His, our laboratory examined the nature of the direct DNA strand scission products formed under the reaction conditions used in earlier investigations of their site-selectivities (10 mM sodium cacodylate, pH 7.5) [86]. By employing duplex substrates with 5′- *or* 3′-^{32}P end-labeling of the *same* DNA restriction fragment strand, cleavage analyses by PAGE permitted identification of the new 3′- and 5′-termini present at a site of metallopeptide-induced DNA fragmentation, respectively. These experiments were performed in parallel with Fe(II)·bleomycin-induced cleavage of the same labeled restriction fragments to allow a direct comparison to be made with the products from an authentic, well-characterized C4′-oxidant [28–32].

The termini identified at the sites of Ni(II)·Gly-Gly-His- and Ni(II)·Lys-Gly-His-induced cleavage, when activated with KHSO$_5$, MMPP, or hydrogen peroxide, were all found to be consistent with deoxyribose C4′ H abstraction and included: (1) 5′-phosphorylated termini and 3′-phosphoroglycolate termini, indicating the intermediacy of a C4′-hydroperoxide lesion and (2) 3′-termini that were phosphorylated, or upon post-reaction work-up with NaOH or NH$_2$NH$_2$ contained deoxyribose-derived fragmentation products characteristic of the breakdown of keto-aldehydic abasic lesions resulting from C4′-hydroxylation of a target nucleotide (Fig. 5.9) [28–32]. In addition to these DNA termini-attached products, monomeric products representing the remaining fragments of a targeted deoxyribose ring released into solution were also identified and quantitated: (1) free nucleobases and nucleobase propenals (detected as thiobarbituric acid-reactive components) were detected and found to correlate well with the extents of keto-aldehydic abasic site and 3′-

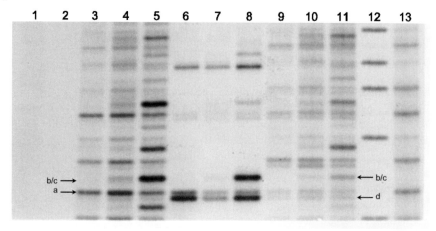

Fig. 5.9. Portion of an autoradiogram of a high-resolution denaturing polyacrylamide gel analyzing the new 3′-termini produced from a 5′-³²P end-labeled restriction fragment cleaved by Ni(II)·Lys-Gly-His (lanes 3–5) or Ni(II)·Gly-Gly-His (Lanes 9–11) + KHSO₅ compared to Fe(II)·bleomycin (lanes 6–8). Lane 1, intact DNA; lane 2, reaction control, metal + KHSO₅ alone; lanes 12 and 13, Maxam-Gilbert G + A, and T + C reactions, respectively. Labeled bands illustrate: (a) 3′-phosphorylated termini; (b/c) closely migrating NaOH-induced 3′-termini and 3′-phosphopyridazine termini (lanes 4/5, 7/8, and 10/11, respectively); and (d) 3′-phosphoglycolate termini (lanes 3–11).

phosphoroglycolate lesion formation, respectively, for a given metallopeptide and (2) the identity of the nucleobases released also paralleled the site-selectivity of the metallopeptides. It should be noted, however, that free guanine was found to be partially degraded by the action of Ni(II)·Xaa-Xaa-His + KHSO₅, making its quantitation difficult in comparison to the remaining three nucleobases [86]. In addition to the above, experiments designed to trap intermediates produced from metallopeptide interactions with deoxyribose sites other than the C4′ H failed to detect their presence [86].

The results described above indicate that Ni(II)·Gly-Gly-His and Ni(II)·Lys-Gly-His (activated with KHSO₅, MMPP, or hydrogen peroxide) associate with the DNA minor groove under the conditions employed and, like Fe(II)·bleomycin, mediate C4′ H abstraction to generate C4′ radicals or C4′ cations leading to two distinct mechanistic pathways to final product formation (Scheme 5.1): (1) C4′-hydro-peroxide lesions are formed upon O₂ addition to a C4′ radical resulting in direct DNA strand scission with the release of 5′-phosphorylated termini, 3′-phosphor-oglycolate termini, and nucleobase propenals and (2) C4′-hydroxylated lesions are generated upon solvent addition to a C4′ cation leading to keto-aldehydic abasic sites and a concomitant loss of free nucleobases. These abasic lesions can subsequently be trapped with NaOH or NH₂NH₂ in a fashion consistent with the characterized alkaline-labile lesion produced by Fe(II)·bleomycin [28–32].

Along with defining the locations of binding and the chemistry of direct DNA strand scission under our conditions, these mechanistic experiments also revealed

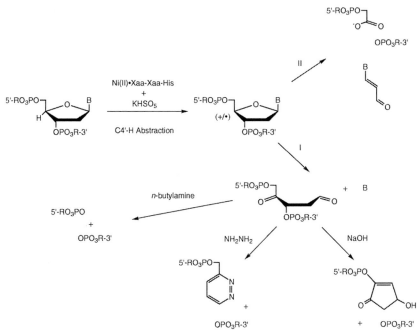

Scheme 5.1. Summary of DNA cleavage products observed and proposed pathways of deoxyribose-based DNA strand scission by Ni(II)·Xaa-Gly-His metallopeptides activated with KHSO$_5$, MMPP, or H$_2$O$_2$. Pathway I involves the intermediate formation of a C4′-OH modified deoxyribose and release of free nucleobases, resulting in keto-aldehydic alkaline-labile lesions that can be further modified with NaOH, hydrazine, or n-butylamine while pathway II occurs through an intermediate C4′-OOH resulting in direct DNA strand scission and the formation of 3′-phosphoroglycolate termini and nucleobase propenals.

key differences between the activities of the individual complexes examined. While Ni(II)·Gly-Gly-His and Ni(II)·Lys-Gly-His generated the same products resulting from C4′ H abstraction, indicating an identical underlying mechanism, they did so with differing ratios of C4′-OH to C4′-OOH lesions (Fig. 5.9). In the case of the charge-neutral complex Ni(II)·Gly-Gly-His, which binds DNA less tightly than the positively charged complexes, the C4′-OOH lesion appeared to a greater extent in comparison to Ni(II)·Lys-Gly-His (as monitored through 3′-phosphoroglycolate and nucleobase propenal formation). This observation suggests that relatively tight metallopeptide-DNA binding can limit the rate of diffusion of O$_2$ to the site of C4′ H abstraction decreasing the formation of 3′-phosphoroglycolate lesions (i.e. direct strand breaks) while with weaker binding metallopeptides, O$_2$ can add readily to the radical lesion formed at the C4′ carbon. Interestingly, these results imply that a lower DNA binding affinity, or careful consideration of association/dissociation rates of complex formation, may actually improve the efficiency of direct strand scission by a cleavage agent that uses similar pathways to final product formation.

Additionally, with Ni(II)·Gly-Gly-His the ratio of C4'-OH to C4'-OOH lesions appears also to be affected by the particular nucleotide sequence cleaved. Thus, as observed previously with Fe(II)·bleomycin [87] and synthetic diimine complexes of Rh(III) [88], the microheterogeneity of the DNA helix, and its influence on the DNA-binding affinity of a targeted agent, alters the partitioning between two available mechanistic pathways to final product formation. The above may also explain the lack of observed 3'-phosphoroglycolate termini in studies of Ni(II)·Gly-Gly-His(Hin 139–190) [52]. The protein portion of this conjugate mediates tight DNA association, limiting, perhaps also in conjunction with the specific nucleotide sequence cleaved, the accessibility of O_2 to the initial C4' radical lesion resulting predominantly in C4' hydroxylation.

5.2.3.3 Nature of the intermediate involved in deoxyribose oxidation

Given the above, it appears that Ni(II)·Xaa-Xaa-His metallopeptides associate selectively with the DNA minor groove resulting, upon activation, in abstraction of the C4' H prominently located there. As stated earlier, activation of the Ni(II)·Xaa-Xaa-His metallopeptides can be supported by $KHSO_5$, the organic peracid MMPP, or hydrogen peroxide (albeit with differing efficiencies, ratios of activating agent to metallopeptide, and reaction times) [86]. Worthy of note is the fact that the final site-selectivities observed, identity and ratios of products released, and the mechanisms of action employed are identical for a given metallopeptide when these three rather structurally diverse activating reagents are used suggesting the formation of a common activated intermediate (Fig. 5.10) [86]. Certainly, the generation of individual activated metallopeptide species that differ with each of these supporting reagents would likely lead to differences in their final site-selectivities or mechanisms/ratios of product formation.

To probe the nature of the activated Ni(II)·Xaa-Xaa-His metallopeptide responsible for C4' H abstraction and direct DNA strand scission, the effects of commonly employed radical scavengers on Ni(II)·Lys-Gly-His-induced conversion of supercoiled form I DNA to relaxed form II DNA were compared upon activation of this metallopeptide with $KHSO_5$, MMPP, and hydrogen peroxide [86]. Under the assay conditions used, which monitored direct DNA strand scission without subsequent chemical workup, the scavengers ethanol, *tert*-butyl alcohol, DMSO, and mannitol had little to no effect on plasmid conversion from form I to form II DNA while parallel control experiments showed substantial inhibition of Fe(II)·EDTA-induced DNA cleavage by hydroxyl radicals.

These data indicate that the "activated" metallopeptide formed by $KHSO_5$, MMPP, and hydrogen peroxide not only produced identical site-selective DNA cleavage and deoxyribose fragmentation products, but also behaved similarly in the presence of four well-established radical scavengers supporting the notion of some common activated intermediate. These results, in total, also indicate that the activated Ni(II)·Xaa-Xaa-His metallopeptide formed does not produce direct, deoxyribose-centered strand scission through the generation of diffusible hydroxyl radicals, as noted by others [66]. Also, these data indicate that the active intermediate involved in direct DNA strand scission via deoxyribose modification is not a diffu-

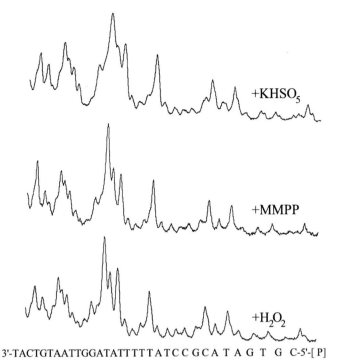

3'-TACTGTAATTGGATATTTTTATCCGCATAGT G C-5'-[P]

Fig. 5.10. Densitometric analyses of autoradiograms from high-resolution denaturing polyacrylamide gels illustrating the similarities in site-selectivities and product (termini) formation that occurs upon metallopeptide activation with KHSO$_5$, MMPP, or hydrogen peroxide.

sible or metal-bound sulfate radical (SO$_4$$^{\bullet-}$) when KHSO$_5$ is employed in activation [86, 89]. While these results apparently rule out the involvement of the above diffusible radical species in the mechanism of deoxyribose-centered strand scission, as noted at the time, they do not preclude their formation and involvement in other pathways of DNA modification such as nucleobase modification that does not result in direct DNA strand scission [90].

In light of the above evidence, it appears that C4' H abstraction of the deoxyribose ring occurs through a common activated metallopeptide formed in the presence of KHSO$_5$, MMPP, or hydrogen peroxide. Given the characteristics of this intermediate, we have proposed [86] that this species may be a high valent, peptide-bound Ni(III)-O· or Ni(IV)=O, effectively a Ni-bound hydroxyl radical equivalent, which may be formed through the heterolytic splitting of the oxygen-oxygen bond contained within KHSO$_5$, MMPP, and hydrogen peroxide (Fig. 5.11). While yet to be proven definitively, indirect support for this intermediate is provided by several lines of evidence: (1) There are observed differences in the relative reactivities of the three activating reagents used (KHSO$_5$ > MMPP ≫ H$_2$O$_2$) which parallel closely the acidity of their leaving groups upon oxygen–oxygen bond het-

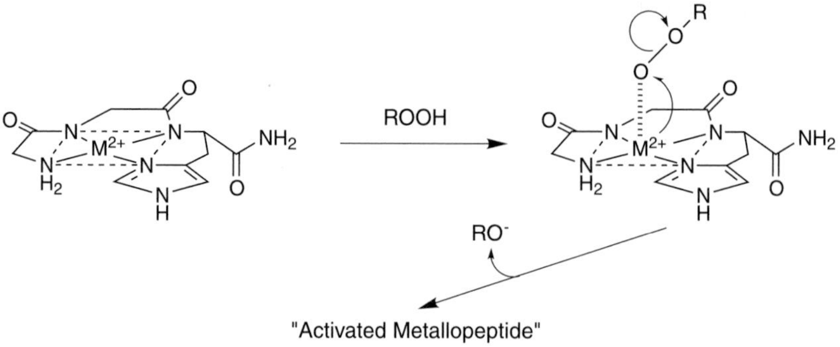

Fig. 5.11. Proposed mechanism of Ni(II)·Xaa-Xaa-His activation through Ni-promoted heterolytic cleavage of a peroxide/peracid oxygen–oxygen bond.

erolysis ($H_2SO_4 > COOHC_6H_4COOH \gg H_2O$), a feature that is known to influence the ability of a peroxide functional group to act as an oxygen atom donor to a metal center [91, 92]. (2) Upon activation of Ni(II)·Lys-Gly-His in the absence of a DNA substrate, a UV–visible absorption shift occurs from 425 nm to a more intense peak at 375 nm indicative of a structural change of the metal complex from square-planar to an oxidized, higher valent Ni(III) or Ni(IV) species of octahedral geometry [93]. (3) Ni(II)·Gly-Gly-His has also been shown to catalyze alkene epoxidations [51, 52], suggesting that, like other Ni complexes where a ligand-bound Ni(III)-O· or Ni(IV)=O has been proposed [94, 95], a similar intermediate may be generated with Ni(II)·Xaa-Xaa-His metallopeptides. Certainly, such an activated intermediate associated with a minor groove-bound metallopeptide would be capable of abstracting the C4′ H, leading to the observed profile of deoxyribose fragmentation products.

5.2.3.4 Generation of minor groove binding combinatorial libraries

Having developed a working model of their minor groove binding, this laboratory is pursuing a systematic examination of Ni(II)·Xaa-Xaa-His metallopeptide–DNA interactions through the development of combinatorial libraries [76]. As a test case, it was determined initially if an "optimized" Ni(II)·Xaa-Xaa-His metallopeptide could be developed to target B-form DNA through the synthesis of positional scanning combinatorial libraries [15–17]. These libraries were designed to examine nearly all possible combinations of naturally occurring L-α-amino acids within the Xaa positions of the Xaa-Xaa-His tripeptide-ligand, excluding Cys and Trp residues to prevent disulfide bond formation, as confirmed [96], and the partial intercalation of DNA [5–7], respectively. Using standard *t*-Boc synthesis protocols [97], two main libraries were generated in which the first and second peptide positions were varied systematically (Fig. 5.12). These libraries were verified by quantitative amino acid analyses to contain a uniform representation of all the amino acids included in the split-and-mix synthesis.

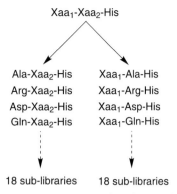

Fig. 5.12. Generation of combinatorial Xaa$_1$-Xaa$_2$-His ligand–peptide libraries.

Initially, the DNA-cleavage efficiencies of the individual members of these libraries were assayed for their ability to convert directly supercoiled, form I plasmid DNA to nicked-circular form II DNA via the chemistry described earlier (using KHSO$_5$ or MMPP as activating agents) [76, 86]. This assay provides a reasonably high-throughput means to determine the relative activities of each member of the synthetic libraries but does not select for any particular site-selectivity. For both libraries it was found that DNA-cleavage efficiency was highly dependent upon the identity of the amino acid found at a given Xaa position (Fig. 5.13). With the library that systematically varied the N-terminal Xaa$_1$ position, it was found that the DNA-

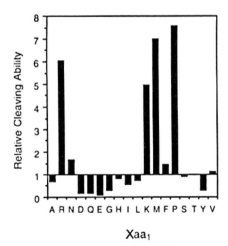

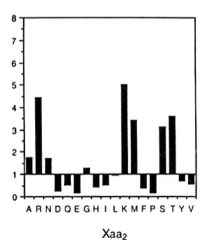

Fig. 5.13. Quantitation of direct DNA strand scission activity by combinatorial libraries of Ni(II)·Xaa$_1$-Xaa$_2$-His metallopeptides relative to Ni(II)·Gly-Gly-His (1 on relative scale). Xaa$_1$ denotes the identity of the amino acid (in single-letter code) present at the N-terminal peptide position, while Xaa$_2$ denotes the identity of the amino acid present at the second peptide position.

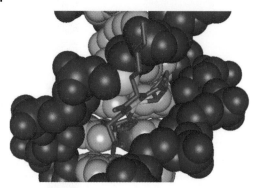

Fig. 5.14. Molecular model of Ni(II)·Pro-Lys-His (red) bound to the minor groove of B-form DNA with the His imidazole H-bonded to the N3 of adenine (green): (or O2 of thymine) and the Ni(II) center (yellow) in close proximity to a C4′ H (yellow).

cleavage efficiency was enhanced between 5- and 8-fold relative to Ni(II)·Gly-Gly-His, which served as a baseline for comparison, when Arg, Lys, Met, or Pro were present (Pro > Met > Arg > Lys). In the case of the second Xaa_2 peptide position, DNA-cleavage efficiency was enhanced between 3- and 5-fold when Arg, Lys, Met, Ser, or Thr were present (Lys > Arg > Met/Ser/Thr). In both the N-terminal and second peptide positions, the remaining amino acids included in the library syntheses exhibited cleavage efficiencies that were either similar to, or slightly reduced in comparison to Ni(II)·Gly-Gly-His, most notably Glu, Asp, Gln, Tyr, and Phe (Fig. 5.13).

The results of the above experiments are important in that they: (1) predict that Ni(II)·Pro-Lys-His, containing the most active amino acids in their respective positions, would substantially increase direct DNA strand scission relative to Ni(II)·Gly-Gly-His and (2) allow a visualization of the contribution of nearly all naturally occurring amino acids and their side chain functionalities within the Ni(II)·Xaa-Xaa-His framework on the cleavage of B-form DNA. In the former case, independent synthesis of Ni(II)·Pro-Lys-His verified its ability to bind metal ions and cleave DNA an order of magnitude better than Ni(II)·Gly-Gly-His, thus indicating that the substitutions had nearly an additive effect [76].

Overall, it appears that for B-form DNA the positively charged amino acids Lys and Arg can favorably influence metallopeptide–DNA binding in either the first or second peptide positions with a preference for Arg in the first position. Lys was, however, most active in the second peptide position. These preferences can be attributed to an increased electrostatic attraction between the metallopeptide and the polyanionic DNA backbone. In addition to electrostatic interactions, however, it was surprising to find that Pro was the most active amino acid in the N-terminal peptide position, increasing the direct DNA strand scission activity of the metallopeptides to an even greater extent than Lys and Arg. As expected, however, Pro in the second position prevented metal binding and exhibited a lack of DNA reactivity. It was found also that Thr and Ser residues, containing side chain alcohol functionalities, contributed to the DNA-cleavage activity of the metallopeptides only when located in the second peptide position while Met residues in the first or second peptide positions also contributed positively to the reactivity of these metallopeptides. In the case of Met residues, under the oxidative conditions employed to activate the metallopeptides, it is likely that Met sulfones or sulfoxides were present.

To probe the possible basis for the increased direct strand scission activity of the combinatorially selected metallopeptides, further experiments were performed [76]. It was found that: (1) the overall metal-binding abilities of individual members of the metallopeptide libraries were equivalent, with 1:1 complexes being formed (except with Xaa-Pro-His peptides); (2) the DNA-binding affinities of the selected metallopeptides were found to be similar within a factor of 2–4 in comparison to Ni(II)·Gly-Gly-His; (3) the cleavage efficiencies of ^{32}P end-labeled restriction fragments by individual library members paralleled their activities in the initial activity screen; (4) while subtle differences between individual active library members were observed, their DNA site-selectivities maintained an underlying selectivity for AT-rich regions, as expected given the nature of the initial activity screen employed and the absence of D-amino acid substitutions [74]; and (5) minor groove binding and the mechanism of direct deoxyribose-centered strand scission via C4′ H abstraction remained unaltered. Given that cleavage enhancements for the selected metallopeptides were up to 10-fold greater than that exhibited by Ni(II)·Gly-Gly-His, these findings suggested that particular structural features of the selected metallopeptides, beyond their simple 2- to 4-fold increase in equilibrium binding affinity, may be responsible for the observed enhancement in cleavage activity. It is possible that the selected combinations of amino acids in the metallopeptide framework result in an increased anchoring/rigid binding interaction in the minor groove or an advantageous positioning relative to the C4′ H position of a target nucleotide, leading to the observed increased in B-form DNA strand scission activity.

To examine the possible structural basis for the increased activity of select members of the metallopeptide libraries, molecular modeling was performed using the metallopeptide with the highest selected activity, Ni(II)·Pro-Lys-His (Fig. 5.14, see p. 106) [76]. In light of modeling and structures [43] described earlier, and given that the common structural element of the metallopeptide that supports AT-selectivity is the His imidazole ring, models were constructed in which the N3 nitrogen of the His imidazole was hydrogen bonded to either the N3 of adenine or the O2 of thymine located on the floor of the minor groove. With this interaction, the β, γ, and δ carbons of the Pro ring are positioned deeply in the minor groove and make van der Waals contacts with the deoxyribose sugars which form the walls of this groove. The narrow structure of AT-rich regions complements well the width of the metallopeptide, suggesting a stabilized interaction that would be lacking with only a Gly substitution [14, 77, 82]. With the insertion of the first and third amino acids into the minor groove, the Lys residue contained within the second peptide position is exposed at the surface of the DNA helix and poised to allow contact between the ε-amino group of its side chain and phosphate groups proximal to the bound metallopeptide. The side chain alcohol functional groups of Ser and Thr would be capable of similar interactions with the DNA when included at this peptide position.

The above models suggest that for the direct strand scission of B-form DNA, Ni(II)·Xaa-Xaa-His metallopeptides, in addition to the role of the His imidazole, can be optimized through the placement of hydrophobic features via a Pro residue in the first peptide position and either positively charged residues (Lys/Arg) or polar residues (Ser/Thr) in the second, exterior-facing, peptide position. As found in earlier models, this overall structural orientation with relationship to a B-form DNA helix allows the metal center to be poised to interact (3–4 Å) with the C4' H of an adjacent deoxyribose ring. Using this model as a guide, ongoing investigations are attempting to refine its accuracy and to examine details of this peptide–DNA interaction through high-resolution methods.

The studies described above have verified the utility of combinatorial methodologies, as applied in other systems [98, 99], toward the discovery and development of Ni(II)·Xaa-Xaa-His metallopeptides with select properties. While this initial study did not attempt to screen for select binding sites, ongoing studies involving expanded libraries containing natural and select unnatural amino acids will screen these libraries for their ability to target particular linear sequence-selectivities and anomalous structural features within DNA. In addition to the above, it is also important to note that the majority of the amino acids selected through the above-described library screening procedure, and found to enhance the minor groove interaction of the metallopeptide framework, are those included in several important classes of minor groove-binding protein motifs. Examples of these proteins include: Hin recombinase [100, 101]; *E. coli* integration host factor [102]; the repeating Ser-Pro-Lys/Arg-Lys/Arg motifs found in sea urchin histones H1 and H2B [103]; the AT-hook motif found in mammalian high mobility group (HMG-I) chromosomal proteins [104, 105]; and DNA architectural proteins [106]. This finding suggests that the activity of Ni(II)·Xaa-Xaa-His metallopeptides is influenced by and can serve to model the fundamental amino acid–minor groove interactions that nature has chosen to use in protein–minor groove recognition. Thus, a systematic examination of Ni(II)·Xaa-Xaa-His systems may provide insight into protein–DNA interactions while also providing information that assists in the development of low-molecular-weight DNA-binding agents.

5.2.3.5 Guanine nucleobase modification/oxidation

Along with their ability to site-selectively recognize and bind to the minor groove and to induce direct DNA strand scission through deoxyribose-centered oxidation, Ni(II)·Xaa-Xaa-His metallopeptides can be induced to oxidize guanine nucleobases in a $KHSO_5$ or sulfite/O_2-dependent reaction [107]. In an initial study of their reactivity in comparison with other Ni(II) complexes that mediate guanine nucleobase modification, however, Ni(II)·Gly-Gly-His (containing a carboxylated terminus) was found to exhibit little DNA modification of a ^{32}P end-labeled oligonucleotide substrate in the presence of 10 mM phosphate, pH 7.0, 100 mM NaCl and

100 μM KHSO$_5$ [75]. This result can be attributed to the overall negative charge of this complex and its tendency to decarboxylate [43] under oxidative conditions.

Subsequent to our report of the direct DNA strand scission activity of Ni(II)·Lys-Gly-His containing a C-terminal amide [74], further studies of this same metallopeptide indicated that Ni(II)·Lys-Gly-His was capable of guanine nucleobase oxidation in the presence of KHSO$_5$ (or through the autooxidation of sulfite) [107] in a reaction reminiscent of other Ni(II) complexes that were studied as probes of DNA structure [24]. The conditions employed in these reactions involved 10 μM Ni(II)·Lys-Gly-His, 100 mM NaCl, and 10 mM sodium phosphate, pH 7.0, activated with 100 μM KHSO$_5$ or 50–100 μM NaSO$_3$. The reactions were incubated for 1 hour prior to a post-reaction chemical work-up (involving 1 M piperidine/90°C/ 30 min) that was necessary to induce DNA polymer strand scission. It was further reported that the DNA modification observed was consistent with the formation of an intermediate guanine radical cation [90] resulting in guanine nucleobase oxidation products and ultimately DNA strand scission upon hot piperidine treatment. The reactivity of guanine residues found within a restriction fragment substrate paralleled their reported sequence-dependent ionization potentials [108]. Importantly, this reaction suggested a potential mechanism for the toxicity associated with sulfite exposure especially in light of the binding of Ni(II) by proteins such as histones [107]. At that time, it was proposed that Ni(II)·Lys-Gly-His catalyzed the formation of a Ni(III)-bound sulfate radical [107]. The redox cycling of this system was examined further [109], supporting the intermediacy of a Ni(III)·peptide species.

Further analysis of the guanine nucleobase modification capabilities of Ni(II)· Lys-Gly-His supported the findings discussed above (Fig. 5.15) [86]. It was found

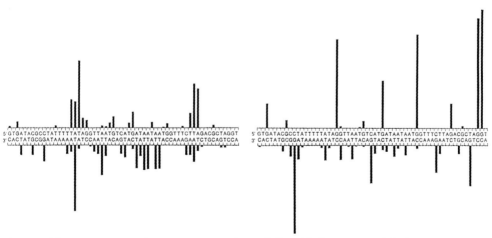

Fig. 5.15. Histograms generated upon densitometric analysis of 3'-^{32}P end-labeled restriction fragment cleavage by Ni(II)·Lys-Gly-His under low ionic strength/low KHSO$_5$ conditions (left) and high ionic strength/high KHSO$_5$ conditions (right). Both cleavage reactions illustrated received a post-reaction work-up with piperidine/90°C.

that Ni(II)·Lys-Gly-His produced guanine nucleobase modification of restriction fragment substrates when the reactions included 100 μM KHSO$_5$, 100 mM NaCl, 10 mM phosphate, pH 7.0, for 30 min followed by a post-reaction work-up involving piperidine/90°C/30 min. Gel analyses of reaction aliquots of these samples prior to piperidine treatment revealed slightly less direct strand scission products at the expected AT-rich regions reported previously [74]. However, when compared in parallel to the reaction conditions initially employed [74] in an examination of Ni(II)·Xaa-Xaa-His sequence-selectivity (10 mM sodium cacodylate, pH 7.5, one equivalent of KHSO$_5$), Ni(II)·Lys-Gly-His induced the expected intensity of direct strand scission of DNA at AT-rich regions with or without a post-reaction piperidine treatment [86]. No appreciable guanine nucleobase modification was observed under these low salt/low KHSO$_5$ conditions.

In order to determine if the guanine nucleobase chemistry induced by Ni(II)·Lys-Gly-His occurred as a function of the activating agent employed, further experiments were carried out using MMPP and hydrogen peroxide as supporting reagents [86]. It was found that Ni(II)·Lys-Gly-His activated with MMPP or hydrogen peroxide in 10 mM sodium cacodylate, pH 7.5, produced the expected pattern of deoxyribose-centered direct strand scission products at AT-rich regions [74]. Importantly, however, even when aliquots of these same reactions were treated with piperidine/90°C, guanine nucleobase modification was not observed [86]. Additionally, it was reported that Ni(II)·Lys-Gly-His could not be activated to produce guanine nucleobase oxidation with MMPP or hydrogen peroxide even under the conditions used to produce this activity with KHSO$_5$ (100 mM NaCl, 10 mM phosphate, pH 7.0, for 30 min followed by a post-reaction work-up involving piperidine/90°C/30 min) [86].

These findings indicate that guanine nucleobase modification [107] by Ni(II)·Lys-Gly-His occurs most readily as a function of metallopeptide activation with KHSO$_5$ under specific conditions (excess KHSO$_5$ relative to metallopeptide, 100 mM NaCl, 10 mM phosphate, pH 7.0, for 30 min followed by a post-reaction work-up involving piperidine/90°C/30 min) while deoxyribose-centered oxidation via C4' H abstraction can occur upon metallopeptide activation with one equivalent of KHSO$_5$ or MMPP/hydrogen peroxide under relatively low ionic strength conditions [74, 86]. These studies suggest that Ni(II)·Lys-Gly-His may be capable of forming two distinctly different "activated" species when KHSO$_5$ is used as a supporting reagent under different reaction conditions: (1) An oxygenated Ni(III)-bound activated species when one equivalent of KHSO$_5$ is used, in common with the activated metallopeptides also produced by MMPP and hydrogen peroxide, that is capable of recognizing and binding to the minor groove of DNA resulting in specific C4' H abstraction under low ionic strength conditions [74, 86]. (2) A sulfate radical-derived species that forms under conditions of excess KHSO$_5$ in the presence of 100 mM NaCl resulting in predominant guanine nucleobase modification via a guanine radical cation intermediate [107]. While the interaction of species similar to the latter with DNA have been postulated as proceeding through the intermediate metallation of the N7 position of the guanine nucleobase [24], recent investigations with other Ni(II) complexes also support the potential for an outer-sphere interaction with the DNA helix [110].

5.2.4
DNA Strand Scission by Co·Xaa-Xaa-His Metallopeptides

5.2.4.1   Activation via ambient O_2 and light

While activation systems involving Ni(II) or Cu(II) complexes of Xaa-Xaa-His peptides have aided our understanding of their DNA recognition, binding, and modification properties, the need for exogenous chemical reductants or oxidants during the course of their activation often limit investigations to systems tolerant of these co-reactants and can alter the observed products of a reaction. Thus, the development of strategies to activate Xaa-Xaa-His-derived metallopeptides with a minimal number of co-reactants may facilitate their study and application in further biochemical analyses.

Towards the above goal, this laboratory has examined the activation and DNA-recognition/cleavage properties of Co(II/III)·Xaa-Xaa-His metallopeptides [111, 112]. The use of cobalt as the metal center in these second-generation systems was prompted by the established reactivity of complexes of this metal with ambient O_2 [113–115] and upon light activation [21, 23, 116]. Indeed, Co(III) derivatives of bleomycin and analogs of this drug exhibit light- and O_2-dependent DNA strand scission [117–119]. In addition to considerations of reactivity, cobalt has been established to bind Xaa-Xaa-His peptides through equatorial coordination of the peptide terminal amine, two deprotonated amides and the His imidazole at alkaline pH values to generate exchange-inert Co(III)·Xaa-Xaa-His metallopeptides similar to those generated through Ni(II) or Cu(II) binding (Fig. 5.16) [120]. Importantly, with regards to the study of metallopeptide–nucleic acid recognition, Co(III)·Xaa-Xaa-His systems increase the dimensionality of the overall metallopeptide in comparison to those containing Ni(II) or Cu(II) by providing axial coordination sites for further nitrogen- or oxygen-based ligands, resulting in an octahedral metallopeptide. This increased dimensionality may provide additional points of contact between the metallopeptide and a target nucleic acid, thus expanding their molecular recognition capabilities.

Initially, it was reported that irradiated admixtures of Co(II) + Gly-Gly-His in 5 mM sodium borate buffer, pH 8.0, resulted in the direct conversion of supercoiled form I DNA to nicked-circular form II DNA [111]. These experiments included a

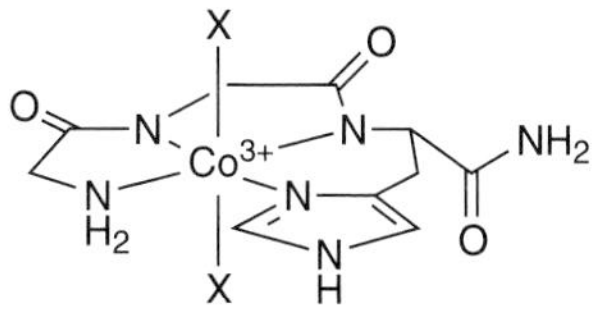

Fig. 5.16.   The structure of Co(III)·Xaa-Xaa-His metallopeptides containing axial (X) nitrogen- or oxygen-bearing ligands.

comparison of: (1) Co(II)·Gly-Gly-His + ambient O_2 + hv; (2) Co(II)·Gly-Gly-His + ambient O_2 without irradiation; and (3) preformed $(NH_3)_2$Co(III)·Gly-Gly-His + hv. Of these reactions, it was found that Co(II)·Gly-Gly-His + ambient O_2 + hv was the most active, being able to totally convert the available form I DNA into form II while the remaining two systems converted <50% of the available substrate. This comparison suggested that DNA cleavage by Co(II)·Gly-Gly-His + ambient O_2 + hv was not simply due to the Fenton-like generation of radicals nor excitation of the ligand field bands of these complexes alone. Curiously, however, this comparison did reveal the activity of Co(II)·Gly-Gly-His + O_2 in the absence of irradiation. Subsequent reactions also examined the activity of Co(II) + Lys-Gly-His under the conditions described above [111]. It was found that this metallopeptide, now containing a positively charged Lys residue, increased the level of DNA cleavage activity in comparison to Co(II)·Gly-Gly-His due to considerations of electrostatics. This observation also suggested the specific delivery of an active metal complex to the DNA as opposed to the generation of radicals in bulk solution.

Given its increased level of DNA cleavage activity, Co(II) + Lys-Gly-His + ambient O_2 + hv was employed in an examination of the mechanism of this reaction [111]. It was reported that: (1) hydroxyl radical scavengers (mannitol and DMSO) diminished DNA cleavage activity by ~10%; (2) the reaction was unenhanced in D_2O, indicating the absence of singlet oxygen sensitization; and (3) DNA cleavage was unaffected by superoxide dismutase. Given these findings, and in the light of the reported activity of Co(III)·bleomycin and its analogs [117–119], it was suggested that DNA cleavage by Co(II) + Xaa-Xaa-His + ambient O_2 + hv may occur through the intermediate formation of an oxygenated Co(III)·peptide that generates hydroxyl radicals in close proximity to a DNA substrate.

In support of an intermediate oxygenation step, control experiments found that anaerobic Co(II) + peptide admixtures were unable to cleave DNA upon irradiation [111]. Futher analysis of the interaction of O_2 with Co(II)·Lys-Gly-His by UV–visible spectroscopy (Fig. 5.17) indicated the rapid formation of μ-peroxo dimers [113–115] of these metallopeptides that decompose over the course of an hour (Fig. 5.18) in a fashion analogous to the oxygenation chemistry of Co(II)·bleomycin [121]. This reactivity results in oxygenated or aquated Co(III)·Xaa-Xaa-His metallopeptides that mimic the DNA cleavage activity of the Co(III)·bleomycin "brown" complex [117, 119]. Importantly, as suggested at the time, these systems decouple peptide metallation and activation thus allowing the purification and storage of a Co(III)-containing metallopeptide (or a protein modified for affinity cleavage) for later use and photochemically induced nucleic acid strand scission.

5.2.4.2 Highly selective DNA cleavage via ambient O_2 activation

As a follow-up to the DNA cleavage observed with Co(II)·Gly-Gly-His + ambient O_2 without irradiation and the ability of Co(II)·Xaa-Xaa-His metallopeptides to form μ-peroxo dimers, our laboratories pursued a further investigation of the aerobic activation of Co(II)-containing metallopeptides in the presence of DNA [112]. In the area of DNA recognition and modification, synthetic compounds that activate O_2 to induce nucleic acid modification are of keen interest due to the physio-

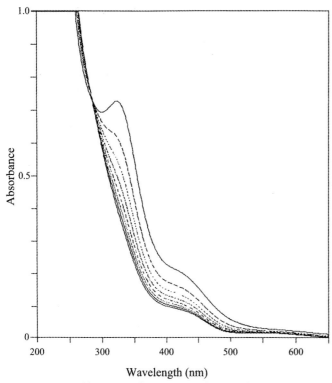

Fig. 5.17. UV-visible spectra of Co(II) + Lys-Gly-His in the presence of ambient dioxygen (top scan, $t = 0$; bottom scan, $t = 10$ min; the spectrum continues to decrease steadily for up to 1 h).

2 Co(II)•Lys-Gly-His

$\downarrow$ O$_2$

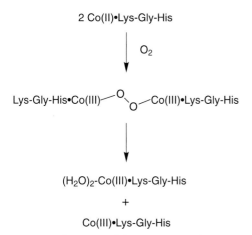

Lys-Gly-His•Co(III)⌐O⌐O⌐Co(III)•Lys-Gly-His

$\downarrow$

(H$_2$O)$_2$-Co(III)•Lys-Gly-His

+

Co(III)•Lys-Gly-His

Fig. 5.18. Illustration of the oxygenation of Co(II) + Lys-Gly-His resulting in the intermediate formation of a μ-peroxo dimer and final monomeric Co(III)·Lys-Gly-His metallopeptides.

logical availability of molecular oxygen. Such systems could lead to agents that function *in vivo* [122, 123].

To examine the above, admixtures of Co(II) in the presence of one equivalent of Lys-Gly-His were allowed to react with ^{32}P end-labeled restriction fragment substrates in the presence or absence of ambient O_2 [112]. It was found that Co(II) + Lys-Gly-His induced the O_2-dependent modification of a single thymine residue within the sequence 5'-AGGTGG (Fig. 5.19). The chemistry of the initial DNA lesion formed was found to be labile to a post-reaction work-up with piperidine (1 M piperidine/90°C/30 min). Mild NaOH treatment (0.1 M NaOH/60°C/5 min) did not induce strand breakage, suggesting that DNA modification proceeds

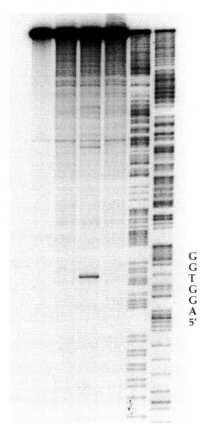

Fig. 5.19. Autoradiogram of a denaturing polyacrylamide gel illustrating the O_2-dependent modification of 5'-AGGTGG sites within ^{32}P end-labeled DNA restriction fragments. All lanes received a post-reaction work-up with piperidine: lane 1, reaction control, intact DNA; lane 2 reaction control, 100 μM Co(II); lane 3, 100 μM Co(II) + 100 μM Lys-Gly-His + O_2; lane 4, 100 μM Co(II) + 100 μM Lys-Gly-His under anaerobic conditions; lanes 5 and 6, Maxam-Gilbert G + A and T + C reactions, respectively.

through a mechanism involving thymine nucleobase modification and not through deoxyribose oxidation [90].

Additional analysis of this reaction revealed the high level of selectivity of Co(II)·Lys-Gly-His for the 5'-AGGTGG site [112]: (1) Other restriction fragments containing this same 5'-AGGTGG sequence were similarly targeted. (2) While a lower level of DNA modification is observed at 5'-A̲TGAGA sites, modification does not occur at nested subsets of the main target sequence nor the very similar sequence 5'-CGGTGG. (3) Restriction fragments bearing ^{32}P end-labeled strands containing the sequence complementary to the targeted 5'-AGGTGG site (i.e. 5'-CCACCT) do not become modified, supporting the involvement of a non-diffusible active species as also suggested by radical scavenger experiments. (4) While other metallopeptides, including Gly-Gly-His, Arg-Gly-His, Gly-Asn-His, and Pro-Lys-His are capable of generating μ-peroxo dimers in the presence of Co(II) + O_2, and mediating the relaxation of supercoiled DNA, they did not target 5'-AGGTGG sites nor other sites contained within the restriction fragments examined. (5) To assess the reactivity and accessibility of the T residue within 5'-AGGTGG sites, restriction fragments containing this sequence were also treated with $KMnO_4$ [124]. This sequence was found to exhibit a low reactivity to this reagent, suggesting that the T residue within the 5'-AGGTGG site was efficiently stacked within the DNA double helix.

Overall, the above evidence supported the notion that a specific binding interaction occurs between the active metallopeptide and the targeted 5'-AGGTGG sequence. To test the potential involvement of the μ-peroxo dimer formed through aerobic oxidation of two Co(II)·Lys-Gly-His metallopeptides, modification of the 5'-AGGTGG sequence was monitored as a function of μ-peroxo dimer decomposition [112]. These experiments revealed that the intensity of T modification in the target sequence diminished in parallel with the decomposition of the μ-peroxo dimer. At early time points in the reaction when the μ-peroxo dimer concentration is high, T nucleobase modification is most intense; as the dimer decomposes, T nucleobase modification also diminishes. This observation supported strongly the notion that the μ-peroxo dimer was responsible for T nucleobase modification within the 5'-AGGTGG sequence.

In addition to the above, indirect support for the participation of the μ-peroxo dimer in the targeting of the 5'-AGGTGG site was derived from the examination of molecular models [112]: (1) It was found that the size of the μ-peroxo dimer of two Co(III)·Lys-Gly-His metallopeptides was capable of interacting with most of the six-base-pair recognition site, whereas a metallopeptide monomer would only be capable of interacting with two to three base pairs. (2) Considering that the recognition site contains two 5'-XGG subsequences and the known targeting of the major groove by other Co complexes [125–127], if each metallopeptide half of the μ-peroxo dimer were to interact with one of the available 5'-GG sites, the dimer could deliver its oxidizing peroxyl bridging ligand to the 5–6 double bond of the T residue located in the middle of the recognition sequence. Oxidation of the 5–6 double bond of thymine would result in a piperidine-labile nucleobase lesion [90], as observed.

Overall, edit μ-peroxo dimers of Co(II)·Xaa-Xaa-His metallopeptides may lead to reactive, extended metallopeptide structures capable of recognizing and delivering a reactive oxygen equivalent derived from molecular oxygen to select DNA sites. Further study of this strategy will seek to investigate details of the chemistry of the DNA lesion(s) formed and will also attempt to harness this chemistry, in conjunction with combinatorial peptide–ligand libraries [76], to deliver a reactive O_2 equivalent to alternative DNA sites.

5.3
Recognition and Cleavage of RNA by Ni(II)·Xaa-Xaa-His Metallopeptides

Along with their ability to selectively recognize and induce DNA strand scission, Ni(II)·Xaa-Xaa-His metallopeptides also mediate selective RNA modification and cleavage [128]. As demonstrated for DNA, Ni(II)·Xaa-Xaa-His metallopeptides have the ability to increase our knowledge of fundamental RNA molecular recognition events by providing various amino acid contacts and overall complex architectures to complement the wealth of tertiary structures found in folded RNAs [27].

As an initial test of their abilities, Ni(II)·Gly-Gly-His, Ni(II)·Arg-Gly-His, and Ni(II)·Lys-Gly-His, again containing side chain functional groups often found in RNA–protein interactions [129, 130], were reacted with two unique RNA structures: 3′-^{32}P end-labeled tRNA^{Phe} and the TAR RNA of HIV-1 [128]. As shown (Fig. 5.20), treatment of these RNAs with the metallopeptides listed above followed by the addition of one equivalent of KHSO$_5$ resulted in predominant modification within the loop structures of these RNA substrates upon post-reaction work-up with aniline-acetate: (1) in tRNA^{Phe}, the metallopeptides were observed to modify the D, anticodon, and the TΨC loops and (2) with TAR, RNA modification was observed to occur primarily in the apical loop of this hairpin structure. Unique from their reactivity with DNA, the Ni(II)·Xaa-Gly-His metallopeptides employed in this experiment produced identical patterns of RNA damage regardless of the nature of the terminal Xaa amino acid [128].

Given the requirement for an aniline-acetate post-reaction work-up, it is likely that the metallopeptides target these RNA structures through an initial act of nucleobase modification that may be similar to other reported Ni complexes [24, 131]. However, with their apparent selectivity for loop structures, the variety of nucleobases targeted, and their lack of reactivity with single-stranded regions containing similar nucleobases, it is possible that RNA modification is preceded by an act of RNA recognition and binding.

While their exact mode of RNA recognition and modification remains undefined, it is interesting to note that within the RNA loops targeted, each metallopeptide appeared to produce a pattern of damage mainly toward their 3′-halves, suggesting an RNA-induced orientation of the metallopeptide that directs the modification of accessible nucleobases within each loop. This observation: (1) differentiates the metallopeptides from several other reagents that interact and modify single-stranded RNA, often based on nucleobase modification alone [22, 27,

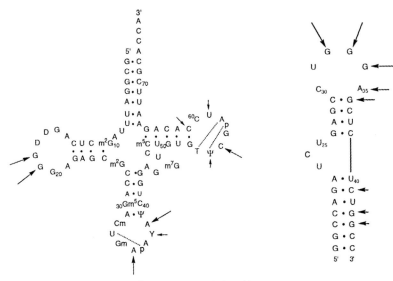

Fig. 5.20. Sites of RNA modification (arrows) induced by
Ni(II)·Xaa-Gly-His metallopeptides activated with KHSO₅. The
secondary structure of tRNA^Phe is shown on the left (dashed
lines indicate intra-loop hydrogen bonding) and the secondary
structure of the TAR RNA of HIV-1 is shown on the right.

132] and (2) suggests that the metallopeptides may behave in a fashion that mimics RNA-binding protein elements that often selectively recognize loop structures [129, 130].

At the time, speculation as to their mode of interaction resulted in the suggestion that RNA loops may create binding pockets that complement the overall structure of Ni(II)·Xaa-Gly-His metallopeptides. Given the similar patterns of RNA loop damage observed for all the metallopeptides tested, it is possible that, as with the minor groove of DNA, the His imidazole may insert into the loop region and sense its local environment, resulting in the oxidative modification of the nucleobases present. Thus, Ni(II)·Xaa-Gly-His metallopeptides may sense intra-loop structural features such as the "uridine turns" present in the anticodon and TΨC loops of tRNA^Phe where the U residues of these loops are hydrogen bonded to a phosphate on the opposite side of the loop [133]. Curiously, examination of the metallopeptide-induced RNA cleavage patterns within these loops suggested that the metallopeptides avoided the nucleotides immediately adjacent to the phosphates involved in these "U-turns". In comparison, with the apical loop of the TAR RNA, which lacks intra-loop hydrogen bonding [134], metallopeptide-induced modification occurs unimpeded throughout its 3′-half.

While the above study of RNA recognition by Ni(II)·Xaa-Gly-His metallopeptides tested only two substrates out of many unique RNA structures, it did indicate the ability of these metallopeptides to selectively modify RNA, allowed insight as to the chemistry of this nucleic acid modification, and provided a basis and incentive for future experiments. As with studies being carried out with DNA, the availability of combinatorial libraries of Ni(II)·Xaa-Xaa-His metallopeptides will allow many more amino acid–RNA interactions to be tested, perhaps allowing the discovery of unique RNA-targeted three-dimensional structures [27, 135].

5.4
Summary

Metallopeptides derived from Ni(II), Cu(II), or Co(III) complexes of Gly-Gly-His-derived peptides create unique and reactive agents that impact our understanding of nucleic acid structure and recognition through the development of affinity cleavage strategies and as stand-alone, low-molecular-weight complexes. As activated metallotripeptides, these complexes have the ability to recognize DNA and RNA structures resulting in selective strand scission and modification thus permitting details of amino acid– and peptide–nucleic acid recognition to be revealed. Information gathered by studies of metallopeptide–nucleic acid recognition and chemistry, through the use of combinatorial peptide–ligand libraries, their available well-defined three-dimensional shapes, and the potential for ambient O_2 activation, may eventually permit the generation of molecules with unique biological properties while also providing fundamental insight into protein–nucleic acid interactions.

Acknowledgements

The authors would like to acknowledge, with gratitude, the contributions of their co-workers and colleagues, whose names appear in the individual literature citations. The work carried out in the laboratory of the authors was supported by the National Institutes of Health, the US Department of Education GAANN Program, and the Purdue Research Foundation.

References

1 DYSON, H. J., WRIGHT, P. E. Defining solution conformations of small linear peptides. Annu. Rev. Biophys. Biophys. Chem. 1991, 20, 519–538.

2 STEITZ, T. A. Structural studies of protein–nucleic acid interaction: The sources of sequence-specific binding. Q. Rev. Biophys. 1990, 23, 205–280.

3 HELENE, C., MAURIZOT, J. C. Interactions of oligopeptides with nucleic acids. CRC Crit. Rev. Biochem. 1981, 10, 213–258.

4 BEHMOARAS, T., TOULME, J. J., HELENE, C. A tryptophan-containing peptide recognizes and cleaves DNA at apurinic sites. Nature (Lond.) 1981, 292, 858–859.

5 BRUN, F., TOULME, J. J., HELENE, C. Interactions of aromatic residues of proteins with nucleic acids. Fluorescence studies of the binding of oligopeptides containing tryptophan and tyrosine residues to polynucleotides. Biochemistry 1975, 14, 558–563.

6 SHEARDY, R. D., WILSON, W. D., KING, H. D. From reporter molecules to peptides – Interactions with nucleic acids, in *Chemistry and Physics of DNA–Ligand Interactions*, ed. N. R. Kallenbach, Adenine Press, Albany, New York, 1989, 175–212.

7 SHEARDY, R. D., GABBAY, E. J. Stereospecific binding of diastereomeric peptides to salmon sperm deoxyribonucleic acid. Further evidence for partial intercalation. *Biochemistry* 1983, *22*, 2061–2067.

8 ZIMMER, C., WAHNERT, U. Nonintercalating DNA-binding ligands: Specificity of the interaction and their use as tools in biophysical, biochemical and biological investigations of the genetic material. *Prog. Biophys. Mol. Biol.* 1986, *47*, 31–112.

9 GEIERSTANGER, B. H., WEMMER, D. E. Complexes of the minor groove of DNA. *Annu. Rev. Biophys. Biomol. Struct.* 1995, *24*, 463–393.

10 BAILLY, C., CHAIRES, J. B. Sequence-specific DNA minor groove binders. Design and synthesis of netropsin and distamycin analogues. *Bioconj. Chem.* 1998, *9*, 513–538.

11 DERVAN, P. B., BURLI, R. W. Sequence-specific DNA recognition by polyamides. *Curr. Opin. Chem. Biol.* 1999, *3*, 688–693.

12 WURTZ, N. R., DERVAN, P. B. Sequence specific alkylation of DNA by hairpin pyrrole-imidazole polyamide conjugates. *Chem. Biol.* 2000, *7*, 153–161.

13 KIELKOPF, C. L., BREMER, R. E., WHITE, S. *et al.* Structural effects of DNA sequence on T·A recognition by hydroxypyrrole/pyrrole pairs in the minor groove. *J. Mol. Biol.* 2000, *295*, 557–567.

14 KIELKOPF, C. L., WHITE, S., SZEWCZYK, J. W. *et al.* A structural basis for recognition of A·T and T·A base pairs in the minor groove of B-DNA. *Science* 1998, *282*, 111–115.

15 DOOLEY, C. T., HOUGHTEN, R. A. The use of positional scanning synthetic peptide combinatorial libraries for the rapid determination of opioid receptor ligands. *Life Sci.* 1993, *52*, 1509–1517.

16 HOUGHTEN, R. A., PINILLA, C., BLONDELLE, S. E., APPEL, J. R., DOOLEY, C. T., CUERVO, J. H. Generation and use of synthetic peptide combinatorial libraries for basic research and drug discovery. *Nature* 1991, *354*, 84–86.

17 PEREZ-PAYA, E., HOUGHTEN, R. A., BLONDELLE, S. E. Functionalized protein-like structures from conformationally defined synthetic combinatorial libraries. *J. Biol. Chem.* 1996, *271*, 4120–4126.

18 CHOREV, M., GOODMAN, M. A dozen years of retro-inverso peptidomimetics. *Acc. Chem. Res.* 1993, *26*, 266–273.

19 LONG, E. C., EASON, P. D., LIANG, Q. Synthetic metallopeptides as probes of protein–DNA interactions. *Met. Ions. Biol. Sys.* 1996, *33*, 427–452.

20 LONG, E. C. Ni(II)·Xaa-Xaa-His metallopeptide-DNA/RNA interactions. *Acc. Chem. Res.* 1999, *32*, 827–836.

21 PYLE, A. M., BARTON, J. K. Probing nucleic acids with transition metal complexes. *Prog. Inorg. Chem.* 1990, *38*, 413–475.

22 SIGMAN, D. S., MAZUMDER, A., PERRIN, D. M. Chemical nucleases. *Chem. Rev.* 1993, *93*, 2295–2316.

23 LONG, E. C. The DNA helical biopolymer: A template for the binding, assembly, and reactivity of metal ions and complexes. *J. Inorg. Organomet. Polym.* 1993, *3*, 3–39.

24 BURROWS, C. J., ROKITA, S. E. Recognition of guanine structure in nucleic acids by nickel complexes. *Acc. Chem. Res.* 1994, *27*, 295–301.

25 *Metal Ions in Biological Systems*: Probing of nucleic acids by metal ion complexes of small molecules, Vol. 33, eds. A. Sigel and H. Sigel, Marcel Dekker, New York, 1996.

26 ERKKILA, K. E., ODOM, D. T., BARTON, J. K. Recognition and reaction of metallointercalators with DNA. *Chem. Rev.* 1999, *99*, 2777–2795.

27 CHOW, C. S., BOGDAN, F. M. A structural basis for RNA–ligand interactions. *Chem. Rev.* 1997, *97*, 1489–1513.

28 BURGER, R. M. Cleavage of nucleic acids by bleomycin. *Chem. Rev.* **1998**, *98*, 1153–1169.

29 CLAUSSEN, C. A., LONG, E. C. Nucleic acid recognition by metal complexes of bleomycin. *Chem. Rev.* **1999**, *99*, 2797–2816.

30 STUBBE, J., KOZARICH, J. W., WU, W., VANDERWALL, D. E. Bleomycins: A structural model for specificity, binding, and double strand cleavage. *Acc. Chem. Res.* **1996**, *29*, 322–330.

31 KANE, S. A., HECHT, S. M. Poly-nucleotide recognition and degra-dation by bleomycin. *Prog. Nucl. Acids Res. Mol. Biol.* **1994**, *49*, 313–352.

32 BOGER, D. L., CAI, H. Bleomycin, syn-thetic and mechanistic studies. *Angew. Chem. Int. Ed.* **1999**, *38*, 448–476.

33 JAMIESON, E. R., LIPPARD, S. J. Structure, recognition, and processing of cisplatin-DNA adducts. *Chem. Rev.* **1999**, *99*, 2467–2498.

34 WONG, E., GIANDOMENICO, C. M. Current status of platinum-based antitumor drugs. *Chem. Rev.* **1999**, *99*, 2451–2466.

35 PABO, C. O., PEISACH, E., GRANT, R. A. Design and selection of novel Cys_2His_2 zinc finger proteins. *Annu. Rev. Biochem.* **2001**, *70*, 313–340.

36 REDDY, B. S. P., SONDHI, S. M., LOWN, J. W. Synthetic DNA minor groove-binding drugs. *Pharmacol. Therapeuts.* **1999**, *84*, 1–111.

37 WANG, L., BAILLY, C., KUMAR, A. *et al.* Specific molecular recognition of mixed nucleic acid sequences: an aromatic dication that binds in the DNA minor groove as a dimer. *Proc. Natl Acad. Sci. USA* **2000**, *97*, 12–16.

38 TAO, Z.-F., FUJIWARA, T., SAITO, I., SUGIYAMA, H. Rational design of sequence-specific DNA alkylating agents based on duocarmycin A and pyrrole-imidazole hairpin polyamides. *J. Am. Chem. Soc.* **1999**, *121*, 4961–4967.

39 SATZ, A. L., BRUICE, T. C. Synthesis of a fluorescent microgonotropen (FMGT-1) and its interactions with the dodecamer d(CCGGAATTCCGG). *Bioorg. Med. Chem. Lett.* **1999**, *9*, 3261–3266.

40 CHIANG, S.-Y., BRUICE, T. C., AZIZKHAN, J. C., GAWRON, L., BEERMAN, T. A. Targeting E2F1-DNA complexes with microgonotropen DNA binding agents. *Proc. Natl. Acad. Sci. USA* **1997**, *94*, 2811–2816.

41 HARFORD, C., SARKAR, B. Amino terminal Cu(II)- and Ni(II)-binding (ATCUN) motif of proteins and peptides: Metal binding, DNA cleavage, and other properties. *Acc. Chem. Res.* **1997**, *30*, 123–130.

42 CAMERMAN, N, CAMERMAN, A, SARKAR, B. Molecular design to mimic the copper(II) transport site of human albumin. The crystal and molecular structure of copper(II)-glycylglycyl-L-histidine-N-methyl amide monoaquo complex. *Can. J. Chem.* **1976**, *54*, 1309–1316.

43 BAL, W., DJURAN, M. I., MARGERUM, D. W. *et al.* Dioxygen-induced decarboxylation and hydroxylation of [Ni^{II}(glycyl-glycyl-L-histidine)] occurs via Ni^{III}: X-ray crystal structure of [Ni^{II}(glycyl-glycyl–hydroxy-D,L-histamine)]·$3H_2O$. *J. Chem. Soc., Chem. Commun.* **1994**, 1889–1890.

44 KOZLOWSKI, H., BAL, W., DYBA, M., KOWALIK-JANKOWSKA, T. Specific structure-stability relations in metallopeptides. *Coord. Chem. Rev.* **1999**, *184*, 319–346.

45 CARTER, B. J., MURTY, V. S., REDDY, K. S., WANG, S.-N., HECHT, S. M. A role for the metal binding domain in determining the DNA sequence selectivity of Fe-bleomycin. *J. Biol. Chem.* **1990**, *265*, 4193–4196.

46 KIMOTO, E., TANAKA, H., GYOTOKU, J., MORISHIGE, F., PAULING, L. Enhancement of antitumor activity of ascorbate against Ehrlich ascites tumor cells by the copper:glycylglycylhistidine complex. *Cancer Res.* **1983**, *43*, 824–828.

47 CHIOU, S.-H. DNA- and protein-scission activities of ascorbate in the presence of copper ion and a copper-peptide complex. *J. Biochem.* **1983**, *94*, 1259–1267.

48 CHIOU, S.-H. DNA-scission activities of ascorbate in the presence of metal chelates. *J. Biochem.* **1984**, *96*, 1307–1310.

49 CHIOU, S.-H., CHANG, W.-C., JOU, Y.-S., CHUNG, H.-M. M., LO, T.-B. Specific cleavages of DNA by ascorbate in the presence of copper ion or copper chelates. *J. Biochem.* **1985**, *98*, 1723–1726.

50 MACK, D. P., IVERSON, B. L., DERVAN, P. B. Design and chemical synthesis of a sequence-specific DNA cleaving protein. *J. Am. Chem. Soc.* **1988**, *110*, 7572–7574.

51 MACK, D. P., DERVAN, P. B. Nickel-mediated sequence-specific oxidative cleavage of DNA by a designed metalloprotein. *J. Am. Chem. Soc.* **1990**, *112*, 4604–4606.

52 MACK, D. P., DERVAN, P. B. Sequence-specific oxidative cleavage of DNA by a designed metalloprotein, Ni(II)·GGH(Hin139–190). *Biochemistry* **1992**, *31*, 9339–9405.

53 NAGAOKA, M., HAGIHARA, M., KUWAHARA, J., SUGIURA, Y. A novel zinc finger-based DNA cutter: Biosynthetic design and highly selective DNA cleavage. *J. Am. Chem. Soc.* **1994**, *116*, 4085–4086.

54 HARFORD, C., NARINDRASORASAK, S., SARKAR, B. The designed protein M(II)-Gly-Lys-His-Fos(138–211) specifically cleaves the AP-1 binding site containing DNA. *Biochemistry* **1996**, *35*, 4271–4278.

55 MCKAY, D. J., RENAUX, B. S., DIXON, G. H. Human sperm protamines. Amino-acid sequences of two forms of protamine P2. *Eur. J. Biochem.* **1986**, *156*, 5–8.

56 BAL, W., JEZOWSKA-BOJCZUK, M., KASPRZAK, K. S. Binding of nickel(II) and copper(II) to the N-terminal sequence of human protamine HP2. *Chem. Res. Toxicol.* **1997**, *10*, 906–914.

57 BAL, W., LUKSZO, J., KASPRZAK, K. S. Mediation of oxidative DNA damage by nickel(II) and copper(II) complexes with the N-terminal sequence of human protamine HP2. *Chem. Res. Toxicol.* **1997**, *10*, 915–921.

58 GROKHOVSKY, S. L., NIKOLAEV, V. A., ZUBAREV, V. E. *et al.* Sequence-specific cleavage of DNA by netropsin analog containing a copper(II)-chelating peptide Gly-Gly-His. *Mol. Biol. (Moscow)* **1992**, *26*, 1274–1297.

59 SHULLENBERGER, D. F., LONG, E. C. Design and synthesis of a DNA-cleaving metallopeptide. *Bioorg. Med. Chem. Lett.* **1993**, *3*, 333–336.

60 MORIER-TEISSIER, E., BOITTE, N., HELBECQUE, N. *et al.* Synthesis and antitumor properties of an anthraquinone bisubstituted by the copper chelating peptide Gly-Gly-L-His. *J. Med. Chem.* **1993**, *36*, 2084–2090.

61 FOOTER, M., EGHOLM, M., KRON, S., COULL, J. M., MATSUDAIRA, P. Biochemical evidence that a D-loop is part of a four-stranded PNA-DNA bundle. Nickel-mediated cleavage of duplex DNA by a Gly-Gly-His Bis-PNA. *Biochemistry* **1996**, *35*, 10673–10679.

62 DE NAPOLI, L., MESSERE, A., MONTESARCHIO, D. *et al.* A new solid-phase synthesis of oligonucleotides 3′-conjugated with peptides. *Bioorg. Med. Chem.* **1999**, *7*, 395–400.

63 TRUFFERT, J.-C., ASSELINE, U., BRACK, A., THUONG, N. T. Synthesis, purification and characterization of two peptide-oligonucleotide conjugates as potential artificial nucleases. *Tetrahedron* **1996**, *52*, 3005–3016.

64 STEULLET, V., DIXON, D. W. Design, synthesis and DNA-cleavage of Gly-Gly-His-naphthalene diimide conjugates. *Bioorg. Med. Chem. Lett.* **1999**, *9*, 2935–2940.

65 CUENOUD, B., TARASOW, T. M., SCHEPARTZ, A. A new strategy for directed protein cleavage. *Tetrahedron Lett.* **1992**, *33*, 895–898.

66 BROWN, K., YANG, S.-H., KODADEK, T. Highly specific oxidative cross-linking of proteins mediated by a nickel-peptide complex. *Biochemistry* **1995**, *34*, 4733–4739.

67 BERTRAND, R., DERANCOURT, J., KASSAB, R. Probing the hydrophobic interactions in the skeletal actomyosin subfragment 1 and its nucleotide complexes by zero-length cross-linking with a nickel-peptide chelate. *Biochemistry* **1997**, *36*, 9703–9714.

68 Brown, K. C., Yu, Z., Burlingame, A. L., Craik, C. S. Determining protein-protein interactions by oxidative cross-linking of a glycine-glycine-histidine fusion protein. *Biochemistry* **1998**, *37*, 4397–4406.

69 Van Dijk, J., Lafont, C., Furch, M., Manstein, D. J., Long, E. C., Chaussepied, P. Myosin isoform dependent crosslinking or cleavage of actin-myosin complexes by combinatorial libraries of Ni(II)·Xaa-Xaa-His metallopeptides *Biophys. J.* **2000**, *78*, 243A.

70 Rokita, S. E., Burrows, C. J. Nickel- and cobalt-dependent oxidation and crosslinking of proteins. *Met. Ions Biol. Syst.* **2001**, *38*, 289–311.

71 Person, M., Brown, K. C., Mahrus, S., Craik, C. S., Burlingame, A. L. Novel inter-protein cross-link identified in the GGH-ecotin D137Y dimer. *Protein Sci.* **2001**, *10*, 1549–1562.

72 Shullenberger, D. F., Eason, P. D., Long, E. C. Design and synthesis of a versatile DNA-cleaving metallopeptide structural domain. *J. Am. Chem. Soc.* **1993**, *115*, 11038–11039.

73 Long, E. C., Eason, P. D., Shullenberger, D. F. Incorporation of square-planar metal binding sites into protein polymeric structures, in *Metal-Containing Polymeric Materials*, eds. C. E. Carraher, M. Zeldin, J. E. Shats, B. M. Culbertson, C. U. Pittman, Plenum Publishing, New York, 1996, 481–489.

74 Liang, Q., Eason, P. D., Long, E. C. Metallopeptide-DNA interactions: Site-selectivity based on amino acid composition and chirality. *J. Am. Chem. Soc.* **1995**, *117*, 9625–9631.

75 Chen, X., Rokita, S. E., Burrows, C. J. DNA modification: Intrinsic selectivity of nickel(II) complexes. *J. Am. Chem. Soc.* **1991**, *113*, 5884–5886.

76 Huang, X., Pieczko, M. E., Long, E. C. Combinatorial optimization of the DNA cleaving Ni(II)·Xaa-Xaa-His metallotripeptide domain. *Biochemistry* **1999**, *38*, 2160–2166.

77 Coll, M., Aymami, J., van der Marel, G. A., van Boom, J. H., Rich, A., Wang, A. H.-J. Molecular structure of the netropsin-d(CGCGATATCGCG) complex: DNA conformation in an alternating AT segment. *Biochemistry* **1989**, *28*, 310–320.

78 Gravert, D. J., Griffin, J. H. Specific DNA cleavage mediated by [salenMn(III)]$^+$. *J. Org. Chem.* **1993**, *58*, 820–822.

79 Raner, G., Ward, B., Dabrowiak, J. C. Interactions of cationic manganese porphyrins with DNA. A binding model. *J. Coord. Chem.* **1988**, *19*, 17–23.

80 Burkoff, A. M., Tullius, T. D. The unusual conformation adopted by the adenine tracts in kinetoplast DNA. *Cell* **1987**, *48*, 935–943.

81 Nagane, R., Koshigoe, T., Chikira, M., Long, E. C. The DNA-bound orientation of Cu(II)·Xaa-Gly-His metallopeptides. *J. Inorg. Biochem.* **2001**, *83*, 17–23.

82 Long, E. C. Fundamentals of nucleic acids, in *Bioorganic Chemistry: Nucleic Acids*, ed. S. M. Hecht, Oxford University Press, New York, 1996, 3–35.

83 Eason, P. D. Design of Ni(II)-metallopeptides and their interactions with DNA. PhD dissertation, Purdue University, **1997**.

84 Sundberg, R. J., Martin, R. B. Interactions of histidine and other imidazole derivatives with transition metal ions in chemical and biological systems. *Chem. Rev.* **1974**, *74*, 471–517.

85 Chikira, M., Sato, T., Antholine, W. E., Petering, D. H. Orientation of non-blue cupric complexes on DNA fibers. *J. Biol. Chem.* **1991**, *266*, 2859–2863.

86 Liang, Q., Ananias, D. C., Long, E. C. Ni(II)·Xaa-Xaa-His induced DNA cleavage: Deoxyribose modification by a common "activated" intermediate derived from KHSO$_5$, MMPP, or H$_2$O$_2$. *J. Am. Chem. Soc.* **1998**, *120*, 248–257.

87 Long, E. C., Hecht, S. M., van der Marel, G. A., van Boom, J. H. Interaction of bleomycin with a methylated DNA oligonucleotide. *J. Am. Chem. Soc.* **1990**, *112*, 5272–5276.

88 SITLANI, A., LONG, E. C., PYLE, A. M., BARTON, J. K. DNA photocleavage by phenanthrenequinone diimine complexes of rhodium(III): Shape-selective recognition and reaction. *J. Am. Chem. Soc.* **1992**, *114*, 2303–2312.

89 MULLER, J. G., ZHENG, P., ROKITA, S. E., BURROWS, C. J. DNA and RNA modification promoted by [Co(H$_2$O)$_6$]Cl$_2$ and KHSO$_5$: guanine selectivity, temperature dependence, and mechanism. *J. Am. Chem. Soc.* **1996**, *118*, 2320–2325.

90 BURROWS, C. J., MULLER, J. G. Oxidative nucleobase modifications leading to strand scission. *Chem. Rev.* **1998**, *98*, 1109–1151.

91 LEE, W. A., BRUICE, T. C. Homolytic and heterolytic oxygen-oxygen bond scissions accompanying oxygen transfer to iron(III) porphyrins by percarboxylic acids and hydroperoxides. A mechanistic criterion for peroxidase and cytochrome P-450. *J. Am. Chem. Soc.* **1985**, *107*, 513–514.

92 MEUNIER, B. Metalloporphyrin-catalysed oxygenation of hydrocarbons., *Bull. Soc. Chim. Fr.* **1986**, *4*, 578–594.

93 SUBAK, E. J., LOYOLA, V. M., MARGERUM, D. W. Substitution and rearrangement reactions of nickel(III) peptide complexes in acid. *Inorg. Chem.* **1985**, *24*, 4350–4356.

94 KOOLA, J. D., KOCHI, J. K. Nickel catalysis of olefin epoxidation. *Inorg. Chem.* **1987**, *26*, 908–916.

95 KINNEARY, J. F., ALBERT, J. S., BURROWS, C. J. Mechanistic studies of alkene epoxidation catalyzed by nickel(II) cyclam complexes. ^{18}O labeling and substituent effects. *J. Am. Chem. Soc.* **1988**, *110*, 6124–6129.

96 ROSS, S. A., BURROWS, C. J. Nickel complexes of cysteine- and cysteine-containing peptides: Spontaneous formation of disulfide-bridged dimers at neutral pH. *Inorg. Chem.* **1998**, *37*, 5358–5363.

97 STEWART, J. M., YOUNG, J. D. *Solid-Phase Peptide Synthesis*, Pierce Chemical Co., Rockville, IL, **1984**.

98 LESCRINIER, T., HENDRIX, C., KERREMANS, L. *et al.* DNA-binding ligands from peptide libraries containing unnatural amino acids. *Chem. Eur. J.* **1998**, *4*, 425–433.

99 ALAM, M. R., MAEDA, M., SASAKI, S. DNA binding peptides searched from the solid-phase combinatorial library with the use of the magnetic beads attaching the target duplex DNA. *Bioorg. Med. Chem.* **2000**, *8*, 465–473.

100 SLUKA, J. P., HORVATH, S. J., GLASGOW, A. C., SIMON, M. I., DERVAN, P. B. Importance of minor-groove contacts for recognition of DNA by the binding domain of Hin recombinase. *Biochemistry* **1990**, *29*, 6551–6561.

101 FENG, J.-A., JOHNSON, R. C., DICKERSON, R. E. Hin recombinase bound to DNA: The origin of specificity in major and minor groove interactions. *Science* **1994**, *263*, 348–355.

102 WANG, S., COSSTICK, R., GARDNER, J. F., GUMPORT, R. I. The specific binding of *Escherichia coli* integration host factor involves both major and minor grooves of DNA. *Biochemistry* **1995**, *34*, 13082–13090.

103 SUZUKI, M. SPKK, a new nucleic acid-binding unit of protein found in histone. *EMBO J.* **1989**, *3*, 797–804.

104 REEVES, R., NISSEN, M. S. The A·T-DNA binding domain of mammalian high mobility group I chromosomal proteins. A novel peptide motif for recognizing DNA structure. *J. Biol. Chem.* **1990**, *265*, 8573–8582.

105 LILLEY, D. M. J. HMG has DNA wrapped up. *Nature* **1992**, *357*, 282–285.

106 BEWLEY, C. A., GRONENBORN, A. M., CLORE, G. M. Minor groove-binding architectural proteins: Structure, function, and DNA recognition. *Annu. Rev. Biophys. Biomol. Struct.* **1998**, *27*, 105–131.

107 MULLER, J. G., HICKERSON, R. P., PEREZ, R. J., BURROWS, C. J. DNA damage from sulfite autoxidation catalyzed by a nickel(II) peptide. *J. Am. Chem. Soc.* **1997**, *119*, 1501–1506.

108 SAITO, I., TAKAYAMA, M., SUGIYAMA, H., NAKATANI, K. Photoinduced DNA cleavage via electron transfer: demonstration that guanine residues located 5' to guanine are the most electron donating sites. *J. Am. Chem. Soc.* **1995**, *117*, 6406–6407.

109 LEPENTSIOTIS, V., DOMAGALA, J., GRGIC, I., VAN ELDIK, R., MULLER, J. G., BURROWS, C. J. Mechanistic information on the redox cycling of nickel (II/III) complexes in the presence of sulfur oxides and oxygen. Correlation with DNA damage experiments. *Inorg. Chem.* **1999**, *38*, 3500–3505.

110 STUART, J. N., GOERGES, A. L., ZALESKI, J. M. Characterization of the Ni(III) intermediate in the reaction of (1,4,8,11-tetraazacyclotetradecane)nickel(II) perchlorate with KHSO$_5$: implications to the mechanism of oxidative DNA modification. *Inorg. Chem.* **2000**, *39*, 5976–5984.

111 ANANIAS, D. C., LONG, E. C. DNA strand scission by dioxygen + light-activated cobalt metallopeptides. *Inorg. Chem.* **1997**, *36*, 2469–2471.

112 ANANIAS, D. C., LONG, E. C. Highly selective DNA modification by ambient O$_2$-activated Co(II)·Lys-Gly-His metallopeptides. *J. Am. Chem. Soc.* **2000**, *122*, 10460–10461.

113 MCLENDON, G., MOTEKAITIS, R. J., MARTELL, A. E. Cobalt complexes of ethylenediamine-N,N'-diacetic acid and ethylenediamine-N,N-diacetic acid. Two-nitrogen oxygen carriers. *Inorg. Chem.* **1975**, *14*, 1993–1996.

114 LEVER, A. B. P., GRAY, H. B. Electronic spectra of metal-dioxygen complexes. *Acc. Chem. Res.* **1978**, *11*, 348–355.

115 JONES, R. D., SUMMERVILLE, D. A., BASOLO, F. Synthetic oxygen carriers related to biological systems. *Chem. Rev.* **1979**, *79*, 139–179.

116 FLEISHER, M. B., WATERMAN, K. C., TURRO, N. J., BARTON, J. K. Light induced cleavage of DNA by metal complexes. *Inorg. Chem.* **1986**, *25*, 3549–3551.

117 CHANG, C.-H., DALLAS, J. L., MEARES, C. F. Identification of a key structural feature of cobalt(III)-bleomycins: an exogenous ligand (e.g., hydroperoxide) bound to cobalt. *Biochem. Biophys. Res. Commun.* **1983**, *110*, 959–966.

118 SAITO, I., MORII, T., SUGIYAMA, H., MATSUURA, T., MEARES, C. F., HECHT, S. M. Photoinduced DNA strand scission by cobalt bleomycin green complex. *J. Am. Chem. Soc.* **1989**, *111*, 2307–2308.

119 TAN, J. D., HUDSON, S. E., BROWN, S. J., OLMSTEAD, M. M., MASCHARAK, P. K. Syntheses, structures, and reactivities of synthetic analogues of the three forms of Co(III)-bleomycin: Proposed mode of light-induced DNA damage by the Co(III) chelate of the drug. *J. Am. Chem. Soc.* **1992**, *114*, 3841–3853.

120 HAWKINS, C. J., MARTIN, J. Cobalt(III) complex of glycylglycyl-histidine: preparation, characterization, and conformation. *Inorg. Chem.* **1983**, *22*, 3879–3883.

121 XU, R. X., ANTHOLINE, W. E., PETERING, D. H. Reaction of Co(II)bleomycin with dioxygen. *J. Biol. Chem.* **1992**, *267*, 944–949.

122 CHENG, C.-C., ROKITA, S. E., BURROWS, C. J. Nickel(III)-promoted DNA cleavage with ambient oxygen. *Angew. Chem. Int. Ed. Engl.* **1993**, *32*, 277–278.

123 BHATTACHARYA, S., MANDAL, S. S. Ambient oxygen activating water soluble cobalt-salen complex for DNA cleavage. *J. Chem. Soc., Chem. Commun.* **1995**, 2489–2490.

124 MCCARTHY, J. G., WILLIAMS, L. D., RICH, A. Chemical reactivity of potassium permanganate & diethylpyrocarbonate with B-DNA: Specific reactivity with short A-tracts. *Biochemistry* **1990**, *29*, 6071–6081.

125 GESSNER, R. V., QUIGLEY, G. J., WANG, A. H.-J., VAN DER MAREL, G. A., VAN BOOM, J. H., RICH, A. Structural basis for stabilization of Z-DNA by cobalt hexammine and magnesium cations. *Biochemistry* **1985**, *24*, 237–240.

126 CALDERONE, D. M., MANTILLA, E. J., HICKS, D. H., MURPHY, W. R., SHEARDY, R. D. Binding of Co(III) to a

DNA oligomer via reaction of $[Co(NH_3)_5(OH_2)]^{3+}$ with $(^{5Me}dC-dG)_4$. *Biochemistry* **1995**, *34*, 13841–13846.

127 NUNN, C. M., NEIDLE, S. The high resolution crystal structure of the DNA decamer d(AGGCATGCCT). *J. Mol. Biol.* **1996**, *256*, 340–351.

128 BRITTAIN, I. J., HUANG, X., LONG, E. C. Selective recognition and cleavage of RNA loop structures by Ni(II)·Xaa-Gly-His metallopeptides. *Biochemistry* **1998**, *37*, 12113–12120.

129 VARANI, G. RNA-protein intermolecular recognition. *Acc. Chem. Res.* **1997**, *30*, 189–195.

130 DRAPER, D. E. Protein-RNA recognition. *Annu. Rev. Biochem.* **1995**, *64*, 593–620.

131 CHEN, X., WOODSON, S. A., BURROWS, C. J., ROKITA, S. E. A highly sensitive probe for guanine N7 in folded structures of RNA: application to

tRNA^{Phe} and *Tetrahymena* group I intron. *Biochemistry* **1993**, *32*, 7610–7616.

132 MAZUMDER, A., CHEN, C.-H. B., GAYNOR, R., SIGMAN, D. S. 1,10-Phenanthroline-copper, a footprinting reagent for single-stranded regions of RNAs. *Biochem. Biophys. Res. Commun.* **1992**, *187*, 1503–1509.

133 RICH, A. Three-dimensional structure and biological function of transfer RNA. *Acc. Chem. Res.* **1977**, *10*, 388–396.

134 PUGLISI, J. D., TAN, R., CALNAN, B. J., FRANKEL, A. D., WILLIAMSON, J. R. Conformation of the TAR-arginine complex by NMR. *Science* **1992**, *257*, 76–80.

135 HERMANN, T. Strategies for the design of drugs targeting RNA and RNA–protein complexes. *Angew. Chem. Int. Ed. Engl.* **2000**, *39*, 1890–1905.

6
Salen–Metal Complexes

S. E. Rokita and C. J. Burrows

6.1
Introduction

The distinct properties and ready availability of metal "salens" have inspired their use in numerous disciplines, including inorganic, organic, and biological chemistry. The salen ligand (bis(salicylaldehyde)ethylenediimine) provides a relatively rigid tetradentate environment that significantly influences the steric and electronic properties of their transition metal complexes in a predictable manner [1, 2]. The two phenolic hydrogens dissociate upon metal binding and help to create a strongly electron-donating environment in conjunction with the two neutral Schiff base imines (Scheme 6.1). Much of the recent interest in these systems originates in part from the efficiency of their manganese and nickel derivatives to catalyze olefin epoxidation in the presence of terminal oxidants such as iodosylbenzene, peracids, and hypochlorite (common bleach) [3–5]. A wide array of metal and salen combinations have since been examined for a variety of chemical and biochemical activities. Two notable examples illustrating the diversity of these efforts include development of a manganese–di(*tert*-butylated)salen that promotes asymmetric epoxidation [6] and a cationic nickel–salen that covalently couples to accessible guanine residues in DNA [7].

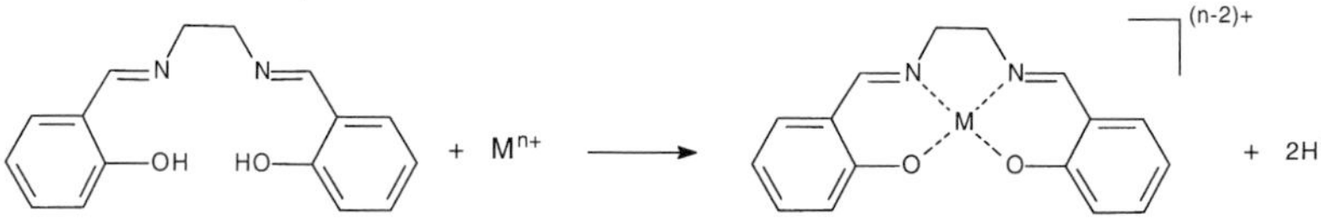

Scheme 6.1. Metal coordination by salen.

Salen derivatives are easily prepared in a single procedure by condensing two equivalents of a salicylaldehyde with a diamine (Scheme 6.2). This assembly usually occurs spontaneously [8] and may also be promoted by addition of a metal or other templates [9, 10]. Alternative stepwise condensation allows for incorporation of non-identical salicylaldehyde components [11]. Most of the early salens were

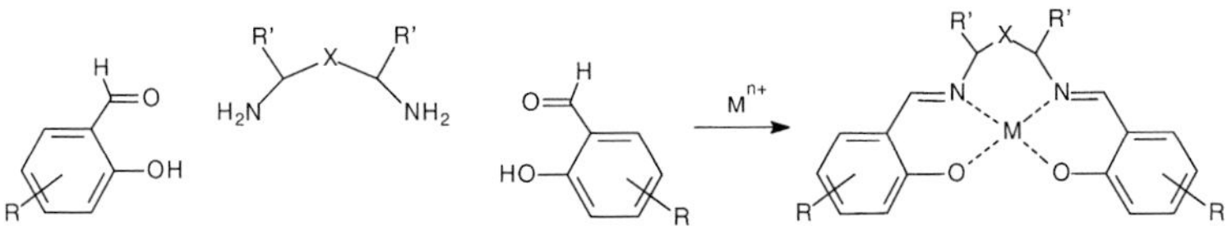

Scheme 6.2. Assembly of a metal–salen complex.

designed to form neutral complexes with dicationic metals for study in organic solvents, but an anionic ligand had also been developed for aqueous application [12]. Neither of these were very appropriate for reaction with polyanionic nucleic acids under physiologically compatible conditions and, in one case, the anionic substituents changed the coordination geometry of the metal [12]. Our laboratories consequently synthesized a salen with two pendent quaternary ammonium groups to provide solubility in water and general electrostatic affinity for nucleic acids [7]. Cationic salen–metal complexes are also now being examined for use in protein [13, 14] and lignin [15] oxidation.

Commercial availability of numerous diamines and salicylaldehydes supports rapid and facile construction of many individual salen derivatives [8, 16] and may be adapted for modern combinatorial methods in the future [17, 18]. By varying the metal and ligand substituents, salen complexes can be shown to either cleave the phosphoribose backbone of DNA or alkylate the nucleobase guanine within DNA and RNA under control of a range of different redox partners and with various structural specificities as described in this chapter. Many of these salen derivatives additionally offer practical alternatives to the most popular probes of nucleic acids that are based on metal complexes such as FeEDTA [19], Cu(1,10-phenanthroline)$_2$ [20], NiCR [21, 22], and those containing rhodium [23] or ruthenium [24]. The ease of functionalizing and assembling salens has similarly expedited their conjugation to biologically active compounds and incorporation into chimeric metal complexes.

6.2
Reversible Binding of Simple Metal–Salen Complexes

Metal–salen complexes are most often adopted for biochemical application based on their ability to modify nucleic acids covalently. Consequently, few studies have focused expressly on their non-covalent interactions, although such studies on binding are necessary if the independent contributions of electrostatics, groove recognition, and intercalation are to be measured. A sequence- or site-specificity determined by the covalent reactivity of a salen should not necessarily be considered synonymous with its binding specificity since only a limited number of orientations may support efficient reaction. Reversible association between salen complexes and nucleic acids are best identified through chemical footprinting [25, 26] and a variety of physical methods [9]. The results of these studies can often be used to rationalize existing data on reaction specificity and help when designing subsequent generations of salen-based reagents.

Both the metal and the ligand affect the interactions of their complex with nucleic acids by defining the net charge, coordination number, and solubility, as well as steric and hydrophobic properties. For example, Cr(III) and Mn(III) form water-soluble complexes with the parent salen ligand (Scheme 6.1) because of their net (+1) charge but their interactions with DNA can vary due to the square-pyramidal geometry of the pentadentate manganese complex with one axially ligated water and the octahedral geometry of the hexadentate chromium complex with two axially waters [27]. Ni(II) and Cu(II) form neutral complexes with the parent salen ligand and thus require charged substituents on the ligands to achieve water solubility. The M(II) oxidation state of these metals also form square-planar geometries and do not readily gain axially ligands unless converted to their M(III) states. Only one study has directly addressed the competing influence of steric bulk and hydrophobicity, and this relied only on a series of Cu(II)–salen derivatives (Scheme 6.3) [28]. Sterics dominated this example since the affinity for DNA decreased as the size of the ligand increased (R, Me $\Rightarrow$ Bu).

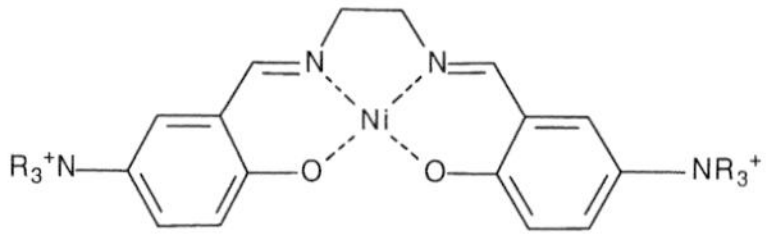

R = Me, Et, *n*-Pr, *n*-Bu

Scheme 6.3. Systematic variation of the steric and hydrophobic properties of a Cu(II) salen.

Most other characteristics of binding between metal–salens and nucleic acids also derive from studies on DNA and cationic copper complexes. No doubt the net positive charge of these derivatives confers a favorable electrostatic affinity to the polyanionic phosphoribose backbone of DNA. Whether further association is based on groove binding or intercalation is somewhat more controversial. ESR spectroscopy indicated that the salen illustrated in Scheme 6.3 (R = Me) orients with respect to DNA fibers in a manner most consistent with binding in the major groove [29]. In contrast, replacing the ethylene diimine with a phenylene diimine unit shifted the salen to an almost perpendicular orientation with respect to the fiber as expected for an intercalative mode of binding [29]. Interestingly, the ethylene derivative ($K_a = 1.0 \times 10^3$ M) bound approximately 50-fold more tightly than the phenylene derivative ($K_a = 53 \times 10^3$ M) to calf thymus DNA [28].

Another salen analog containing a cationic alkylamine-substituted ethylene bridge readily displaced ethidium from calf thymus DNA and appeared by electric linear dichroism to bind perpendicular to the helical axis of DNA much like the previous phenylene derivative [9]. However, this copper–salen did not induce helical unwinding as expected for intercalation and consequently major groove binding was proposed as an alternative to be considered equally [9]. A preference for binding at AT base pairs or AT-rich regions was also detected by its greater affinity for poly(dA-dT) versus poly(dG-dC) [9]. This result suggests that coordination between guanine N7, the strongest DNA ligand for copper [30, 31], and the salen complex did not control binding although it does control the reaction selectivity of nickel–salen as described in Section 6.4. Perhaps the wider major groove formed by AT base pairs relative to GC base pairs stabilizes interactions with Cu(II)–salen. Interestingly, a series of Mn(III)–salen derivatives expressed a reaction specificity for regions of DNA with high AT content in parallel to the binding selectivity expressed by the copper derivatives. But in this case, target selectivity was ascribed to the minor groove characteristics of AT-rich regions [32]. Further experiments will be needed to distinguish between these possibilities since sequence-dependent deviations from the canonical Watson–Crick duplex induce commensurate changes in both major and minor groove dimensions [33]. Regardless of the mode by which metal–salens bind to DNA, their selectivity can be easily manipulated by conjugating them to drugs such as distamycin and ellipticine with known affinity for DNA [25, 26].

Reversible binding of nickel–salens to nucleic acids has not been examined although these complexes covalently couple to unpaired guanines of DNA and RNA with high specificity in the presence of a peracid such as monoperoxyphthalic acid (MMPP) or monoperoxysulfate (HSO_5^-) [7, 34–37]. This ability to detect unusual secondary structures in nucleic acids is unique to the nickel derivatives and may originate from a transient Ni(III) species that coordinates preferentially to the N7 position of highly accessible guanine residues [7, 21, 35]. Such a possibility was recently demonstrated through distance-dependent effects of a paramagnetic complex of nickel as detected by the proton NMR spectra of duplex DNA containing an unpaired guanine [38, 39].

6.3
Nucleic Acid Strand Scission Induced by Simple Metal–Salen Complexes

Metal–salen complexes screened for reactivity with nucleic acids have typically been found to induce oxidative scission of the phosphoribose backbone leading to direct strand fragmentation. This net process can involve a variety of different reactive intermediates and sites of phosphoribose oxidation [40]. A limited number of complexes have the ability to react spontaneously but most require the presence of an oxygen donor and, in certain cases, a reductant. The nature of both the ligand and bound metal govern which conditions are necessary to support covalent reaction of nucleic acids. These requirements in turn delineate the range of applications suitable for each derivative. The reductive activation required by copper–salens is particularly attractive since it avoids use of strong oxidants and has the potential for use *in vivo*. Most other metal–salens require oxidants ranging from hydrogen peroxide to peracids, although some need only molecular oxygen. Each system that is competent at direct strand scission is reviewed in more detail below (Sections 6.3.1–6.3.4). Typical assays used to screen for this type of activity include nicking of supercoiled plasmid DNA and scission of polynucleotide restriction fragments (see below for example, Figs. 6.1 and 6.2). Unfortunately, other types of nucleic acid modification are not detected by these techniques and certain modes of reaction such as covalent adduct formation can proceed without detection unless alternative methods are used for product analysis.

6.3.1
Metal–Salens Activated by Reductants for Strand Scission

In analogy to Cu(1,10-phenanthroline)$_2$ [20], copper–salens are also easily reduced from their Cu(II) to Cu(I) forms. These in turn react readily with ambient levels of molecular oxygen to generate intermediates which drive substrate oxidation [9, 41]. For example, a copper–salen containing a single alkylamine substituent (50 µM) converted a plasmid (1 hour, 37°C) entirely from its supercoiled (form I) to nicked (form II) conformation by strand cleavage in the presence of 2-mercaptopropionic acid (MPA, 2.5 mM) and air-saturated buffer (Scheme 6.4, Fig. 6.1) [9]. The efficiency of plasmid nicking increased moderately after the DNA-binding agents anthracenedione and ellipticine were conjugated through the alkylamine of the copper–salen [25, 41]. Also, MPA was observed to be more effective than dithio-

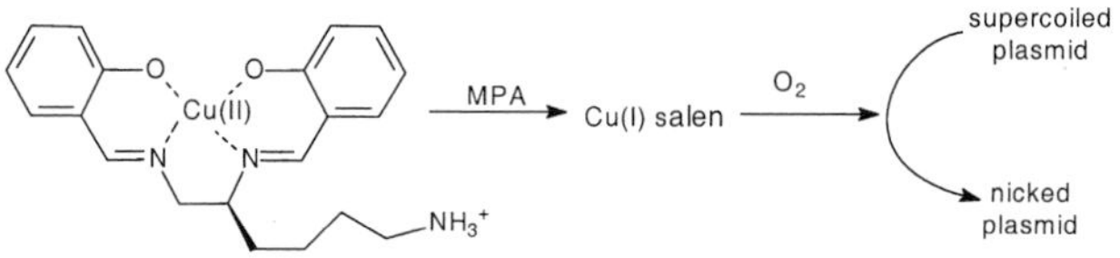

Scheme 6.4. Plasmid strand scission mediated by a copper–salen complex.

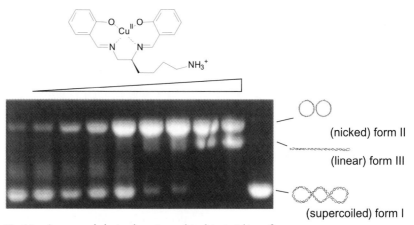

Fig. 6.1. Agarose gel electrophoresis used to detect nicking of supercoiled plasmid DNA in the presence of MPA and varying concentrations of the copper–salen illustrated in Scheme 6.4 under aerobic conditions. Adapted from Ref. [9].

threitol or hydrogen peroxide at promoting these processes. Surprisingly, an equivalent distamycin conjugate acted on a restriction fragment without sequence selectivity despite the preferential binding of distamycin to AT-rich regions [26]. The origins of this result are not yet clear since the ultimate oxidant for at least $Cu(1,10$-phenanthroline$)_2$ and other copper-containing complexes is thought to be non-diffusible and hence be retained within the site of DNA recognition [40, 42]. Even the diffusible hydroxyl radical generated by iron–EDTA affinity conjugates primarily acts within the local vicinity of its DNA target [43, 44].

6.3.2
Metal–Salens Activated by Peracids for Strand Scission

The majority of metal–salens rely on the presence of an added oxidant to initiate DNA scission or nicking although exact requirements depend on the nature of the salen and its bound metal. Some derivatives effect scission only in the presence of strong oxidants such as a peracid (MMPP or $KHSO_5$) while others need only molecular oxygen. One of the very first studies reported a reaction between a salen derivative and plasmid DNA involving a Ni(II) complex of the parent salen, bis(salicylaldehyde)ethyl enediimine, along with oxidants MMPP and iodosylbenzene alternatively [45]. Both conditions supported strand scission but their efficiencies could not be easily determined due to the low solubility of the neutral Ni(II)–salen and iodosylbenzene. In addition, the high molecular weight of the polynucleotide products prevented detection of the unique ability of nickel–salen to couple covalently to its target sites of DNA [7].

A Mn(III) complex of the same salen parent has the advantage of a net $(+1)$ charge that enhanced its solubility in aqueous media and provided electrostatic affinity to DNA. Reaction of this complex in the presence of MMPP was characterized in part by direct strand scission of a 517-base-pair restriction fragment (Fig. 6.2) [32]. This allowed for product identification at nucleotide resolution and revealed a selectivity for AT-rich regions [32]. The observed selectivity was attributed to preferential binding and phosphoribose oxidation in the minor groove of DNA by a Mn(V)oxo–salen intermediate or a species with equivalent activity. In this case, association in the minor groove was implied by a characteristic 3′ offset in the profile of scission products formed by the complementary strand. These results were also reminiscent of the reaction and target selectivity of a cationic Mn(III) porphyrin derivative in the presence of $KHSO_5$ [46]. An extensive series of Mn(III) –salens were subsequently characterized to establish a structural basis for reaction efficiency and target recognition [16]. Despite considerable perturbation of the aromatic ring's electronic and steric properties and the diimine linker's composition, the position of DNA scission remained relatively constant. In contrast, reaction efficiency varied by over 10-fold. Significant differences in efficiency and even some variation in product profiles were also detected for enantiomers created by asymmetric substitution on the diimine linker [16].

Alternative salen derivatives based on the presence of two pendant quarternary amines were developed to improve the solubility properties of otherwise neutral

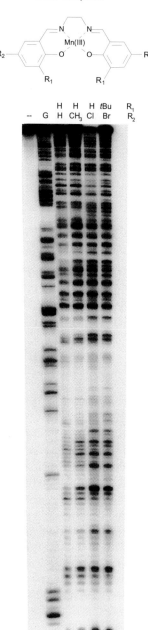

Fig. 6.2. Denaturing polyacryamide gel electrophoresis used to detect direct strand scission of DNA by a series of manganese salens in the presence of MMPP. Adapted from Ref. [32].

M(II)–salens and to create an electrostatic attraction for nucleic acids [7, 47]. One of these ligand systems was then used to compare the activity of Mn(III), Ni(II), Cu(II) , Co(II) and Cr(III) derivatives in the presence of MMPP (Scheme 6.5). Plasmid cleavage assays confirmed the expected high reactivity of the Mn(III)– and Ni(II)–salens (1–50 μM) and also revealed a lack of reactivity of Cu(II)–, Co(II)– and Cr(III)–salens [47, 48]. The importance of electrostatics in designing salens was concurrently illustrated by the inability of a Ni(II)–salen complex containing an anionic appendage (Scheme 6.5) to either oxidize or reversibly bind DNA in a detectable manner. The need to match reaction conditions with each type of salen complex was also highlighted by the ability of the Co(II)–salen to nick plasmid DNA in the absence, rather than presence, of MMPP (see discussion below on spontaneous activation of molecular oxygen, Section 6.3.4). Similarly, the copper–salen complex was unable to induce strand scission in the presence of an oxidant such as MMPP but was quite proficient at scission in the presence of molecular oxygen and a reductant such as MPA [9].

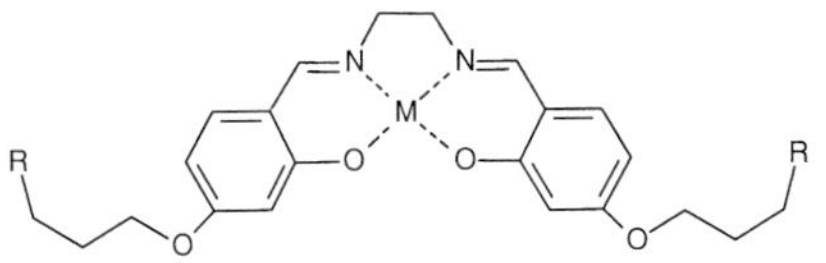

M = Mn(III), Ni(II), Cu(II), Cr(III)

R = N(CH$_3$)$_3$$^+$, CH$_2CH_2COO^-$

Scheme 6.5. Cationic and anionic metal salens.

Primer extension assays were used in further studies to compare the same series of metal–salens and did indeed verify the reactivity of the Ni(II)– and Mn(III)–salens in a qualitative manner. This assay relies on the inability of DNA polymerase to replicate past sites of nucleobase or phosphoribose damage [49]. Unlike the plasmid-nicking assay, primer extension is very useful for identifying the exact sites of nucleotide modification and hence the sequence selectivity of reaction. However, this assay cannot distinguish between strand scission and many types of base modification. Thus, reaction efficiencies measured by the plasmid-nicking assay should be considered independently from those of primer extension. Similarly, the observable end points of the two assays are quite different since 100% nicking of plasmid DNA is equivalent to approximately one oxidation event per >8600 nucleotides (pBR322), whereas primer extension may detect only one event per ~200 nucleotides. If oligonucleotide systems are used, even more abundant reaction might be necessary for its detection [7]. Whatever type of modification was ultimately detected with the Ni(II)– and Mn(III)–salens, the former expressed a preference for GC sequences and the latter expressed a preference for AT sequences [47, 48]. These results are generally consistent with the nickel–salens described in Section 6.4 on nucleic acid adduct formation as well as with other nickel and manganese complexes not based on salens [16, 21, 32, 46, 49, 50].

6.3.3
Metal Salens Activated by Hydrogen Peroxide for Strand Scission

Metal salens that are active with hydrogen peroxide rather than a stronger oxidant such as MMPP offer milder conditions for nucleic acid reaction. However, these types of systems have not been widely explored nor applied in biochemical studies. Their modes of action are often difficult to characterize since hydrogen peroxide has the potential to act as a radical precursor, a reductant and an oxidant. A Mn(III)–salen had originally been shown to promote double-strand cleavage of plasmid DNA with the same specificity, whether MMPP or hydrogen peroxide was present [32]. Manganese derivatives also exhibit the unusual ability to react under diverse conditions ranging from ascorbate to $KHSO_5$ [51]. Hydrogen peroxide only very weakly supports reaction of a copper–salen–anthracenedione conjugate in comparison to its reaction in the presence of the reductant MPA [41] and would not be expected to support reaction of the nickel–salens. In contrast, hydrogen peroxide-dependent reactions based on iron are ubiquitous, and thus efficient plasmid nicking by hydrogen peroxide and an iron–salen complex was not surprising [52]. An Ru(III)–salen derivative containing two pendant cationic groups also promotes hydrogen peroxide-dependent oxidation of DNA [53]. The resulting products of this reaction did not undergo spontaneous strand scission but instead required subsequent treatment with hot piperidine for scission. All nucleotides were susceptible to modification and modest selectivity for non-helical DNA, including bulge and hairpin structures, was observed. The basis for selectivity of the Ru(III)–salen has not yet been identified, but intercalation of the salen complex appears unlikely [53].

6.3.4
Metal–Salens Activated by Molecular Oxygen for Strand Scission

Recent efforts have begun to focus on metal–salen complexes that may act spontaneously in the absence of any co-reactant or at least in the mere presence of molecular oxygen. Such complexes have the potential for broad application and, in particular, may be used for characterizing nucleic acid conformation and dynamics *in vivo* [54]. The ability of Co(II)–salens to bind molecular oxygen reversibly has been known for over six decades and has already been subject to intensive investigation [8, 55, 56] A Co(II)–salen containing two cationic appendages (Scheme 6.5) has also been shown capable of nicking plasmid DNA under standard aerobic conditions [57]. Reaction was inhibited by the presence of the oxidant MMPP but greatly stimulated by the reductant dithiothreitol (DTT) (Scheme 6.6). Together, these and other data suggest formation of a complex between molecular oxygen and Co(II)–salen which serves as a crucial intermediate for nucleic acid oxidation. Primer extension analysis of duplex DNA treated with the Co(II)–salen in the presence and absence of added DTT revealed a slight preference for reaction at guanine residues but the exact products formed under these conditions have not yet been identified [57].

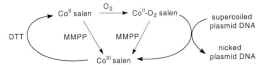

Scheme 6.6. Oxygen activation by Co(II)–salen.

A Ni(II)–salen complex equivalent to the Co(II) complex above was incapable of inducing plasmid scission in the presence of only molecular oxygen [57]. Model studies in organic solvents indicate that standard Ni(II)– and Cu(II)–salens neither bind nor activate molecular oxygen under ambient conditions [58, 59]. Reduction of the salen diimine to a diamine, however, generates a ligand that still coordinates to Ni(II) and now enables the resulting complex to undergo oxidation in the presence of molecular oxygen [58, 59]. Derivatives of Ni(II)–tetrahydrosalen have since been constructed for aqueous solubility but not yet applied to nucleic acid analysis [60]. In these cases, the reduced ligand might serve as the sacrificial reductant of molecule oxygen. Sulfite (SO_3^{2-}) has similarly been used in this manner during a process involving nickel–salen-dependent reduction of molecular oxygen (see Section 6.5).

Unlike the Ni(II) derivatives, a Cu(II)–tetrahydrosalen remained inert to molecular oxygen [59]. Copper(II)–dioxygen chemistry was alternatively stimulated by use of salen derivatives containing hydroxy groups on the aromatic rings [61]. A derivative containing bis(*para*-hydroxy) substituents most readily formed oxygen radicals in the presence of molecular oxygen and similarly expressed greatest proficiency at nicking plasmid DNA when compared to its bis(*ortho*-hydroxy) analog [61]. In contrast, the meta isomer did not detectably produce such radicals and was unable to cause plasmid nicking.

A similar trend was observed for the iron complexes of these same bishydroxy-substituted salens [62]. The *para*-substituted derivative (50 μM) nicked 75% of the available supercoiled DNA in 3.5 h under aerobic conditions (37°C), whereas the *ortho* and *meta* derivatives nicked no more than 6% of the DNA. Iron complexes of the parent salen lacking the hydroxy groups and the reduced tetrahydrosalen exhibited little or no reaction. The *para* derivative also bound reversibly to DNA with surprisingly much greater affinity than any of its analogs as indicated by its ability to stabilize helical DNA and significantly increase the T_m values for poly(dA-dT) and calf thymus DNA [62]. The basis for this strong association has not yet been identified and does not influence or guide reaction to any particular nucleotide sequence. Direct polynucleotide fragmentation is generated by this iron–salen without sequence specificity in a manner reminiscent of iron–EDTA and, like iron–ETDA, the iron–salen has been applied to mapping DNA–drug interactions [62].

Finally, Cr(V)–salens may yet serve as the epitome of a self-activated reagent since no additional oxidant or reductant was necessary to achieve plasmid nicking [63] and base oxidation [64].

6.4
Covalent Coupling between Simple Nickel–Salen Complexes and Nucleic Acids

Each assay frequently used to identify and define nucleic acid modification is typically capable of observing only one type of product and hence provides only a limited perspective on the overall reaction. As described above, plasmid nicking is exquisitely sensitive to strand fragmentation but offers no information on the sites of fragmentation or on other types of reaction. Primer extension alternatively detects the sites of reaction but cannot distinguish between base modification and strand fragmentation. Examining the full range of products from a particular set of reaction conditions then necessitates a number of complementary approaches. Without a series of oligonucleotide-based studies, the unique ability of nickel–salen derivatives to attach covalently to nucleic acids in a nucleotide- and conformation-specific pattern would not have been readily apparent [7, 35]. This activity has since provided an important alternative to the more commonly observed strand scission and may ultimately allow for examining systems of great complexity.

The efficiency and type of oxidation promoted by nickel complexes was shown very early to be highly sensitive to its ligand environment. Tetrazamacrocycles forming square-planar complexes with nickel (see, for example, NiCR, Fig. 6.3) were most proficient at inducing oxidation of guanine residues exhibiting high solvent accessibility using a peracid such as MMPP or $KHSO_5$ as oxidant [21, 65, 66]. In contrast, a pentaazamacrocyclic complex of nickel required only molecular oxygen and appeared to oxidize both guanine and adenine residues [21, 67]. Both nickel complexes were capable of nicking plasmid DNA, and nucleobase oxidation was evident after hot piperidine treatment induced the diagnostic strand scission (see, for example, lane 7, Fig. 6.3). When a cationic nickel–salen, Ni(II)–TMAPES, was tested in the presence of $KHSO_5$, a distinct DNA product migrating more slowly than the parent oligonucleotide was detected by gel electrophoresis (lane 2, Fig. 6.3) [7]. This product was unique to the salen complex and was not observed under the same conditions with the tetraazamacrocyclic complexes of nickel nor a diimine complex in which salicylaldehyde was replaced by the non-aromatic acetylacetone (NiTMAPAA, lane 8, Fig. 6.3). This slowly migrating product was first proposed to be a covalent adduct between DNA and Ni(II)–TMAPES, and this was later confirmed by mass spectroscopy [7, 35]. Some heterogeneity in the type of salen–DNA linkage was also originally considered since not all of the slowly migrating product fragmented under standard hot piperidine conditions (30 min) (lane 4, Fig. 6.3). However, extended treatment with piperidine fully cleaved the modified DNA and no heterogeneity of reaction sites is now suspected [35, 49].

The mechanism of Ni(II)–TMAPES coupling to DNA most likely involves formation of a ligand center that couples to the C8 position of guanine, a typical site for radical addition (Scheme 6.7) [68–70]. Such a proposal is consistent with the inhibition of coupling observed for derivatives containing methyl and chloro substituents at the ortho and para positions of the phenolic ring that sterically hinder the centers with most radical character [35]. Salen ligand radicals were also previously invoked to explain oxidative polymerization of nickel–salens during elec-

NiTMAPES NiCR NiTMAPAA

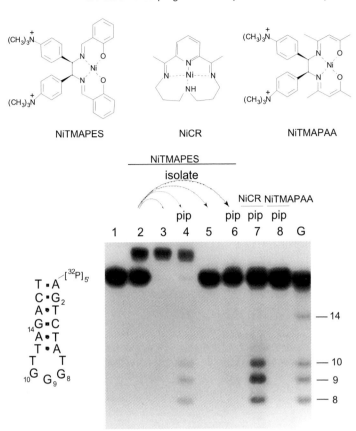

Fig. 6.3. Denaturing polyacryamide gel electrophoresis used to detect covalent coupling and piperidine-induced strand scission of a hairpin forming oligonucleotide in the presence of $KHSO_5$ and NiTMAPES versus other nickel complexes. Adapted from Ref. [7].

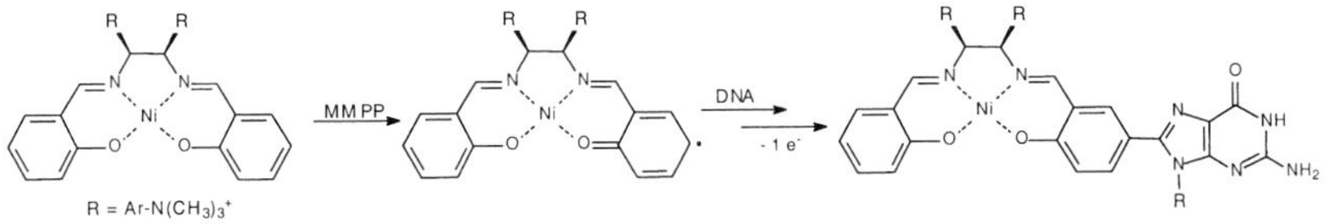

Scheme 6.7. Proposed mechanism of coupling between NiTMAPES and an accessible guanine residue.

trochemical analysis [71, 72]. Certain nickel protein and peptide complexes similarly appear to generate related tyrosyl radicals in the presence of a peracid and induce a coupling reaction as indicated by subsequent protein–protein crosslinking [73, 74].

The conformational selectivity of Ni(II)–TMAPES was first demonstrated by its coupling to any of three extrahelical guanines within a hairpin loop (lane 4, Fig. 6.3) [7]. No reaction was evident for the two helical guanines even though one is only a single base pair from the helix terminus. The conformational dependence of modification is equivalent to that previously described for a macrocyclic complex of nickel NiCR [21, 22] (lane 7, Fig. 6.3) and subsequently expanded to include another nickel–salen [37]. Target recognition for at least NiCR seems to correlate with access to guanine N7, an intrinsic site for nickel coordination [21, 38, 39, 75] and likely the other complexes act similarly. However, an advantage of the Ni(II)–salen derivatives is their ability to add bulky substituents rather than oxidatively degrade the reactive guanine residues. This became apparent from the strong signals detected by primer extension when mapping RNA and DNA structure with Ni(II)–TMAPES [34, 35]. Ni(II)–TMAPES has since been used to characterize folding intermediates of self-splicing RNA [76] and junctions of B- and Z-helical DNA [77].

Widespread application of Ni(II)–TMAPES has been limited in part by its difficult preparation although a relatively efficient synthesis has recently been published [7, 78, 79]. Still, alternative salens retaining both their solubility under aqueous conditions and their ability to couple with nucleic acids can now be prepared in three steps from commercially available starting materials (Scheme 6.8) [35, 48]. Many analogous derivatives may also be envisioned to perform equally well as structural probes of DNA and RNA.

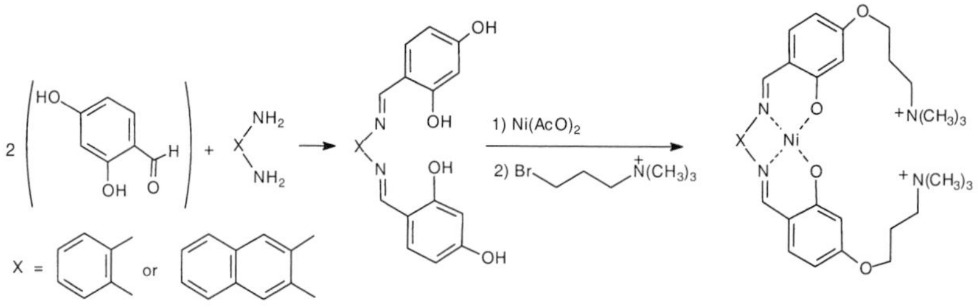

Scheme 6.8. Efficient synthesis of cationic salens.

An especially versatile salen ligand was constructed with an alkylamine appendage that has the ability to either confer water solubility to an otherwise neutral M^{2+} complex or act as a convenient attachment for DNA-directing reagents (Scheme 6.4) [9, 25, 26, 41]. An analogous nickel–salen–biotin conjugate has since been generated from combining components of two complementary salen ligands, the alkylamine containing ethylenediamine derivative illustrated in Scheme 6.4 and the quarternary amine containing salicylaldehyde derivative illustrated in Scheme 6.5 [36]. The resulting salen maintained its selectivity for coupling to guanine residues and provided an expedient method of biotinylating DNA in a conformation selective manner (Scheme 6.9). This reagent may now provide entry into the myriad of biotin-derived kits available for molecular biology and has already supported isolation of modified DNA by avidin-based affinity chromatography and its chemiluminescent detection by streptavidin–alkaline phosphatase conjugates.

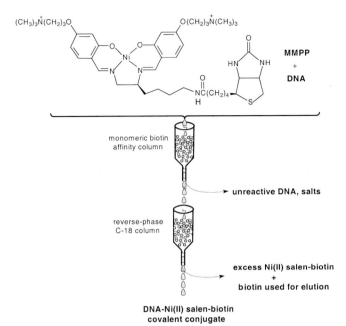

Scheme 6.9. A conjugate of Ni(II) salen and biotin allows for selective reaction and product isolation.

6.5
Chimeric Metal–Salen Complexes

The simplicity and adaptability of metal ligands based on salen have just begun to inspire design of chimeric (or hybrid) derivatives that combine the beneficial properties of salicylaldimine with other known coordination systems. The breadth of opportunity presented by this strategy has the potential to create many predictable, yet new and significant, activities. Both classical and combinatorial approaches may once again be applied to such development as practiced earlier with salen itself.

1,10-Phenanthroline has garnered much attention in biological chemistry due to extensive use of its copper complex for mapping nucleic acid and protein structure [80]. This ligand controls the redox activity of the bound copper as well as the interaction of the complex with biomolecules. Features of phenanthroline and salen have recently been integrated to form a novel ligand with a metal-dependent affinity for DNA (Scheme 6.10) [51]. The cobalt complex significantly stabilized duplex DNA by enhancing its T_m by as much as 10°C. The corresponding nickel and manganese complexes exhibited weaker stabilization and the copper complex expressed only a minimal effect. The cobalt complex also induced plasmid nicking in the presence of a thiol reductant (MPA) and molecular oxygen in analogy to that observed with the earlier salen-based system. However, the equivalent copper

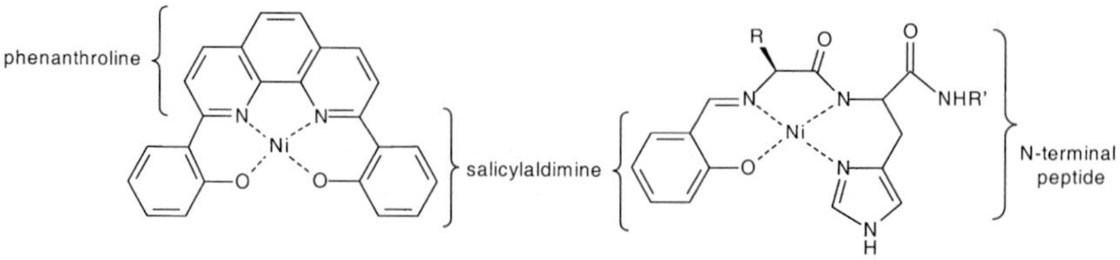

Scheme 6.10. Salen-derived chimeric complexes.

complex was surprisingly unreactive with plasmid DNA under the same conditions that had previously supported nicking of DNA by the copper salen. Reaction of the nickel complex was similarly suppressed and no covalent coupling to DNA was observed [51]. Identifying the source of these divergent characteristics should ultimately help distinguish the relative influence of target association and ligand chemistry on the modes and selectivity of nucleic acid modification.

Peptides and proteins with N-terminal sequences containing histidine at position 3 offer another intriguing ligand system for combination with salen on account of their ability to coordinate and activate metals for protein crosslinking [73, 81] and DNA oxidation [82–84]. Each of these activities has been successfully used to detect protein–protein and protein–DNA interactions. Replacing the terminal amino acid with salicylaldehyde creates an alternative chimera (Scheme 6.10) that may share the reactivity of salen but exhibit target selectivity directed by an attached (poly)peptide. Initial investigations on a related Cu(II)–salicylideneglycyl-glycine complex suggested possible coordination to cytidine as indicated by paramagnetic line broadening of its ^{1}H NMR signals [85]. More extensive analysis has been published on a Ni(II)–salicylidene diamino acid complex containing an arginine for water solubility and a histidine to support tetradentate coordination (Scheme 6.10) [86]. Reaction of the peptide–salen chimera was initiated by addition of KHSO$_5$ or alternatively by autoxidation of sulfite. This additional activity appears to generate HSO$_5^-$ *in situ* through a process that is catalyzed by nickel–peptide complexes including that formed by the salen–peptide hybrid [87]. Consistent with the nickel and salen components of the chimera, covalent coupling between a guanine residue and the complex was observed and apparently formed through an intermediate phenolic radical reminiscent of that generated by Ni(II)–TMAPES [86]. When desired, the DNA–metal adduct could be partially dissembled by treatment with EDTA. Removal of the nickel induced hydrolysis of the ligand's imine linkage and released the peptide component. The net effect of this chimera consequently allows for delivery of a salicylaldehyde equivalent to DNA and hence the introduction of a unique aldehyde group that is available for subsequent conjugation and detection.

6.6
Conclusion

By the appropriate choice of metal and salen components, coordination complexes may be constructed to promote most every type of nucleic acid oxidation that has been achieved collectively from a broad range of individual metal systems including (1) direct strand scission via oxidation of sugar residues, (2) base modification leading to piperidine-sensitive cleavage, and (3) oxidative adduct formation. The conformational selectivity exhibited by certain nickel–salens provides useful probes of nucleic acid structure while the lack of selectivity exhibited by certain iron and manganese derivatives provides reagents for footprinting nucleic acid interactions with proteins and small molecules. Metal–salen complexes have also been designed to function under reductive, oxidative and ambient conditions, thus supporting numerous applications including the potential for study *in vivo*. Simultaneous control of both target recognition and type of modification has been rather challenging but may become more common with designs based on salen conjugates and chimeras. Recent use of DNA as a template for assembling a salen–oligonucleotide conjugate illustrates a particularly exciting strategy for tailoring the function of a salen derivative directly to its intended target [10].

References

1 D. CHEN, A. E. MARTELL, Y. SUN. New synthetic cobalt schiff base complexes as oxygen carriers. *Inorg. Chem.* **1989**, *28*, 2647–2652.

2 A. BÖTTCHER, E. R. BIRNBAUM, M. W. DAY, H. B. GRAY, M. W. GRINSTAFF, J. A. LABINGER. How do electronegative substituents make metal complexes better catalysts for the oxidation of hydrocarbons by dioxygen. *J. Mol. Cat.* **1997**, *117*, 229–242.

3 K. SRINIVASAN, P. MICHAUD, J. K. KOCHI. Epoxidation of olefins with cationic (salen)Mn^{III} complexes. The modulation of catalytic activity by substituents. *J. Am. Chem. Soc.* **1986**, *108*, 2309–2320.

4 H. YOON, T. R. WAGLER, K. J. O'CONNOR, C. J. BURROWS. High turnover rates in pH-dependent alkene epoxidation using NaOCl and square-planar nickel(II) catalysts. *J. Am. Chem. Soc.* **1990**, *112*, 4568–4570.

5 W. ZHANG, J. L. LOEBACH, S. R. WILSON, E. N. JACOBSEN. Enantioselective epoxidation of unfunctionalized olefins catalyzed by (salen)manganese complexes. *J. Am. Chem. Soc.* **1990**, *112*, 2801–2803.

6 E. N. JACOBSEN, W. ZHANG, M. L. GULER. Electronic tuning of asymmetric catalysts, *J. Am. Chem. Soc.* **1991**, *113*, 6703–6704.

7 J. G. MULLER, S. J. PAIKOFF, S. E. ROKITA, C. J. BURROWS. DNA modification promoted by water-soluble nickel(II) salen complexes: a switch to DNA alkylation. *J. Inorg. Biochem.* **1994**, *54*, 199–206.

8 R. H. BAILES, M. CALVIN. The oxygen-carrying synthetic chelate compounds. VII. Preparation. *J. Am. Chem. Soc.* **1947**, *69*, 1886–1893.

9 S. ROUTIER, J.-L. BERNIER, M. J. WARING, P. COLSON, C. HOUSSIER, C. BAILLY. Synthesis of a functionalized salen-copper complex and its interactions with DNA. *J. Org. Chem.* **1996**, *61*, 2326–2331.

10 J. L. CZLAPINSKI, T. L. SHEPPARD. Nucleic acid template-directed assembly of metallosalen-DNA conjugates. *J. Am. Chem. Soc.* **2001**, *123*, 8618–8619.

11 J. LOPEZ, S. LIANG, X. R. BU. Unsymmetric chiral salen Schiff bases: a new chiral ligand pool from bis-Schiff bases containing two different salicylaldehyde units. *Tetrahedron Lett.* **1998**, *39*, 4199–4202.

12 A. K. MUKHERJEE, P. RÂY. Metal chelate complexes of sulfosalicylaldehyde. *J. Indian Chem. Soc.* **1955**, *32*, 632–643.

13 H. Y. SHRIVASTAVA, B. U. NAIR. Cleavage of human orosomucoid by a chromium(V) species: relevance in

biotoxicity of chromium. *Biochem. Biophys. Res. Commun.* **2000**, *279*, 980–983.

14 H. Y. SHRIVASTAVA, B. U. NAIR. Chromium(III)-mediated structural modification of glycoprotein: impact of the ligand and the oxidants. *Biochem. Biophys. Res. Commun.* **2001**, *285*, 915–920.

15 A. HAIKARAINEN, J. SIPILÄ, P. PIETIKÄINEN, A. PAJUNEN, I. MUTIKAINEN. Salen complexes with bulky substituents as useful tools for biomimetic phenol oxidation research. *Bioorg. Med. Chem.* **2001**, *9*, 1633–1638.

16 D. J. GRAVERT, J. H. GRIFFIN. Steric and electronic effects, enantio-specificity, and reactive orientation in DNA binding/cleaving by substituted derivatives [salenMn^{III}]$^+$. *Inorg. Chem.* **1996**, *35*, 4837–4847.

17 K. D. SHIMIZU, M. L. SNAPPER, A. H. HOVEYDA. *Combinatorial Approaches in Comprehensive Asymmetric Catalysis* Vol. III, eds E. N. Jacobsen, A. Pfaltz, H. Yamamoto, Springer-Verlag, New York, 1999, 1389–1399.

18 M. B. FRANCIS, E. N. JACOBSEN, Discovery of novel catalysts for alkene epoxidation from metal-binding combinatorial libraries. *Angew. Chem. Int. Ed. Engl.* **1999**, *38*, 937–941.

19 W. J. DIXON, J. J. HAYES, J. R. LEVIN, M. F. WEIDNER, B. A. DOMBROSKI, T. D. TULLIUS. Hydroxyl radical footprinting. *Methods Enzymol.* **1991**, *208*, 380–413.

20 D. S. SIGMAN. Nuclease activity of 1,10-phenanthroline–copper ion. *Acc. Chem. Res.* **1986**, *19*, 180–186.

21 C. J. BURROWS, S. E. ROKITA. Probing guanine structure in nucleic acid folding using nickel complexes. *Acc. Chem. Res.* **1994**, *27*, 295–301.

22 S. E. ROKITA, C. J. BURROWS. Structural studies of nucleic acids using nickel and cobalt based reagents, in *Current Protocols in Nucleic Acid Chemistry*, ed. G. Glick, Wiley, New York, 2000, 6.4.1–6.4.7.

23 C. S. CHOW, L. S. BEHLEN, O. C. UHLENBECK, J. K. BARTON. Recognition of tertiary structure in

tRNAs by Rh(phen)$_2$phi^{3+}, a new reagent for RNA structure-function mapping. *Biochemistry* **1992**, *31*, 972–982.

24 P. J. CARTER, C.-C. CHENG, H. H. THORP. Oxidation of DNA hairpins by oxoruthenium(IV): effects of sterics and secondary structure. *Inorg. Chem.* **1996**, *35*, 3348–3354.

25 S. ROUTIER, J.-L. BERNIER, J.-P. CATTEAU, P. COLSON, C. HOUSSIER, C. RIVALLE, E. BISAGNI, C. BAILLY. Synthesis, DNA binding and cleaving properties of an ellipticine-salen-copper conjugate. *Bioconj. Chem.* **1997**, *8*, 7890–7792.

26 S. ROUTIER, J.-L. BERNIER, J.-P. CATTEAU, C. BAILLY. Recognition and cleavage of DNA by a distamycin-salen-copper conjugate. *Bioorg. Med. Chem. Lett.* **1997**, *7*, 1729–1732.

27 S.-W. LEE, S. CHANG, D. KOSSAKOVSKI, H. COX, J. L. BEAUCHAMP. Slow evaporation of water from hydrated salen transition metal complexes in the gas phase reveals details of metal ligand interactions. *J. Am. Chem. Soc.* **1999**, *121*, 10152–10156.

28 T. TANAKA, K. TSURUTANI, A. KOMATSU *et al.* Synthesis of new cationic Schiff base complexes of copper(II) and their selective binding with DNA. *Bull. Chem. Soc. Jpn* **1997**, *70*, 615–629.

29 K. SATO, M. CHIKIRA, Y. FUJII, A. KOMATSU. Stereospecific binding of chemically modified salen-type Schiff base complexes of copper(II) with DNA. *J. Chem. Soc. Chem. Commun.* **1994**, 625–626.

30 R. B. MARTIN. Dichotomy of metal ion binding to N1 and N7 of purines, in *Metal Ions in Biological Systems*, Vol. 32, eds A. Sigel and H. Sigel, Marcel Dekker, New York, 1996, 61–89.

31 B. H. GEIERSTANGER, T. F. KAGAWA, S.-L. CHEN, G. J. QUIGELY, P. S. HO. Base-specific binding of copper(II) to Z-DNA. *J. Biol. Chem.* **1991**, *266*, 20185–20191.

32 D. J. GRAVERT, J. H. GRIFFIN. Specific DNA cleavage mediated by [salenMn(III)]$^+$. *J. Org. Chem.* **1993**, *58*, 820–822.

33 D. M. Crothers, T. E. Haran, J. G. Nadeau. Intrinsically bent DNA. *J. Biol. Chem.* **1990**, *265*, 7093–7096.

34 S. A. Woodson, J. G. Muller, C. J. Burrows, S. E. Rokita. A primer extension assay for modification of guanine by Ni(II) complexes. *Nucleic Acids Res.* **1993**, *21*, 5524–5525.

35 J. G. Muller, L. A. Kayser, S. J. Paikoff *et al.* Formation of DNA adducts using nickel(II) complexes of redox-active ligands: a comparison of salen and peptide complexes. *Coord. Chem. Rev.* **1999**, *185–186*, 761–774.

36 X. Zhou, J. M. Shearer, S. E. Rokita. A Ni(salen)-biotin conjugate for rapid isolation of accessible DNA. *J. Am. Chem. Soc.* **2000**, *122*, 9046–9047.

37 S. Routier, J.-L. Bernier, J.-P. Catteau, C. Bailly. Highly preferential cleavage of unpaired guanines in DNA by a functionalized salen-nickel complex. *Bioorg. Med. Chem. Lett.* **1997**, *7*, 63–66.

38 H.-C. Shih, N. Tang, C. J. Burrows, S. E. Rokita. Nickel-based probes of nucleic acid structure bind to guanine but do not perturb a dynamic equilibrium of extrahelical guanine residues. *J. Am. Chem. Soc.* **1998**, *120*, 3284–3288.

39 H.-C. Shih, H. Kassahun, C. J. Burrows, S. E. Rokita. Selective association between a macrocyclic nickel complex and extrahelical guanine residue. *Biochemistry* **1999**, *38*, 15034–15042.

40 W. K. Pogozelski, T. D. Tullius. Oxidative strand scission of nucleic acids: routes initiated by hydrogen abstraction from the sugar moiety. *Chem. Rev.* **1998**, *98*, 1089–1107.

41 S. Routier, N. Cotelle, J.-P. Catteau *et al.* Salen-anthraquinone conjugates. Synthesis, DNA-binding and cleaving properties, effects on topoisomerases and cytotoxicity. *Bioorg. Med. Chem.* **1996**, *4*, 1185–1196.

42 D. S. Sigman, A. Mazumder, D. M. Perrin. Chemical nucleases. *Chem. Rev.* **1983**, *93*, 2295–2316.

43 P. G. Schultz, J. S. Taylor, P. B. Dervan. Design and synthesis of a sequence-specific DNA cleaving molecule (Distamycin-EDTA)iron(II). *J. Am. Chem. Soc.* **1982**, *104*, 6861–6863.

44 R. S. Youngquist, P. B. Dervan. Sequence-specific recognition of B-DNA by oligo(*N*-methylpyrrole-carboxamide)s. *Proc. Natl Acad. Sci. USA* **1985**, *82*, 2565–2569.

45 J. R. Morrow, K. A. Kolasa. Cleavage of DNA by nickel complexes. *Inorg. Chim. Acta* **1992**, *195*, 245–248.

46 M. Pitié, G. Pratviel, J. Bernadou, B. Meunier. Preferential hydroxylation by the chemical nuclease meso-tetrakis-(4-*N*-methylpyridiniumyl) porphyrinatomanganese[III] pentaacetate/KHSO$_5$ at the 5′ carbon of deoxyribose on both 3′ sides of three contiguous A-T base pairs in short double-stranded oligonucleotides. *Proc. Natl Acad. Sci. USA* **1992**, *89*, 3967–3971.

47 S. S. Mandal, N. V. Kumar, U. Varshney, S. Bhattacharya. Metal-ion-dependent oxidative DNA cleavage by transition metal complexes of a new water-soluble salen derivative. *J. Inorg. Biochem.* **1996**, *63*, 265–272.

48 S. S. Mandal, U. Varshney, S. Bhattacharya. Role of the central metal ion and ligand charge in the DNA binding and modification by metallosalen complexes. *Bioconj. Chem.* **1997**, *8*, 798–812.

49 C. J. Burrows, J. G. Muller. Oxidative nucleobase modifications leading to strand scission. *Chem. Rev.* **1998**, *98*, 1109–1151.

50 C. J. Burrows, J. G. Muller, G. T. Poulter, S. E. Rokita. Nickel-catalyzed oxidations: from hydrocarbons to DNA. *Acta Chem. Scand.* **1996**, *50*, 337–344.

51 S. Routier, V. Joanny, A. Zaparucha *et al.* Synthesis of metal complexes of 2,9-bis(2-hydroxyphenyl)-1,10-phenanthroline and their DNA binding and cleaving activities. *J. Chem. Soc. Perk. Trans II* **1998**, 863–868.

52 A. S. Kumbhar, S. G. Damle, S. T. Dasgupta, S. Y. Rane, A. S. Kumbhar. Nuclease activity of oxo-bridged diiron complexes. *J. Chem. Res.* **1999**, 98–99.

53 C.-C. CHENG, Y.-L. LU. Novel water-soluble 4,4-disubstituted ruthenium(III)-salen complexes in DNA stranded scission. *J. Chinese Chem. Soc.* **1998**, *45*, 611–617.

54 S. E. ROKITA. Chemical reagents for investigating the major groove of DNA in Current Protocols in *Nucleic Acid Chemistry*, ed. G. Glick, Wiley, New York, 2001, 6.6.1–6.6.16.

55 T. TSUMAKI. Nebenvalenzringver-bindungen. IV. Über einige inner-komplexe kobaltsalze der oxyaldimine. *Bull. Chem. Soc. Jpn* **1938**, *13*, 252–260.

56 R. D. JONES, D. A. SUMMERVILLE, F. BASOLO. Synthetic oxygen carriers related to biological systems. *Chem. Rev.* **1979**, *79*, 139–176.

57 S. BHATTACHARYA, S. S. MANDAL. Ambient oxygen activating water soluble cobalt-salen complex for DNA cleavage. *J. Chem. Soc. Chem. Commun.* **1995**, 2489–2490.

58 A. BÖTTCHER, H. ELIAS, L. MÜLLER, H. PAULUS. Oxygen activation of nickel(II)tetrahydrosalen complexes with the formation of nickel(II)dihydrosalen complexes. *Angew. Chem. Int. Ed. Engl.* **1992**, *31*, 623–625.

59 A. BÖTTCHER, H. ELIAS, E.-G. JÄGER *et al.* Comparative study on the coordination chemistry of cobalt(II), nickel(II) and copper(II) with derivatives of salen and tetrahydro-salen: metal-catalyzed oxidative dehydrogenation of the C-N bond in coordinated tetrahydrosalen. *Inorg. Chem.* **1993**, *32*, 4131–4138.

60 R. F. ZABINSKI. Synthesis and characterization of water soluble tetrahydrosalen ligands and their nickel complexes: studies toward O_2 activation and DNA reactivity. M.S., SUNY, Stony Brook, 1995.

61 E. LAMOUR, S. ROUTIER, J.-L. BERNIER, J.-P. CATTEAU, C. BAILLY, H. VEZIN, Oxidation of Cu^{II} to Cu^{III}, free radical production, and DNA cleavage by hydroxy-salen-copper complexes. Isomer effects studied by ESR and electrochemistry. *J. Am. Chem. Soc.* **1999**, *121*, 1862–1869.

62 S. ROUTIER, H. VEZIN, E. LAMOUR, J.-L. BERNIER, J.-P. CATTEAU, C. BAILLY. DNA cleavage by hydroxy-salicylidene-ethylendiamine-iron complexes. *Nucleic Acids Res.* **1999**, *27*, 4160–4166.

63 C. T. DILLION, P. A. LAY, A. M. BONIN, N. E. DIXON, Y. SULFAB. DNA interactions and bacterial mutagenicity of some chromium(III) imine complexes and their chromium(V) analogues. Evidence for chromium(V) intermediates in the genotoxicity of chromium(III). *Aust. J. Chem.* **2000**, *53*, 411–424.

64 K. D. SUGDEN, C. K. CAMPO, B. D. MARTIN. Direct oxidation of guanine and 7,8-dihyhdro-6-oxoguanine in DNA by a high-valent chromium complex: a possible mechanism for chromate genotoxicity. *Chem. Res. Toxicol.* **2001**, *14*, 1315–1322.

65 X. CHEN, S. E. ROKITA, C. J. BURROWS. DNA modification: intrinsic selectivity of nickel(II) complexes. *J. Am. Chem. Soc.* **1991**, *113*, 5884–5886.

66 J. G. MULLER, X. CHEN, A. C. DADIZ, S. E. ROKITA, C. J. BURROWS. Ligand effects associated with the intrinsic selectivity of DNA oxidation promoted by nickel(II) macrocyclic complexes. *J. Am. Chem. Soc.* **1992**, *114*, 6407–6411.

67 C.-C. CHENG, S. E. ROKITA, C. J. BURROWS. Nickel(III)-promoted DNA scission using ambient dioxygen. *Angew. Chem. Int. Ed. Engl.* **1993**, *32*, 227–278.

68 N. V. S. RAMAKRISHNA, E. L. CAVALIERI, E. G. ROGAN *et al.* Synthesis and structure determination of the adducts of the potent carcinogen 7,12-dimethylbenz[a]anthracene and deoxyriboucleosides formed by electrochemical oxidation: models for metabolic activation by one-electron oxidation. *J. Am. Chem. Soc.* **1992**, *114*, 1863–1874.

69 S. HIX, M. D. S. MORAIS, O. AUGUSTO. DNA Methylation by *tert*-butyl hydroperoxide-iron(II). *Free Rad. Biol. Med.* **1995**, *19*, 293–301.

70 A. AKANNI, Y. J. ABUL-HAJJ. Estrogen-nucleic acid adducts: reaction of 3,4-estrone-o-quinone radical anion with deoxyribonucleosides. *Chem. Res. Toxicol.* **1997**, *10*, 760–766.

71 K. A. GOLDSBY, J. K. BLAHO, L. A. HOFERKAMP. Oxidation of nickel(II) bis(salicylaldimine) complexes: solvent control of the ultimate redox site. *Polyhedron* **1989**, *8*, 113–115.

72 K. A. GOLDSBY. Symmetric and unsymmetric nickel(II) Schiff base complexes; metal-localized versus ligand-localized oxidation. *J. Coord. Chem.* **1988**, *19*, 83–90.

73 K. C. BROWN, T. KODADEK. Protein cross-linking mediated by metal ion complexes, Ch. 12 in *Metal Ions in Biological Systems*, Vol. 38, ed. H. Sigel, Marcel Dekker, New York, 2001, 351–384.

74 S. E. ROKITA, C. J. BURROWS. Nickel- and cobalt-dependent oxidation and cross-linking of proteins, Ch. 10 in *Metal Ions in Biological Systems*, Vol. 38, ed. H. Sigel, Marcel Dekker, New York, 2001, 289–311.

75 X. CHEN, S. A. WOODSON, C. J. BURROWS, S. E. ROKITA. A highly sensitive probe for guanine N7 in folded structures of RNA: application to tRNA[Phe] and *Tetrahymena* group I intron. *Biochemistry* **1993**, *32*, 7610–7616.

76 J. PAN, S. A. WOODSON. Folding intermediates of a self-splicing RNA: mispairing of the catalytic core. *J. Mol. Biol.* **1998**, *280*, 597–609.

77 N. TANG, J. G. MULLER, C. J. BURROWS, S. E. ROKITA. Nickel and cobalt reagents promote selective oxidation of Z-DNA. *Biochemistry* **1999**, *38*, 16648–16654.

78 F. VÖGTLE, E. GOLDSCHMITT. Die Diaza-Cope-Umlagerung. *Chem. Ber.* **1976**, *109*, 1–40.

79 J. M. SHEARER, S. E. ROKITA. Diamine preparation for synthesis of a water soluble Ni(II) salen complex. *Bioorg. Med. Chem. Lett.* **1999**, *9*, 501–504.

80 D. S. SIGMAN. Site specific oxidative scission of nucleic acids and proteins, in *DNA and RNA Cleavers and Chemotherapy of Cancer and Viral Diseases*, ed. B. Meunier, Kluwer Academic Publishers, Netherlands, 1996, 119–132.

81 K. C. BROWN, S.-H. YANG, T. KODADEK. Highly specific oxidative cross-linking of proteins mediated by a nickel-peptide complex. *Biochemistry* **1995**, *34*, 4733–4739.

82 D. P. MACK, P. B. DERVAN. Sequence-specific oxidative cleavage of DNA by a designed metalloprotein Ni(II)-GGH(Hin139–190). *Biochemistry* **1992**, *31*, 9399–9405.

83 D. F. SHULLENBERGER, P. D. EASON, E. C. LONG. Design and synthesis of a versatile DNA-cleaving metallopeptide structural domain. *J. Am. Chem. Soc.* **1993**, *115*, 11038–11039.

84 M. NAGAOKA, M. HAGIHARA, J. KUWAHARA, Y. SUGIURA. A novel zinc finger-based DNA cutter. Biosynthetic design and highly selective DNA cleavage. *J. Am. Chem. Soc.* **1994**, *116*, 4085–4086.

85 M. PALANIANDAVAR. Models for enzyme-copper-nucleic acid interaction. *Biol. Trace Element Res.* **1989**, *21*, 41–48.

86 A. J. STEMMLER, C. J. BURROWS. The Sal-XH motif for metal-mediated oxidative DNA-peptide cross-linking. *J. Am. Chem. Soc.* **1999**, *121*, 6956–6957.

87 J. G. MULLER, R. P. HICKERSON, R. J. PEREZ, C. J. BURROWS. DNA damage from sulfite autoxidation catalyzed by a nickel(II) peptide. *J. Am. Chem. Soc.* **1997**, *119*, 1501–1506.

7
Charge Transport in DNA

Tashica T. Williams and Jacqueline K. Barton

7.1
Introduction

Double-helical DNA contains within its interior an extended π-stacked array of aromatic, heterocyclic base-pairs. As such, the DNA double helix represents, almost uniquely, a well-defined molecular π-stack. Solid-state π-stacked materials, particularly when doped, can be effective conductors and semiconductors [1, 2]. In the years since the delineation of the DNA structure by Watson and Crick [3], by analogy, it has been asked whether double-helical DNA might similarly display properties associated with molecular conductivity [4, 5]. Indeed, whether double-helical DNA, owing to its π-stacked structure, might be an effective conduit for charge transport (CT) has intrigued physicists, chemists, and biologists for over 40 years.

Early studies of DNA charge transport generated significant interest and extensive debate. Physicists carried out varied measurements of electron transport in DNA, resulting in DNA being described by some as an insulator and others as a quantum wire [6–8]. Based upon pulse radiolysis studies, radiation biologists suggested charge could migrate over as short a distance as 3 base pairs [9] or as long a distance as 200 base pairs [10]. More recent studies applied more sophisticated techniques in measurements of DNA conductivity. Differential conductivity depending upon DNA orientation was seen in studies of aligned DNA films [11], a large current that increased linearly with applied voltage was found when the DNA molecules were oriented perpendicular to the electrode, and in contrast, no current was detected with a parallel alignment. Direct conduction measurements were carried out on small collections of DNA duplexes arranged into 600-nm-long DNA ropes, and these studies revealed semiconductor behavior, with resistivity values that were comparable to those of other conducting or semiconducting materials [12].

Furthermore, single-molecule conduction measurements on dry poly(G)-poly(C) showed no conduction at low voltages, but at higher voltages, DNA seemed to support large currents [13]. While some consensus may now have been reached that DNA possesses properties of a wide band gap semiconductor [14], varied conductivity measurements on dry DNA samples at low temperatures continue to appear consistent with DNA being everything from an insulator to a superconductor [15]. Likely the variability in DNA structure and integrity in these physical measurements adds to the variability in conclusions obtained.

Chemists first focused on CT in DNA in the context of earlier studies of protein electron transfer [16], and *analogously* to protein systems, well-defined oligonucleotide duplexes containing pendant donors and acceptors were constructed so as to measure electron transfer rates and yields as a function of distance on DNA duplexes in solution. The ability of the DNA π-stack to serve as a medium for long-range electron transport was quantified utilizing β, which, from Marcus theory, is a measurement of the exponential decay with distance of electronic coupling of the medium [17]. For σ-bonded systems, including proteins, the value of β was found to be ~ 1.0 Å^{-1}; for comparison, a β value of ~ 0.1 Å^{-1} is found for conjugated polymers [18]. In measurements of electron transfer on DNA assemblies, β was found to range from $= 0.2$ Å^{-1} [19, 20] to ~ 0.6–0.7 Å^{-1} [21], to as high as 1.4 Å^{-1} [22]. Recent experiments have demonstrated the complexity of experimental evaluation of β for DNA, and have suggested that the apparent disparity in CT distance dependence may arise because different experimental assemblies might operate within different regimes of a mechanistic continuum between one-step superexchange and multi-step hopping [23]. Hence β, in the context of non-adiabatic tunneling models, may not be the most appropriate parameter to characterize DNA-mediated CT.

Studies in our research group have focused on using DNA assemblies with intercalators or modified bases serving as donors and acceptors. These studies using intercalators as direct probes of the π-stack have revealed particularly shallow distance dependences in CT, underscoring the importance of stacking to effective CT in DNA.

Here we focus on these studies of DNA CT using well-stacked probes of the DNA duplex. This discussion is therefore intended not as an exhaustive review of the literature but instead as an illustrative description of what we have learned through a mix of biochemical, spectroscopic, and electrochemical studies in our laboratory. We have found CT in DNA to be exquisitely sensitive to π-stacking of the donor, the acceptor, and the intervening base pair array. As a result, CT may be most usefully applied as a probe of nucleic acid structure and structural dynamics.

7.2
DNA Metallointercalators

In characterizing the π-stack as a medium for DNA-mediated CT, it is important to have the probes directly coupled into the π-stack. Therefore, our laboratory has utilized a number of metallointercalators to study long-range electron transfer in DNA, particularly, dipyridophenazine (dppz) complexes of ruthenium(II) (e.g. [Ru(1,10-phenanthroline (phen))(4′-methylbipyridine-4-butyric acid (bpy′))(dppz)]$^{2+}$ and 9,10-phenanthrenequinone diimine (phi) complexes of rhodium(III) (e.g. [Rh(phi)$_2$bpy′]$^{3+}$) [24]. These ligands afford the octahedral ruthenium and rhodium metal complexes tight DNA binding via intercalation ($K \geq 10^6$ M^{-1}), and the DNA interactions of these intercalators have been extensively characterized [25].

7.2.1
Phenanthrenequinone Diimine Complexes of Rhodium

Studies of phi complexes of rhodium have shown that these complexes bind avidly to DNA ($K \geq 10^6$ M^{-1}) by intercalation of the phi ligand [26, 27]. The sequence specificity of these complexes can be varied by altering the ancillary, non-intercalating ligands. High-resolution NMR studies initially provided insight into the binding mode of these phi complexes [28]. The studies revealed that the complexes bound DNA from the major groove, and that intercalation provided a means by which the functional groups on the ancillary ligands and those contained in the major groove could interact. In fact, NMR studies of the complex, Δ-α-[Rh(R,R-dimethyltrien)phi]$^{3+}$ (where R,R-dimethyltrien is $2R,9R$-diamino-4,7-diazadecane), which was designed to recognize the sequence 5′-TGCA-3′, revealed site-specific binding of the complex to a decamer that centrally contained a 5′-TGCA-3′ sequence and delineated specific, rationally designed hydrogen bonding and methyl–methyl contacts [29]. Recently, a 1.2-Å crystal structure was obtained for the Δ-α-[Rh(R,R-dimethyltrien)phi]$^{3+}$ complex, intercalated into a DNA octamer (Fig. 7.1) [30]. The structure revealed that intercalation of the complex occurs via the major groove, and, based upon five independent views of the intercalator, minimal perturbation of the π-stack is observed; the phi ligand inserts into and essentially serves as an additional base pair within the π-stack. In the context of our studies of DNA CT, this intercalation does not appear to perturb the base stack either globally or locally.

The rich photochemistry of phi complexes of rhodium enables demarcation of binding sites of the metal complexes in addition to providing potent, intercalating photooxidants [27]. Hence, irradiation of the complexes bound to DNA at different wavelengths induces different photochemical reactions [31]. Irradiation of the phi complexes with ultraviolet light ($\lambda = 313$ nm) leads to direct strand scission of the DNA sugar–phosphate backbone with products consistent with abstraction of the C3′ hydrogen from the sugar that is near the activated phi ligand at the intercalation site. This chemistry then marks the site of binding. If, instead, the phi complexes are irradiated with visible light bound to DNA ($\lambda \geq 365$ nm), oxidative damage to the DNA bases results. These phi complexes of rhodium, when photoexcited, appear to be sufficiently potent to promote oxidation of all of the nucleotides.

7.2.2
Dipyridophenazine Complexes of Ruthenium

The dipyridophenazine complexes of ruthenium(II) have been shown to possess remarkable photophysical properties when bound to DNA [32, 33]. Irradiation of these complexes with visible light leads to a metal-to-ligand charge transfer excited

(a)   (b)

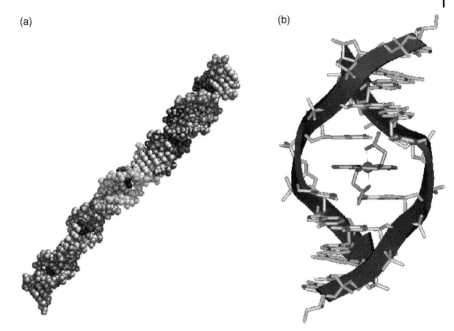

Fig. 7.1. (a) Five Δ-α-[Rh(*R,R*-dimethyltrien) phi]³⁺-DNA octamer complexes stacked end to end in the asymmetric cell of the crystal. (b) A major groove view of intercalation of Δ-α-[Rh(*R,R*-dimethyltrien)phi]³⁺ bound to 5'-G-dIU-TGCAAC-3'. As shown in the figure, minor perturbation of the π-stack is observed. The intercalator is inserted as an additional base-pair step. Adapted from [30].

state, which is localized on the dppz ligand. This localization of the charge on the dppz ligand has been particularly valuable in probing DNA-mediated CT, because excitation of intercalated dipyridophenazine complexes directs the charge transfer directly into the π-stack, not onto an ancillary ligand [34].

Perhaps more important is the interesting differential luminescence quenching seen with the complex when free in aqueous solution compared with bound to DNA. Luminescence of the dppz complex of ruthenium is easily detected in organic solvents; however, none is observed in aqueous solution. This quenching by water has been attributed to proton transfer from the solvent to the nitrogen atoms of the phenazine [35]. Upon intercalation of the dppz into DNA, however, the luminescence is maintained, for the DNA π-stack protects the phenazine nitrogen atoms from the solvent. Hence dppz complexes of ruthenium serve as sensitive "light switches" for DNA.

This "light switch" characteristic of the dipyridophenazine complexes has proven to be a valuable spectroscopic tool in delineating how the intercalated dppz ligand is bound to DNA [36]. NMR studies of partially deuterated [Δ-Ru(phen)₂dppz]²⁺ bound to a hexamer revealed that the dppz ligand prefers to intercalate via the major groove and that two binding orientations (e.g. symmetrical and asymmetrical) were identified [37]. In the symmetrical mode, the phenazine nitrogens are protected from the solvent, whereas in the asymmetrical mode, the phenazine nitrogens are more solvent exposed. Other studies have suggested that some dppz complexes may bind to DNA also from the minor groove side [38]. A DNA intercalator that possesses both major and minor groove orientations has not been established previously. It would therefore be valuable to establish whether both binding orientations occur through intercalation and what truly determines access of the dppz complex to the base pair stack.

7.3
Photophysical Studies of Electron Transport in DNA

7.3.1
Electron Transport between Ethidium and a Rhodium Intercalator

Our early finding of long-range electron transfer quenching in an assembly containing tethered metallointercalators provided an intriguing introduction to studies of CT through DNA [19]. These early studies underscored clear differences between CT in DNA versus protein systems [39]. To further probe the distance dependence of CT in DNA and to begin to delineate the importance of π-stacking to the CT process, the classic organic intercalator, ethidium, was employed as the photoexcited electron donor and $[\text{Rh(phi)}_2\text{bpy}']^{3+}$ as the acceptor (Fig. 7.2) [40]. Both the donor and acceptor molecules were tethered to duplexes of variable length

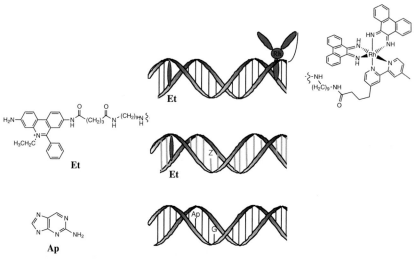

Fig. 7.2. Schematic illustration of some DNA assemblies used to probe photoinduced charge transfer. Shown (top to bottom) are duplexes containing donors and acceptors as two pendant intercalators [20], an intercalator and modified base [23, 41], and two bases [42, 44].

to permit systematic investigation of the electron transfer over defined donor–acceptor distances. Photoinduced quenching of the ethidium excited state was monitored for a series of duplexes which were 10–14 base pairs in length and had donor–acceptor separations that ranged from 24 to 38 Å. Fluorescence quenching provided a measure of electron transfer efficiency. Interestingly, a shallow distance dependence in the quenching yield was observed for this system. Moreover, although the quenching variations appeared to reflect an attenuation in the yield of electron transfer, time-resolved measurements showed that dramatic variations in the electron transfer rate were not apparent. Thus we proposed that electron transfer through the base pair stack was faster than our ability to detect it ($>10^9$ s^{-1}) and that electron transfer through DNA was sensitive to π-stacking, so that the distance dependence in quenching yield observed reflected the increased probability of a stacking defect with increasing donor–acceptor distance, not a significant change in rate.

To further probe the sensitivity of the electron transfer process to π–stacking, a CA mismatch, which is known to cause local perturbations in base-stacking but not global structural distortions, and a GA mismatch, which continues to be well stacked in the DNA due to increased aromatic surface area, were introduced into the DNA assemblies. Substantial electron transfer quenching was still evident in the duplex containing the GA mismatch, which had a quenching yield similar to that of the Watson–Crick paired duplex. However, a significant reduction in quenching yield was observed in the presence of the CA mismatch. This result provided a clear demonstration of the sensitivity of the electron transport process to perturbations in the intervening base pair stack. Moreover this experiment established unambiguously that the path of CT is through π-stacked bases.

7.3.2
Ultrafast Charge Transport in DNA: Ethidium and 7-Deazaguanine

Having observed photoinduced long-range electron transport between metallo-intercalators, we also wanted to examine assemblies in which a DNA base could serve as the reactant. The guanine base analog 7-deazaguanine possesses an oxidation potential that is $\sim$300 mV below that of guanine. Hence, in assemblies containing tethered ethidium as the photoexcited donor, selective oxidative quenching by this analog but not guanine is possible within a mixed DNA sequence [41].

Fluorescence measurements of the ethidium-deazaguanine system (Fig. 7.2) first revealed that electron transport could occur over a range of donor–acceptor separations (6–24 Å) and that the transport occurred on a subnanosecond time scale, with a quenching yield that also exhibited a shallow distance dependence. Furthermore, as in the earlier study, upon the incorporation of mismatches into the ethidium/7-deazaguanine duplexes, the quenching yield was significantly diminished. Hence, results were consistent with fast electron transport between the intercalated ethidium and deazaguanine mediated by the DNA π-stack.

The critical importance of DNA dynamics in attenuating the CT process was then underscored in time-resolved studies on the femtosecond time scale of the photooxidation of deazaguanine in a series of tethered ethidium duplexes [23]. Again the measurements indicated that it was primarily the yield of electron transfer that varied with distance rather than the rate. Transient absorption measurements revealed a 5-ps component in the electron transfer decay, which we assigned to direct electron transfer from deazaguanine, and a 70-ps component, which corresponded to the time scale for motion of the ethidium within its binding site. With an increase in donor–acceptor separation, we observed that these components decreased in yield but not significantly in their decay times. Hence, we understood these results in the context of a model in which the initial population of duplexes that were properly stacked and aligned to permit effective coupling were represented by the 5-ps component, yielding fast long-range CT, while CT on the longer 70-ps time scale was gated by the motion of the ethidium within its intercalation site, requiring motion and realignment of the ethidium to permit long-range CT. Interestingly, these studies provided the first direct observation of electron transfer rates through DNA and showed the sensitivity of electron transfer to dynamical motions within the base pair stack.

7.3.3
Base–Base Charge Transport

While metallointercalators have been useful in probing the DNA-mediated CT process, we were also interested in developing systems through which we might examine base–base electron transfer directly. This was accomplished by exploiting the fluorescence properties of 2-aminopurine and ethenoadenine, analogs of adenine, as well as 7-deazaguanine.

A systematic investigation of DNA-mediated photooxidation of guanine by ethenoadenine and 2-aminopurine was first performed using 12-base-pair duplexes with donor–acceptor separations ranging from 3.4 Å to 13.6 Å (Fig. 7.2) [42]. The studies revealed fast CT over a range of distances in duplexes containing guanine and 2-aminopurine; the rates for interstrand base–base electron transfer were $\sim 10^8$ s^{-1} with $\beta = 0.1$ Å^{-1}. These rates indicated efficient coupling among the donor, acceptor, and intervening bases. Notably, however, in duplexes containing ethenoadenine, a bulkier fluorescent analog of adenine, and guanine, a steeper distance dependence was observed ($\beta = 1.0$ Å^{-1}). Analogously, slow electron transfer kinetics and steep distance dependences were also observed in a system using stilbene-bridged DNA hairpins ($\beta = 0.6$ Å^{-1}) [21, 43].

The base–base electron transfer studies also allowed us to differentiate between interstrand and intrastrand electron transfer [42]. In B-form duplexes containing 2-aminopurine, the electron transport kinetics for the intrastrand reactions were monitored on the femtosecond time scale and were seen to be $>10^3$ times faster than the interstrand reactions [44]. This difference again is consistent with the path for charge transport being through the stacked bases. In the case of the interstrand reaction. However, because the B-form duplex maintains primarily intrastrand stacking, interstrand electron transfer requires transfer across a hydrogen-bonded base pair.

Thus, coupling of the donor, acceptor, and intervening base moieties into the π-stack is paramount for effective CT. In systems that have a well-stacked π-way, lower β values are expected, whereas in systems that do not have significant base stacking, β values approaching those observed in proteins ($\beta \sim 1.0$ A^{-1}) are expected. From these studies, it can be concluded that fast CT is mediated by an intrastrand, π-stacked pathway.

7.4
DNA-mediated Electron Transport on Surfaces

7.4.1
Characterization of DNA-modified Surfaces

The remarkable ability of DNA to form self-assembled monolayers on gold surfaces by way of aliphatic alkanethiol linkers has permitted us to utilize also electrochemical techniques to study DNA-mediated charge transport processes that involve ground state reactants [45]. Our laboratory has constructed and extensively characterized the assembly of DNA films on gold surfaces (Fig. 7.3) [46]. To im-

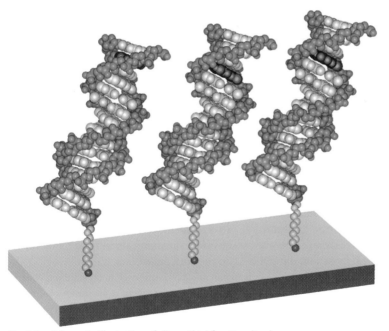

Fig. 7.3. Schematic illustration of alkane thiol-functionalized DNA duplexes, immobilized on a gold electrode. An intercalator (shown in blue), bound near the top of the DNA monolayer, is reduced by electron transfer from the gold surface through the DNA π-stack.

mobilize a high density of DNA duplexes on the surface (e.g. close-packing), the DNA is deposited on the surface in the presence of high concentrations of magnesium ion. Direct quantitation of the surface density is achieved by ^{32}P radioactive labeling. Utilizing atomic force microscopy (AFM), a 45° orientation of the duplexes relative to the gold surface was also determined. Moreover, in AFM studies as a function of applied potential, we found that application of a positive potential caused the DNA, if not closely packed, to lie down upon the surface, hence compressing the monolayer (~20 Å), whereas application of a negative potential induced an increase in monolayer thickness (~50 Å), consistent with the DNA 15-mer duplexes standing upright, perpendicular to the surface.

7.4.2
Electrochemical Probe of Redox Reactions of Intercalators

Early electrochemical experiments performed in our laboratory involved methylene blue, an aromatic heterocycle that binds to DNA by way of intercalation [47]. The reduction of methylene blue at micromolar concentrations could be easily monitored on an electrode modified to contain densely packed DNA duplexes. The binding affinity of the methylene blue to the DNA film was found to be comparable to that seen with DNA in solution; the binding stoichiometry, however, was significantly lower, consistent with methylene blue having access only to sites near the top of the film. Because of the large intervening distance expected between the gold and the film-bound methylene blue, these first studies suggested fast rates of DNA-mediated electron transport. As with other methods of analysis, however, well-defined systems were needed to remove the ambiguity of intercalator binding within the DNA-modified surfaces.

A redox active antitumor intercalator, daunomycin, was chosen as a covalently bound intercalator in studies designed to investigate systematically CT as a function of distance on the DNA-modified surfaces [48]. Daunomycin also undergoes a reversible reduction within the reduction window of the monolayer. Additionally, upon treatment with formaldehyde, the intercalator can crosslink to the 2-amino group of guanine [49] to form an adduct which has been crystallographically characterized [50]. Hence, thiol-derivatized duplexes were constructed containing a single guanine–cytosine base step to probe electrochemically the effects of distance on long-range electron transport [48]. Structural characterization of the DNA films did not reveal a difference between those films containing crosslinked daunomycin duplexes from those that did not contain the crosslinked moiety.

Interestingly, regardless of the position of the daunomycin crosslinked within the 15-mer duplex, reduction of DNA-bound daunomycin was observed and at a comparable rate. Rates of electron transfer through the film could be estimated in measurements where the scan rate was varied, and these showed slow rates of 10^2 s^{-1}, irrespective of the position of the intercalator; in fact similar rates were seen with other redox-active probes simply attached to the alkane-thiol linker we employed. These observations suggested to us that the rate-limiting step of the process was tunneling through the alkane-thiol linkage; in this context, then, CT

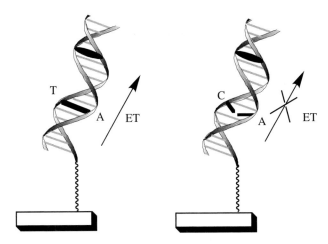

Effective Electron Transport (ET)   Attenuated Electron Transport (ET)

Fig. 7.4. Illustration of the effect of a CA mismatch on electron transport through DNA films. As indicated by the schematic, the introduction of a mismatch, intervening between the gold electrode and intercalator bound near the top of the film, (right) interrupts base-pair stacking locally and effectively attenuates reduction of the intercalator. In the well-stacked duplex (left), in contrast, reduction of the intercalator is efficient.

through 40-Å distances in DNA was necessarily faster. Again to determine explicitly the path of CT in these films, a CA mismatch was introduced at a position between the daunomycin and the gold (Fig. 7.4). With this intervening mismatch, we found no detectable reduction of the daunomycin. The path for CT is therefore through the base stack, and interruptions in the stack within the films also sensitively attenuate CT.

7.4.3
Sensing Mismatches in DNA

While the CA-mismatch experiment represented an important control in our first studies, establishing the path for CT, it also provided a demonstration that a single-base mismatch in DNA could be detected electrochemically. Current techniques for mismatch discrimination commonly rely upon differential hybridization, thermodynamic measurements of the small and sequence-dependent differences between well-matched and mismatched duplexes. Hence, we considered that the sensitivity of long-range CT to stacking might also be exploited in the development of a new class of DNA-based diagnostic tools [51, 52].

To amplify the sensitivity of mismatch detection using this strategy, an electrocatalytic cycle was coupled to the electron transport in well-packed DNA-modified films, magnifying signal differences that arise from small stacking perturbations. The cycle begins with the reduction of the methylene blue intercalator, noncovalently bound near the top of the DNA monolayer. The reduced methylene blue, in turn, reduces ferricyanide, that is contained in solution and repelled by the anionic DNA-modified electrode, hence allowing for further reduction of methylene blue. The catalysis is limited only by the amount of ferricyanide contained in the solution. Using chronocoulometry, integrating the current at the methylene blue potential, all of the possible single-base mismatches (e.g. GA, CA, etc.), including the difficult purine–purine mismatches, in addition to naturally occurring DNA lesions (e.g. A_{ox}:T), could be detected using electrocatalysis [52]. We have also found, perhaps not surprisingly, that mismatches are also sensitively detected in DNA/RNA hybrids (E.M. Boon, unpublished results). Moreover, this technology can be applied in probing how nucleic acid analogs, such as those utilized for antisense applications, affect duplex stacking [53].

We have also applied this assay in tests of physiologically important sequences where mutations cause disease [52]. For example we examined hotspots for mutations in the *p53* gene. These mutations have been implicated in a variety of human cancers. We have also fabricated DNA chips, prepared using gold surfaces of variable size on silicon wafers. We found a linear response in charge accumulation with decreasing electrode size (30–500 μm) and with high signal-to-noise. Given the dimensions of the gold surfaces, assuming close-packing of the DNA, this corresponds to detection of 10^8 DNA molecules on the chip. The technology, therefore, offers promise as a sensitive, new diagnostic tool to detect mutations and single-nucleotide polymorphisms.

7.5
Long-range Oxidative Damage to DNA

Within the cell, radical damage to DNA is known to occur. Given our studies demonstrating long-range CT through DNA, we considered whether CT processes might be an issue physiologically. In other words, can damage to a DNA site arise from a distance through DNA-mediated CT?

7.5.1
Long-range Oxidative Damage at 5'-GG-3' Sites by a Rhodium Intercalator

Of the four DNA bases, guanine has been shown to have the lowest oxidation potential ($E° \sim +1.3$ V versus normal hydrogen electrode (NHE)) [54]. Experimental studies and *ab initio* molecular orbital calculations have revealed that the 5'-G of 5'-GG-3' doublets is preferentially oxidized because the highest occupied molecular orbital (HOMO) lies on the 5'-G of a 5'GG-3' doublet [55, 56]. The oxidation of guanine in DNA leads to piperidine-sensitive base lesions, which are revealed as strand breaks [57]. Hence, this damage at the 5'-G of 5'-GG-3' doublets has become a hallmark for electron transfer chemistry.

Oxidative damage to DNA at a distance was first demonstrated in DNA assemblies containing the tethered, intercalating photooxidant $[Rh(phi)_2bpy']^{3+}$, spatially separated from two 5'-GG-3' sites (Fig. 7.5) [58]. With binding non-covalently

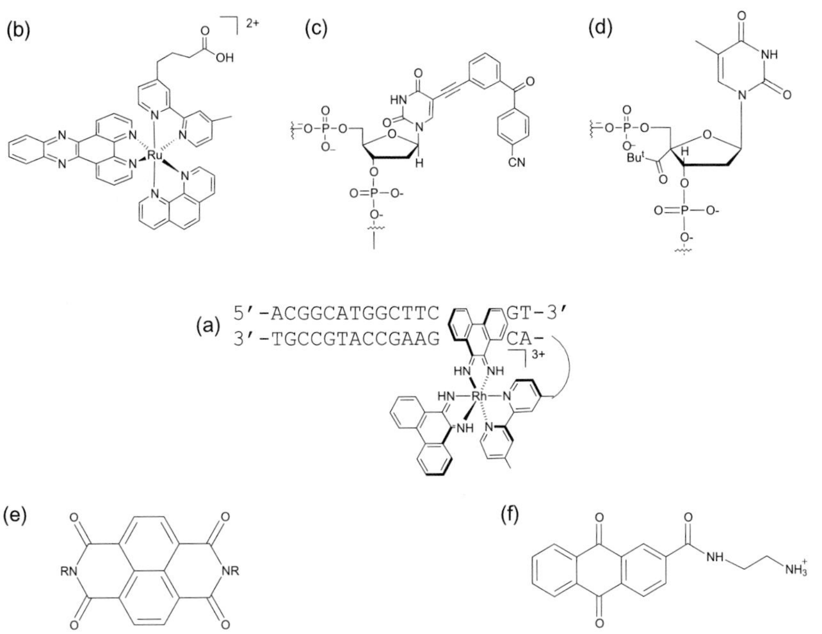

Fig. 7.5. Photooxidants used to probe long-range oxidative damage to DNA. (a) DNA assembly, first used to probe long-range oxidative damage to DNA, containing [Rh(phi)$_2$bpy′]$^{3+}$ spatially separated from two 5′-GG-3′ doublets; (b) [Ru(phen)(dppz)(bpy′)]$^{2+}$; (c) cyanobenzophenone-modified deoxyuridine; (d) 4-pivavoyl-modified deoxythymine; (e) naphthalene diimide; (f) a substituted anthraquinone.

of the sequence-neutral intercalator, [Rh(phi)$_2$ (DMB)]$^{3+}$ (DMB = 4,4′-dimethyl bpy) and irradiation at 365 nm, oxidation of the 5′-G of 5′-GG-3′ was evident; irradiation at higher energy (313 nm) leads instead to direct strand scission, marking the site(s) of metal complex binding. When the rhodium complex was covalently tethered to the duplex, photolysis at 313 nm yielded direct strand breaks near the terminus to which the rhodium complex was tethered. However, with irradiation instead at 365 nm and piperidine treatment, it was found that the two 5′-GG-3′ sites were damaged at distances 17 Å and 34 Å away from the bound metallointercalator.

To examine systematically the distance dependence of long-range oxidative damage, a series of 28 base pair duplexes was constructed that contained proximal and distal 5′-GG-3′ sites [59]. The proximal guanine doublet remained at a fixed distance from the rhodium intercalator, while the distal guanine doublet was marched out in two base-pair increments relative to the position of the intercalator. This incremental increase in distal location allowed for exploration of the importance of helical phasing (e.g. stacking of the intercalator and guanine doublet on the same or opposite side of the helix) in charge transport in addition to the distance dependence of the process. The ratio in yield of oxidative damage at the 5′-G of 5′-GG-3′ doublets provides a measure of the relative efficiency of the charge transport process.

Paralleling our spectroscopic studies, the yield of oxidative damage was not found to be significantly attenuated over a distance of 75 Å. Moreover, the helical phasing of metal complex and the distal guanine site did not appear to be a pertinent parameter. Also, similar to spectroscopic experiments, the long-range CT chemistry was found to be quite sensitive to stacking, both of the oxidant and of the intervening base pairs. Introduction of base bulges (e.g. 5′-ATA-3′) in a DNA assembly containing tethered rhodium between 5′-GG-3′ sites positioned distal and proximal to the rhodium, for example, caused a dramatic diminution in the distal/proximal ratio of oxidative damage [60].

To probe further the distance dependence of the electron transport process, 63 base-pair duplexes containing either tethered $[Rh(phi)_2bpy']^{3+}$ or $[Ru(phen)(dppz)(bpy')]^{2+}$ and six 5′-GG-3′ sites, located 31–197 Å from the metallointercalator, were constructed [59]. Ruthenium(III) complexes of dppz, generated *in situ* with quenching of the ruthenium(II) excited state by a non-intercalating quencher (e.g. methyl viologen) had also been shown to promote oxidative damage to DNA from a distance [61]. Remarkably, in the large assemblies, both excited Rh(III) and ground state Ru(III) intercalating oxidants were found to substantially damage DNA over a distance of 200 Å. Therefore, these studies demonstrated charge migration through DNA to effect damage over a biologically significant distance regime.

Oxidative damage to DNA from a distance does not require a metallointercalator as oxidant. Since our first study establishing oxidative damage from a distance, a variety of oxidants have been used to carry out this long-range chemistry. Figure 7.5 illustrates some of the different photooxidants employed. These include ethidium [62], a cyanobenzophenone-modified deoxyuridine [63], 4-pivaloyl-modified deoxythymine [64], naphthalene diimide [65, 66], and anthraquinone [67]. Indeed studies with the anthraquinone moiety bound at a discrete site on long oligonucleotide duplexes confirmed that long-range oxidative damage can be generated from a 200-Å distance [68]. Many of these photooxidants are currently being employed in experiments to explore DNA CT mechanistically.

7.5.2
Models for Long-range DNA Charge Transport

Once it became clear that oxidative damage to DNA from a distance through CT was general and not a unique property associated with metallointercalators, new theoretical models were proposed to account for the long-range chemistry that had been delineated. These models focused on descriptions of long-range CT through a combination, to varying extents, of hopping and tunneling mechanisms. In a tunneling mechanism, the orbitals of the donor and acceptor molecules are significantly below that of the DNA bridge. Therefore, the charge is considered to tunnel through the DNA bridge without forming transient species along the path. The DNA base-pair stack forms a "virtual" bridge of communication between donor and acceptor, and the rate of CT is expected to show an exponential dependence on the distance separating donor and acceptor. At the other extreme, in the hopping mechanism, the donor and acceptor orbitals are energetically close to or above the bridge states. Hence charge is considered to transiently occupy the bridge. If the rate of charge migration through the DNA bridge is faster than irreversible trapping of the transient radical, then the charge would be able to migrate long distances with a quite shallow distance dependence [69, 70].

Based upon measurements of oxidative yield, Geise, Jortner, and co-workers proposed that CT through DNA occurs by a mixture of hopping and tunneling mechanisms [71]. Specifically, they proposed that hopping occurs primarily among guanine sites, the lower sites energetically, while tunneling is preferred through TA steps. This proposal was based first upon the observation of little oxidative damage when a 5′-TATA-3′ intervened between guanine sites [63, 71]. Schuster also described CT through DNA in the context of a hopping model, but one involving phonon-assisted polaron formation [68, 72]. In this model, he considered that charge might be delocalized over a polaron within the DNA bridge, formed in response to the electrostatic perturbations associated with transport of the charge; propagation of the polaron through the DNA would be assisted by phonon interaction. In reconciling our own photophysical and oxidative damage results, a hopping mechanism seemed appealing to consider. The probability of all the bases being properly stacked and aligned to permit charge transfer over 200 Å was vanishingly small. Thus we considered a "domain hopping" model, in which charge hopping proceeds among domains, where delocalization within a given domain depends upon sequence-dependent dynamics and structure [59].

It is likely, however, that the description is far more complicated still. In the case of photophysical studies with intercalators bound to DNA, the donor and acceptor orbitals energetially can lie close to the bridge states and are extremely well coupled with the base-pair stack. Conversely, in the case of our electrochemical studies, the redox intercalator potential lies far below that of the DNA bridge. Do these different experiments lie at different ends of a mechanistic continuum or can they be considered together? Heller and co-workers introduced the importance of longitudinal polarizability as a factor in considering these results [73]. If the DNA is polarized along the direction of the helix axis, CT is expected to be facilitated. Photoexcitation or the presence of an applied potential within the base-pair stack could provide the means to generate a polarized stack.

Additional mechanistic insights will clearly be needed to reconcile all these results. Irrespective of the mechanism, however, it is clear that stacking within the DNA duplex is a critical factor governing and distinguishing long-range CT.

7.5.3
Sequence Dependence of DNA Charge Transport

To systematically investigate the sequence dependence of charge transport, a series of 21 base-pair oligonucleotides, functionalized with the metallointercalator, $[Rh(phi)_2 bpy']^{3+}$, was designed containing guanine doublets located proximal and distal to the photooxidant [74]. The surrounding sequence of the guanine doublet sites remained constant, while the intervening sequence, containing adenines and thymines, was varied. This family of assemblies was prepared to test directly the proposal of guanine hopping and AT tunneling.

We found that having adenines intervening between the guanine doublet sites led to consistently higher yields of oxidative damage as compared to having either thymines or alternating adenine and thymines between the G-doublet sites (Tab. 7.1). The assemblies containing runs of adenines, with their extensive stacking overlap, would be expected to have the most efficient charge transport based upon simple stacking arguments. Similarly, having thymines in the intervening medium between the guanine doublets would be expected to provide the poorest conduits for charge transport, with AT-tracts yielding an intermediate level of damage. Importantly, and in contrast, based upon an AT tunneling proposal [71], poor transport, if any, would be expected for the assemblies containing a short run of adenines and thymines, with an exponentially decreasing oxidative yield expected as the length of the AT run increased. In fact, increasing the number of adenines or thymines between guanine doublets did not result in marked decreases in long-range oxidative damage [74]. Instead, we observed an *increase* in oxidative yield for all assemblies with increasing oligonucleotide length. Also, inconsistent with guanine hopping, we found that the insertion of a guanine-cytosine step into the otherwise AT sequence actually *decreased* the efficiency of charge transport to the distal guanine site (Tab. 7.1).

These studies certainly illustrate a sequence-dependence in DNA CT, but not with as simple a picture as had been proposed. For example, studies have shown that A-tracts, once nucleated by a sufficient stretch of adenines, readily adopt conformations that differ from that of canonical B-form DNA [75], and this nucleation was used to explain the increase in oxidative damage yield with increasing length. Similarly, many studies have indicated a significant flexibility associated with the sequence 5′-TATA-3′ [76], and this can be used to explain the low efficiency of CT across a 5′-TATA-3′ site. Hence we consider CT in the context of hopping among domains defined by their sequence-dependent conformations [59, 74]. The results, therefore, underscore and reflect the sensitivity of CT to dynamical stacking and to sequence-dependent structure and flexibility.

7.5.4
The Effects of Ion Distribution on Long-Range Charge Transport

Charge transport through well-stacked DNA appears to be much faster than the trapping of the resultant guanine radical by O_2 and H_2O. Hence, one might expect the oxidative yields at guanine doublets contained on the same strand of a duplex to be similar if their thermodynamic potentials are equal. However, using our rhodium(III) and ruthenium(II) intercalators, we observed distal/proximal oxidative damage ratios that were significantly higher than one. A possible explanation for this observation was that the high cationic charge on the metallointercalator

Tab. 7.1. DNA assemblies used to investigate long-range oxidative damage to DNA

Sequence	Distal 5'-G/proximal 5'-G oxidation ratio[a]
Rh — AC GAGCCGTTTTGCCGTAT-3' 3'-TG CTCGGCAAAACGGCATA-5' *[b]	2.5 (2)[c]
AC Rh GAGCCGTTTTTTTTTTGCCGTAT-3' 3'-TG CTCGGCAAAAAAAAAACGGCATA-5' *[b]	1.2 (1)[c]
Rh — AC GAGCCGAAAAGCCGTAT-3' 3'-TG CTCGGCTTTTCGGCATA-5' *[b]	0.9 (1)[c]
Rh — AC GAGCCGAAAAAAAAAAGCCGTAT-3' 3'-TG CTCGGCTTTTTTTTTTCGGCATA-5' *[b]	0.4 (3)[c]
Rh — AC GAGCCGTATAGCTATAGCCGTAT-3' 3'-TG CTCGGCATATCGATATCGGCATA-5' *[b]	0.6 (1)[c]
AC Rh GAGCCGTTTTTTGCCGTAT-3' 3'-TG CTCGGCAAAAAACGGCATA-5' *[b]	5.2 (4)[c,d]
Rh — AC GAGCCGTTTTTTGCCGTAT-3' [b]*3'-TG CTCGGCAAAAAACGGCATA-5'	0.4 (1)[d]
AC G Rh ATGCCGACAGTGTGCCGAT-3' 3'-TGC TACGGCTGTGACACGGCTA-5' *[b]	1.7 (2)[e]
AC G Rh ATGCCGACATTGTGCCGAT-3' 3'-TGC TACGGCTGTTACACGGCTA-5' *[b]	0.2 (1)[e]

[a] Measured damage yields at the 5'-G of 5'-GG-3' sites. The statistical error in the last digit is listed in parentheses.
[b] *indicates location of ^{32}P end-labeling.
[c] Data taken from study of the dependence of CT on sequence.
[d] Data taken from study of the dependence of CT on charge distribution.
[e] Data taken from study of the effects of intervening mismatches.

bound near the terminus of the duplex was sufficient to attenuate the potential of the guanine doublet proximal to the intercalator.

To examine how ion distribution on the DNA helix affects charge transport, we assessed oxidative damage yields upon photooxidation of assemblies covalently tethered with $[Rh(phi)_2bpy']^{3+}$ and ^{32}P-end-labeled either at the 5' end, which introduces two negative charges at the terminus or at the 3' end, which introduces one negative charge [77]. Using an assembly containing an intervening A_6-tract, we found that changing the ^{32}P-label from the 5' end of the duplex to the 3' end resulted in a dramatic decrease in charge transport to the distal guanine site (Tab. 7.1). Furthermore, upon the introduction of an unlabeled phosphate at the 5' end of the duplex and maintenance of the $3'$-^{32}P end-label, an intermediate increase in oxidative yield was observed. Analogous behavior was also seen with mixed DNA sequences.

We have proposed that these results reflect a change in the oxidation potential at the distal guanine site relative to the proximal site due to change in charges at the termini of the oligomer. Coarse calculations of the internal dielectric of DNA based upon these data suggested particularly high values, an indication of the high polarizability of the DNA π-stack. Such results certainly require consideration in developing mechanisms for DNA CT. Indeed the high longitudinal polarizability of DNA offers an attractive means to reconcile photophysical, biochemical, and electrochemical measurements that have been carried out [75].

7.5.5
Mismatch Influence on Long-range Oxidative Damage to DNA

Just as we had seen both in photophysical and electrochemical experiments that contain intervening mismatches, which locally perturbed the DNA base-pair stack and sensitively modulated DNA CT, we were also interested in determining the effect of intervening mismatches on long-range oxidative damage [78]. To investigate this systematically, 22 base-pair duplexes, containing intervening mismatches between guanine doublet sites and functionalized with the metallointercalator, $[Ru(phen)(dppz)(bpy')]^{2+}$, were assembled. This resulted in a total of 16 possible base-pair and mismatch combinations. As in other experiments, the efficiency of charge transport was quantified through measurements of the ratio in yield of 5'-G oxidative damage at guanine doublets located distal versus proximal to the intercalating oxidant. ^1H NMR measurements of base-pair opening lifetimes were also performed to probe mismatch dynamics.

Overall, the trend in oxidative damage yields varied in the order GC $\sim$ GG $\sim$ GT $\sim$ GA > AA > CC $\sim$ TT $\sim$ CA $\sim$ CT. Generally the purine–purine mismatches did little to attenuate the amount of oxidative damage observed at the distal guanine site, while introduction of a pyrimidine–pyrimidine mismatch resulted in much lower oxidative yields (Tab. 7.1). The extent of distal/proximal guanine oxidation in different mismatch-containing duplexes was compared with the helical stability of the duplexes, electrochemical data for intercalator reduction on different mismatch-containing DNA films, and base-pair lifetimes for oligomers containing the different mismatches derived from ^{1}H NMR measurements of the imino proton exchange rates. While a clear correlation was evident both with helix stability and the electrochemical data, monitoring reduction of an intercalator through DNA films, it was interesting to observe that damage ratios correlated most closely with base-pair lifetimes. Competitive hole trapping at the mismatch site, which had been proposed by others [79], did not appear to be a key factor governing the efficiency of transport through the mismatch. Because of the close correlation with dynamical measurements of DNA, these results served once again to underscore the importance of base dynamics in modulating long-range CT through the DNA base-pair stack [78]. Indeed, it has become increasingly clear that measurements of DNA CT may offer a new and sensitive route to assess sequence-dependent base dynamics within the DNA duplex.

7.6
Using Charge Transport to Probe DNA–Protein Interactions and DNA Repair

Given the efficiency of DNA CT that was documented over biologically relevant distance regimes, it became important to consider roles and consequences of DNA CT within the cell. With that aim in mind, we have begun to explore how proteins modulate DNA CT, how nucleosomal packaging affects DNA CT damage, how DNA repair can be promoted through CT, and even whether long-range CT through DNA occurs within the nucleus of the cell.

7.6.1
DNA-Binding Proteins as Modulators of Oxidative Damage from a Distance

Proteins that bind DNA do so utilizing a range of structural motifs. Some modes of DNA binding involve significant perturbations in base-pair structure. Given the sensitivity of DNA CT to such structural perturbations, we were interested in determining how protein binding through different motifs would affect long-range DNA CT.

DNA assemblies were constructed that incorporated specific protein binding sites between two 5'-GG-3' sites that were spatially separated from the 5'-tethered photooxidant $[Rh(phi)_2bpy']^{3+}$ [80]. Again, we assessed the efficiency of DNA charge transport as a function of protein binding through measurements of the ratio of damage yield at the 5'-GG-3' site distal to the photooxidant versus that at the proximal 5'-GG-3' site. The proteins utilized included methyltransferase *Hha* I, the TATA-binding protein (TBP), the restriction endonuclease *Pvu*II, and the Antennapedia homeodomain protein (ANTP), proteins which bind DNA through different, structurally well-characterized modes.

We focused first on perturbations in long-range oxidative damage as a result of binding the base-flipping enzyme methyltransferase *Hha* I [81]. This enzyme recognizes the sequence 5'-G*CGC-3' and methylates the target cytosine (*C) after completely flipping the base out of the duplex stack [82]. The π-cavity that is created is filled by insertion of a glutamine side chain (Gln237) from the enzyme; the glutamine essentially serves as a "bookmark" to hold the place for the cytosine. From the perspective of CT chemistry, the binding of methyltransferase *Hha* I to DNA provided an ideal system to study how the insertion of a σ-bonded residue within the π-stack would affect long-range DNA CT. What we observed was that the yield of oxidative damage at the distal 5'-GG-3' site decreased significantly with protein binding. We also tested the effects of binding of a mutant methyltransferase *Hha* I which inserts a π-stacking tryptophan residue (Q237W) into the cavity; in this case, oxidative damage to the distal site was restored.

Studies of long-range oxidative damage in the presence of other DNA-binding proteins yielded consistent results [80]. For example, crystallographic studies show that DNA binding of TBP, which plays an important role in transcription initiation, to the sequence 5'-TATAAA-3', induces a 90° kink at either end of the already flexible binding site [83, 84]. With this DNA substrate, owing to the flexible 5'-TATAAA-3' site, a low damage ratio was evident even without protein. In the presence of TBP, however, a significant decrease in charge transport to the distal guanine site was also observed as a function of increasing protein concentration. In general, then, it was observed that yields of long-range oxidative damage correlated directly with the nature of the nucleoprotein interaction. Interactions that disturb the DNA π-stack inhibit DNA CT. Alternatively, interactions that promote no helix distortion, such as *Pvu*II and ANTP, but, as a result of tight packing, may rigidify the π-stack, serve instead to enhance the ability of the DNA base stack to provide a conduit for CT. Thus, we observed that protein binding to DNA modulates long-range CT both negatively and positively, depending upon the specific protein/DNA interactions in play.

7.6.2
Detection of Transient Radicals in Protein/DNA Charge Transport

We were interested also in determining if proteins can participate directly in DNA CT chemistry. Can electrons or electron holes be transferred effectively between the DNA and protein within the macromolecular assembly? Both spectroscopic and biochemical techniques were employed to monitor transient radical intermediates as a result of CT through DNA–protein assemblies [85]. To facilitate spectroscopic studies, assemblies containing a methyltransferase *Hha* I-binding site were constructed with the tethered, intercalator [Ru(phen)(dppz)(bpy')]$^{2+}$, and the flash-quench technique [61] was employed to generate the Ru(III) oxidant *in situ*, triggering the long-range oxidation chemistry. Because we could monitor the same assemblies using both biochemical measurements of oxidative damage yield and transient absorption measurements of rates of formation of radical intermediates in the CT process, we could correlate the two directly.

Upon irradiation of an assembly containing the mutant methyltransferase *Hha* I (Q237W)-bound 14 base pairs from the site of ruthenium intercalation, extensive oxidative damage was observed to the guanine located 5' to the tryptophan intercalation site; no similar damage was evident with binding of the wild-type protein

[85]. Parallel transient absorption experiments also revealed the formation of a radical species. Interestingly, the full transient absorption spectrum of this intermediate contained features characteristic of both the guanine and tryptophan radicals. It should be noted that the oxidation potential of the tryptophan is ∼1.0 V versus NHE [86]. Hence hole transport proceeded through DNA to the inserted tryptophan from the protein and to the 5′-guanine, where irreversible oxidative damage occurred.

To examine the distance dependence of the rate of formation of the tryptophan and guanine radicals, a series of ruthenium-tethered duplexes was prepared containing methyltransferase *Hha* I-binding sites, located at distances varying from 24 to 51 Å away from the ruthenium intercalator [85]. The yield of oxidative damage observed at the 5′-position to the inserted tryptophan diminished little with increasing distance. Transient absorption measurements also showed no variation in the rate of radical formation with increasing distance; the rate constant for radical formation was $>10^6$ s^{-1} in each assembly. Hole transport through DNA over this 50-Å distance regime, therefore, was not rate limiting. In fact these kinetic results complemented well our earlier findings of base-base electron transfer rates on the order of 10^{10} s^{-1} [44]. More recently, we have carried out analogous experiments using a methylindole moiety incorporated directly into the DNA as an artificial base [87]. These studies of CT yielded comparable results and, in fact, set the lower limit for CT through the stack as $>10^7$ s^{-1}, coincident with formation of the Ru(III) oxidant through diffusional quenching.

7.6.3
Electrical Detection of DNA–Protein Interactions

Our electrochemical assay on DNA-modified electrodes provided another means to probe protein/DNA interactions by monitoring the effect of protein binding on the DNA base stack [88]. This assay yielded results wholly complementary to our measurements of oxidative damage, despite the fact that here reduction of a bound intercalator was being measured, while in the biochemical experiments photo-induced DNA oxidation chemistry was being probed. Practically, however, in the electrochemistry experiment, the measurement could be made in real time and without gel electrophoresis.

Functionalized with a thiol-terminated linker and with daunomycin crosslinked to a guanine residue near the duplex terminus, DNA assemblies were constructed that contained the binding sites for either methyltransferase *Hha* I, uracil-DNA glycosylase (UDG), TBP, or *Pvu*II. Measures were taken to prevent enzyme turnover (e.g. using 2′-fluorouracil as opposed to uracil in binding UDG), to insure that binding of the native enzyme to DNA could be studied. The self-assembly of the functionalized DNA monolayer on the gold surface was performed in the absence of magnesium ion, to allow for loose packing of the film and binding of protein within the monolayer. Any remaining exposed gold surface was then back-filled with mercaptohexanol. Tab. 7.2 provides a summary of the electrochemical responses obtained by chronocoulometry in the presence of the different proteins.

Tab. 7.2. Electrochemical detection of DNA–protein interactions using tethered daunomycin[a]

Sequence[b]	Assay	DNA association	Integrated charge[c]
SH-5'-AIAIAT*ICIC*AIATCC◖T-3' 3'-TCTCTA*CICI*TCTAGG◗A-5'	Without protein		High
	M. *Hha* I	Base-flipping, inserts Q	Low
	M. *Hha* I (Q237W)	Base-flipping, inserts W	High
SH-5'-AIAIAT*ICIC*AIATCC◖T-3' 3'-TCTCTAC *CI*TCTAGG◗A-5'	Without protein		High
	M. *Hha* I	Base-flipping, inserts Q	Low
	M. *Hha* I (Q237W)	Base-flipping, inserts W	High
SH-5'-AICT*I*AATCAITCC◖T-3' 3'-TCIACT*U*AITCAGG◗A-5'	Without protein		High
	UDG	Base-flipping	Low
SH-5'-IAIATATAAAICACC◖T-3' 3'-CTCTATATTT*CI*TGG◗A-5'	Without protein		High
	TBP	Kinks DNA by 90°	Low
SH-5'-TCTT*CAI*CT*I*AIACC◖T-3' 3'-AIAA*ITMI*ACTCTGG◗A-5'	Without protein		High
	R. *Pvu*II	No structural perturbation	High

[a] Summary of results taken from [88].
[b] Sequence used to form DNA film.
[c] Based on chronocoulometry.
M. *Hha* I, methyltransferase *Hha* I; UDG, uracil-DNA glycosylase; TBP, TATA-
binding protein; R. *Pvu*II, restriction endonuclease *Pvu*II.

As in the biochemical experiments, electrochemistry revealed that binding of native methyltransferase *Hha* I, the base-flipping enzyme that inserts a glutamine into the π-stack, significantly attenuated the electrochemical response. Consistent with transport through a π-way, binding of the mutant protein methyltransferase *Hha* I (Q237W), which inserts a tryptophan into the π-stack, showed little attenuation of electrochemical response. We carried out a parallel series of experiments on DNA-modified electrodes containing an abasic site within the protein recognition site. In this case, only a small electrochemical signal was evident either without protein or with native protein. Interestingly, however, binding of the mutant here as well filled the π-stack and thus yielded a high electrochemical response. UDG, a base excision repair enzyme that extrudes uracil upon binding, was also probed using this assay. Analogous to the results found upon binding of the native methyltransferase *Hha* I, UDG also showed a marked decrease in electrochemical response upon protein binding. Furthermore, introducing a gross structural distortion, as that found upon binding of TBP, which significantly kinks DNA, also led to an attenuation in the electrochemical signal.

To determine whether this novel assay could also be useful in monitoring enzymatic reactions with DNA, a DNA film was prepared to study the endonuclease activity of *Pvu*II. As a function of incubation with the protein, a decrease in signal was evident, and the kinetics, monitored by this signal reduction, paralleled closely that determined by gel assay of restriction fragmentation. We are currently interested in determining whether these assays might be sufficiently sensitive to allow us to follow the real time fluctuations in base-pair motions associated with protein binding and reaction.

These data further illustrate the exquisite sensitivity of DNA electron transport to stacking perturbations and offer a novel assay through which to structurally probe how different proteins interact with DNA. This assay also offers a new opportunity to apply CT chemistry in directly probing DNA–protein interactions in real time.

7.6.4
Repair of Thymine Dimers

In addition to probing oxidative damage to DNA from a distance, we became interested in exploring other reactions on DNA that might be promoted at long-range through CT chemistry. We determined that DNA CT can also be utilized to promote the repair of thymine dimers in DNA [89]. Thymine dimers, the most common photochemical lesions in DNA, result from the photoinduced [2 + 2] cycloaddition reaction between adjacent thymines contained on the same polynucleotide strand to form the cyclobutane dimer [90]. This lesion is removed in eukaryotic cells by the excision of the dimer. However, in bacteria, the photolyase enzyme promotes repair using electron transfer chemistry; with irradiation in the visible region, a reduced flavin cofactor adds an electron to the cyclobutane dimer, triggering repair of the lesion and release of the electron back to the enzyme. In model systems, repair of the cyclobutane dimer has been triggered both oxidatively and reductively [91, 92]. Hence, with our potent rhodium intercalating photooxidant ($E° \sim +2.0$ V versus NHE), we considered that thymine dimer repair might be promoted within the DNA duplex triggered by oxidation [89].

To perform these studies, 16 base-pair duplexes containing the tethered photooxidant $[Rh(phi)_2bpy']^{3+}$ and a thymine dimer located 16–26 Å away from the metallointercalator, were constructed. Upon irradiating the duplexes at 400 nm, a high-performance liquid chromatography assay revealed a substantial amount of thymine dimer repair with increasing irradiation time (Tab. 7.3). Analogous to our studies of oxidative damage, we also found that this repair was insensitive to increasing the distance between the metallointercalator and the thymine dimer (16–26 Å). Furthermore, the repair efficiency was also found to be sensitive to π-stacking perturbations; a decrease in repair efficiency was seen with the introduction of base bulges between the site of rhodium intercalation and the thymine dimer. Hence, not only is DNA able to promote long-range oxidative damage to DNA from a remote site through CT chemistry, but CT chemistry with a similar sensitivity to the intervening base stack can also be harnessed to trigger thymine dimer repair from a distance.

The long-range repair of thymine dimers is not a unique property of the rhodium intercalator. Other potent photooxidants can also promote the repair of thymine dimers [93]. A series of duplexes were constructed containing the thymine dimer lesion and various intercalating photooxidants, including anthraquinone derivatives, naphthalene diimide, $[Rh(phi)_2bpy']^{3+}$, and N-8-glycyl ethidium, were probed. Analogous to our rhodium intercalator, substantial repair from a distance

Tab. 7.3. Long-range thymine dimer repair using the tethered photooxidant [Rh(phi)$_2$(bpy')]$^{3+}$

DNA assembly	Distance (Å)[a]	Repair (%)
Rh A–C–G T–C–T–C–A–A–C–T–C–A–C–G–T T–G–C A–G–A–G–**T**–**T**–G–A–G–T–G–C–A–' 5	19	79[b]
Rh A–C–G T–C–T–C–A–A–C–T–C–A–C–G–T T–G–C A–G–A–G–**T**–**T**–G–A–G–T–G–C–A–5'	26	100[b]
Rh A–C–G T–C–T–C–A–A–C–T–C–A–C–G–T ^{c} T–G–C A–G A G–**T**–**T**–G A G–T–G–C–A–5' C T T C	–	47[b]
Rh A–C–G T–A–T–G–C–C–G–T–C–A–A–C–T T–G–C A–T–A–C–**G**–**G**–C–A–G–**T**–**T**–G–A–5'	36	70[d]

[a] Distances between the rhodium intercalator, as shown schematically, and the cyclobutane ring of the dimer, which are based upon 3.4 Å base-pair separation.
[b] Repair is expressed as the percentage of thymine dimer repaired, as determined by HPLC analysis, after 6 h of irradiation at 365 nm. Data taken from [89].
[c] The thymine dimer is shown in bold with a ∧. The extruding bases schematically show the position of the base bulges.
[d] Data taken from [95] and describes the amount of repaired dimer after 3 h of irradiation at 365 nm.

was observed using the organic intercalator naphthalene diimide, which has a reduction potential of +1.9 V versus NHE. However, repair was not observed using either ethidium or anthraquinone derivatives as non-covalently bound intercalators. In the case of ethidium, it is understandable that the organic intercalator cannot perform repair, since the reduction potential of the ethidium is sufficiently below that of the dimer. Surprisingly, the anthraquinone was also unable to repair the dimer [94]. The anthraquinone derivatives have a reduction potential of ~2.0 V versus NHE from the excited triplet state and therefore should have sufficient driving force for repair. We have attributed the lack of repair with the anthraquinone moiety to the fact that its excited singlet state is particularly short-lived [93]. In fact, anthraquinone derivatives that do not undergo rapid conversion to the excited triplet state can promote thymine dimer repair. Interaction with the singlet state may therefore be critical for effective repair.

Studies were also performed to examine the competition between oxidative damage and thymine dimer repair, since oxidation of guanine sites should be favored thermodynamically [95]. Duplexes were constructed containing either a 5'-GG-3' site, a thymine dimer, or both moieties, and the duplexes were covalently modified with a metallointercalating oxidant (either $[Rh(phi)_2bpy']^{3+}$ or $[Ru(phen)(dppz)(bpy')]^{2+}$). Both thymine dimer repair and guanine oxidation were then systematically examined. Interestingly, the data revealed that thymine dimer repair proceeds efficiently in the presence of potential thermodynamic traps provided by guanine doublets (Tab. 7.3). Although it is thermodynamically favorable to perform guanine oxidation, it appears kinetically preferable to repair thymine dimers. This competition is useful to consider in the context of generally assessing reactions that may proceed within the cell based upon CT chemistry. Indeed other DNA-mediated CT reactions, perhaps initiated by proteins bound to DNA, could conceivably be carried out along the duplex without irreversible reaction at guanines as long as these reactions are similarly kinetically favored.

7.6.5
Oxidative Damage to DNA in Nucleosomes

In eukaryotic cells, DNA is packaged within nucleosome core particles [96]. In this unit, $\sim$150 base pairs of DNA are wrapped around an octamer of histone proteins, to which the DNA is associated by non-specific, electrostatic interactions. It has been thought that this packaging of nucleosomal DNA serves to protect it from damage. The DNA is dynamically restricted and less accessible to solution-borne damaging agents. However, oxidative damage through CT chemistry does not require solution accessibility. On the other hand, the restricted motion and overall bending of the DNA within the nucleosome would not be expected to favor long-range CT. The question of whether DNA CT proceeds through the nucleosome core particle was therefore interesting to consider.

To investigate long-range oxidative damage in nucleosomes, several 146 base-pair palindromic duplexes were constructed, containing 14 5'-GG-3' sites [97]. These duplexes were modeled after that crystallographically characterized within the nucleosome core particle [96]. Nucleosomes were then formed using these duplexes and their structural integrity was confirmed using biochemical methods [97].

First damage within the nucleosome was probed using non-covalently bound $[Rh(phi)_2DMB]^{3+}$. Upon irradiation of the rhodium-bound assemblies, damage was evident at the 5'-G of 5'-GG-3' sites. Little specific binding of the rhodium complex was evident. Perhaps not surprisingly, the nucleosomal DNA cannot unwind to accommodate the intercalator, and hence, intercalator binding is inhibited. Nonetheless, despite the poor binding, significant oxidative damage was observed. Given the poor binding, the results suggested that this damage had arisen through long-range CT within the nucleosome.

To establish long-range CT within the nucleosome definitively and to probe the distance over which charge can migrate, the metallointercalator $[Rh(phi)_2bpy']^{3+}$ was covalently tethered to the 5' ends of the palindromic 146 base-pair duplex and nucleosomes were formed. Upon irradiation of the duplex, selective 5'-G oxidative damage at 5'-GG-3' sites located 8–24 base pairs away from the site of intercalation was revealed (Fig. 7.6, see p. 171). Notably, guanine oxidation was not observed at sites located further away from the site of rhodium intercalation either in the core particle or even in the absence of histones. It is likely that the highly bent DNA structure utilized to obtain consistent nucleosome phasing for these experiments also interfered with long-range CT. Nonetheless, over a significant distance regime (>75 Å), long-range CT was established within the nucleosome core particle.

These data hold implications for damage to and repair of the genome *in vivo*, where much of the DNA is bound within nucleosome core particles. It has been considered that histones, in addition to packaging and regulating DNA, also function to protect DNA, since the histones reduce binding by a variety of potentially dangerous small molecules. In fact, binding of DNA to histone proteins to form nucleosome core particles does reduce the ability even of the rhodium complex to intercalate into the DNA. However, and importantly, packaging of DNA as nucleosomes does not protect it from long-range damage through CT through the base-pair stack. Indeed, such damage generated by long range CT within the nucleosome may persist preferentially and lead to the formation of permanent mutations.

7.6.6
DNA Charge Transport within the Nucleus

Many of our studies lead us to consider the physiological ramifications of DNA-mediated CT. Since DNA is the genetic material encoding all of the information within the cell, it becomes important to ask whether in the cell there are mechanisms to protect DNA from oxidative damage at a distance and additionally perhaps also to exploit DNA CT chemistry. A first step in exploring these issues involves establishing whether, indeed, DNA CT can proceed within the cell nucleus.

Towards that end, our rhodium intercalators were incubated with HeLa cell nuclei and irradiated to promote photooxidation [98]. After isolation of the DNA, and treatment with base excision repair enzymes to reveal base damage through strand breaks, the damaged DNA was amplified using the ligation-mediated polymerase chain reaction. Using this protocol, we probed exon 5 of the *p53* gene as well as a transcriptionally active promoter within the phosphoglycerate kinase gene (*PGK1*). Oxidative damage studies on exon 5 of the *p53* gene revealed clearly the preferential damage at the 5'-G of 5'-GGG-3', 5'-GG-3', and 5'-GA-3' sites, a signature of electron transfer damage to DNA. Thus the rhodium complex, with photoactivation, can promote oxidative damage to DNA within the nucleus. More interesting still were the experiments in which damage on the PGK promoter was probed. Here, oxidative damage was found at protein-bound sites which are inaccessible to rhodium. Thus, on transcriptionally active DNA within the cell nucleus, DNA-mediated CT can promote base damage from a distance. Importantly, then, even within the nucleus, direct interaction of an oxidant is not necessary to generate a base lesion at a specific site. These observations set the stage for studies directed at probing *in vivo* applications of DNA CT. Certainly these data require consideration in understanding cellular mechanisms for DNA damage and repair.

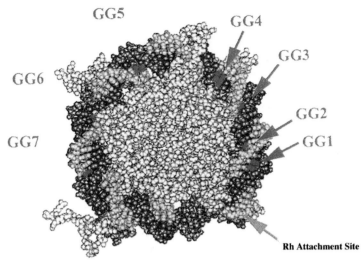

GG5

GG4

GG3

GG6

GG2

GG7

GG1

Rh Attachment Site

Fig. 7.6. Schematic illustration of long-range charge transport in a nucleosome core particle. The nucleosome was assembled containing [Rh(phi)$_2$bpy']$^{3+}$ covalently tethered to the terminus (indicated by the purple arrow). Efficacious charge transport leading to oxida- tive damage at a site 24 base pairs away (GG4) from the site of rhodium intercalation was evident [97]. The nucleosome structure shown here is based upon that determined crystallographically [96] and shows DNA in blue and histones in gray.

7.7
Conclusions

The range of studies described here, including spectroscopic, electrochemical, and biochemical measurements, all have pointed to a remarkable chemistry that depends upon the π-stacked structure of DNA. It is this dependence on stacking that distinguishes CT in nucleic acids from that through other macromolecular assemblies. We continue to be surprised by the exquisite sensitivity of DNA CT to the dynamical π-stacked structure of nucleic acids.

We can now begin to harness this sensitivity to π-stacking in a variety of applications. Electrochemistry on DNA films offers a novel and sensitive route for the diagnosis of mutations in DNA. These electrochemical methods may offer also strategies to analyze protein–DNA interactions and to screen for new drugs that inhibit such interactions. Moreover, CT chemistry may provide a completely new means to probe and characterize the sequence-dependent dynamics and flexibilities of nucleic acids. From the experiments already carried out, it is clear that the dynamical motions within DNA, dependent on sequence, local conformation, and protein interaction, modulate sensitively CT through the base-pair stack.

Perhaps most intriguing to consider is whether this chemistry is harnessed within the cell. CT through DNA proceeds over biologically significant distances. Perhaps then some sequences in DNA must be insulated and protected from long-range CT damage. Other sites within the genome may instead be hot spots to which CT damage is funneled. The modulation of CT by sequence, conformation, and DNA-binding proteins clearly also offers a means of physiological control and regulation within the cell. Indeed, given that proteins can participate in DNA CT chemistry, perhaps they also utilize the DNA stack in long-range signaling or in repair. The remarkable sensitivity we have observed in detecting mutations and lesions in DNA could conceivably be exploited within the cell.

Our studies of DNA CT chemistry has therefore certainly provided a rich treasure of questions and ideas to explore. These experiments have provided us with a new perspective through which to view double-helical DNA. CT chemistry through DNA offers new challenges to exploit and unravel.

Acknowledgements

We are grateful to the NIH, NSF, UNCF-Merck Science Initiative, and NFCR for their continued support of the work described here. Most importantly, this research depended upon the persistence, creativity, and curiosity of our co-workers and collaborators, cited throughout the text, and we thank them for their efforts.

References

1 KITTEL, C. *Introduction to Solid State Physics*, Wiley, New York, 1976.

2 MARKS, T. J. Electrically conductive metallomacrocyclic assemblies. *Science* 1985, *227*, 881–889.

3 WATSON, J. D., CRICK, F. H. C. Molecular structure of nucleic acids: a structure for deoxyribose nucleic acid. *Nature* 1953, *171*, 737–738.

4 SZENT-GYÖRGI, A. The study of energy-levels in biochemistry. *Nature* 1941, *148*, 157–159.

5 ELEY, D. D., SPIVEY, D. I. Semi-conductivity of organic substances. *J. Chem. Soc., Faraday Trans.* 1962, *58*, 411–415.

6 WARMAN, J. M., DE HAAS, M. P., RUPPRECHT, A. DNA: a molecular wire? *Chem. Phys. Lett.* 1996, *249*, 319–322.

7 SNART, R. S. Photoelectric of DNA. *Biopolymers* 1968, *6*, 293–297.

8 LIANG, C. Y., SCALCO, E. G. Electrical conduction of a highly polymerized sample of sodium salt of deoxyribonucleic acid. *J. Chem. Phys.* 1964, *40*, 919–922.

9 ANDERSON, R. F., PATEL, K. B., WILSON, W. R. Pulse-radiolysis studies of electron migration in DNA from DNA base-radical anions to nitro-acridine intercalators in aqueous solution. *J. Chem. Soc., Faraday Trans.* 1991, *87*, 3739–3746.

10 CULLIS, P. M., McCLYMONT, J. D., SYMONS, M. C. R. Electron conduction and trapping in DNA- an electron spin resonance study. *J. Chem. Soc., Faraday Trans.* 1990, *86*, 591–592.

11 OKAHATA, Y., KOBAYASHI, T., TANAKA, K., SHIMOMURA, M. Anisotropic electric conductivity in an aligned DNA cast film. *J. Am. Chem. Soc.* 1998, *120*, 6165–6166.

12 FINK, H.-W., SCHÖNENBERGER, C. Electrical conduction through DNA molecules. *Nature* 1999, *398*, 407–410.

13 PORATH, D., BEZRYADIN, A., DE VRIES, S., DEKKER, C. Direct measurement of electrical transport through DNA molecules. *Nature* 2000, *403*, 635–638.

14 DEKKER, C., RATNER, M. A. Electronic properties of DNA. *Physics World* 2001, 29–33.

15 KASUMOV, A. Y., KOCIAK, M., GUÉRON, S., REULET, B., VOLKOV, V. T., KLINOV, D. V., BOUCHIAT, H. Proximity-induced superconductivity in DNA. *Science* **2001**, *291*, 280–282.

16 GRAY, H. B., WINKLER, J. R. Electron transfer in proteins. *Annu. Rev. Biochem.* **1996**, *65*, 537–561.

17 MARCUS, R. A., SUTIN, N. Electron transfers in chemistry and biology. *Biochim. Biophys. Acta* **1985**, *811*, 265–322.

18 WOITELLIER, S., LAUNAY, J. P., SPANGLER, C. W. Intervalence transfer in pentaammineruthenium complexes of alpha, omega-dipyridyl polyenes. *Inorg. Chem.* **1989**, *28*, 758–762.

19 MURPHY, C. J., ARKIN, M. R., JENKINS, Y., GHATLIA, N. D., BOSSMANN, S. H., TURRO, N. J., BARTON, J. K. Long-range photoinduced electron transfer through a DNA helix. *Science* **1993**, *262*, 1025–1029.

20 KELLEY, S. O., HOLMLIN, R. E., STEMP, E. D. A., BARTON, J. K. Photoinduced electron transfer in ethidium-modified DNA duplexes: dependence on distance and base stacking. *J. Am. Chem. Soc.* **1997**, *119*, 9861–9870.

21 LEWIS, F. D., WU, T., ZHANG, Y., LETSINGER, R. L., GREENFIELD, S. R., WASIELEWSKI, M. R. Distance-dependent electron transfer in DNA hairpins. *Science* **1997**, *277*, 673–676.

22 FUKUI, K., TANAKA, K. Distance dependence of photoinduced electron transfer in DNA. *Angew. Chem. Int. Ed.* **1998**, *37*, 158–161.

23 WAN, C., FIEBIG, T., KELLEY, S. O., TREADWAY, C. R., BARTON, J. K., ZEWAIL, A. H. Femtosecond dynamics of DNA-mediated electron transfer. *Proc. Natl Acad. Sci. USA* **1999**, *96*, 6014–6019.

24 NÚÑEZ, M. E., BARTON, J. K. Probing DNA charge transport with metallo-intercalators. *Curr. Opin. Chem. Biol.* **2000**, *4*, 199–206.

25 ERKKILA, K. E., ODOM, D. T., BARTON, J. K. Recognition and reaction of metallointercalators with DNA. *Chem. Rev.* **1999**, *99*, 2777–2795.

26 SITLANI, A., BARTON, J. K. Sequence-specific recognition of DNA by

phenanthrenequinone diimine complexes of rhodium(III): importance of steric and van der waals interactions. *Biochemistry* **1994**, *33*, 12100–12108.

27 SITLANI, A., LONG, E. C., PYLE, A. M., BARTON, J. K. DNA photocleavage by phenanthrenequinone diimine complexes of rhodium(III): shape-selective recognition and reaction. *J. Am. Chem. Soc.* **1992**, *114*, 2303–2312.

28 DAVID, S. S., BARTON, J. K. NMR Evidence for Specific Intercalation of Δ–Rh(phen)$_2$phi^{3+} in [d(GTCGAC)]$_2$. *J. Am. Chem. Soc.* **1993**, *115*, 2984–2985.

29 HUDSON, B. P., BARTON, J. K. Solution structure of a metallointercalator bound site specifically to DNA. *J. Am. Chem. Soc.* **1998**, *120*, 6877–6888.

30 KIELKOPF, C. L., ERKKILA, K. E., HUDSON, B. P., BARTON, J. K., REES, D. C. Structure of a photoactive rhodium complex intercalated into DNA. *Nature Struct. Biol.* **2000**, *7*, 117–121.

31 TURRO, C., HALL, D. B., CHEN, W., ZUILHOF, H., BARTON, J. K., TURRO, N. J. Solution photoreactivity of phenanthrenequinone diimine complexes of rhodium and correlations with DNA photocleavage and photooxidation. *J. Phys. Chem. A* **1998**, *102*, 5708–5715.

32 FRIEDMAN, A. E, CHAMBRON, J.-C., SAUVAGE, J.-P., TURRO, N. J., BARTON, J. K. Molecular "light switch" for DNA: Ru(bpy)$_2$(dppz)$^{2+}$. *J. Am. Chem. Soc.* **1990**, *112*, 4960–4962.

33 JENKINS, Y., FRIEDMAN, A. E., TURRO, N. J., BARTON, J. K. Characterization of dipyridophenanzine complexes of ruthenium(II): the light switch effect as a function of nucleic acid sequence and conformation. *Biochemistry* **1992**, *31*, 10809–10816.

34 STEMP, E. D. A., HOLMLIN, R. E., BARTON, J. K. Electron transfer between metal complexes bound to DNA: variations in sequence, donor, and metal binding mode. *Inorg. Chim. Acta.* **2000**, *297*, 88–97.

35 OLSON, E. J. C., HU, D., HÖRMANN,

A., JONKMAN, A. M., ARKIN, M. R.,
STEMP, E. D. A., BARTON, J. K.,
BARBARA, P. F. First observation of the
key intermediate in the "light switch"
mechanism of [Ru(phen)₂(dppz)]²⁺.
J. Am. Chem. Soc. **1997**, *119*, 11458–
11467.

36 HARTSHORN, R. M., BARTON, J. K.
Novel dipyridophenazine complexes of
ruthenium(II): exploring luminescent
reporters of DNA. *J. Am. Chem. Soc.*
1992, *114*, 5919–5925.

37 DUPUREUR, C. M., BARTON, J. K. Use
of selective deuteration and ¹H NMR
in demonstrating major groove
binding of Δ-[Ru(phen)₂(dppz)]²⁺ to
d(GTCGAC)₂. *J. Am. Chem. Soc.* **1994**,
116, 10286–10287.

38 TUITE, E., LINCOLN, P., NORDÉN, B.
Photophysical evidence that Δ- and
Λ-[Ru(phen)₂(dppz)]²⁺ intercalate
DNA from the minor groove. *J. Am.
Chem. Soc.* **1997**, *119*, 239–240.

39 ARKIN, M. R., STEMP, E. D. A.,
HOLMLIN, R. E., BARTON, J. K.,
HÖRMANN, A., OLSON, E. J. C.,
BARBARA, P. F. Rates of DNA-mediated
electron transfer between metallointercalators. *Science* **1996**, *273*, 475–
480.

40 KELLEY, S. O., HOLMLIN, R. E., STEMP,
E. D. A., BARTON, J. K. Photoinduced
electron transfer in ethidium-modified
DNA duplexes: dependence on dis-
tance and base stacking. *J. Am. Chem.
Soc.* **1997**, *119*, 9861–9870.

41 KELLEY, S. O., BARTON, J. K. DNA-
mediated electron transfer from a
modified base to ethidium: π-stacking
as a modulator of reactivity. *Chem.
Biol.* **1998**, *5*, 413–425.

42 KELLEY, S. O., BARTON, J. K. Electron
transfer between bases in double
helical DNA. *Science* **1999**, *283*, 375–
381.

43 LEWIS, F. D., LETSINGER, R. L.,
WASIELEWSKI, M. R. Dynamics of
photoinduced charge transfer and hole
transport in synthetic DNA hairpins.
Acc. Chem. Res. **2001**, *34*, 159–170.

44 WAN, C., FIEBIG, T., SCHIEMANN, O.,
BARTON, J. K., ZEWAIL, A. H. Femtosecond direct observation of charge
transfer between bases in DNA. *Proc.*

Natl Acad. Sci. USA **2000**, *97*, 14052–
14055.

45 KELLEY, S. O., BARTON, J. K. In *Metal
Ions in Biological Systems*, Vol. 36, eds
A. Sigel and H. Sigel, Dekker, New
York, 1999, 211–249.

46 SAM, M., BOON, E. M., BARTON, J. K.,
HILL, M. G., SPAIN, E. M. Morphology
of 15-mer duplexes tethered to
Au(111) probed using scanning probe
microscopy. *Langmuir* **2001**, *17*, 5727–
5730.

47 KELLEY, S. O., BARTON, J. K., JACKSON,
N. M., HILL, M. G. Electrochemistry
of methylene blue bound to a DNA-
modified electrode. *Bioconj. Chem.*
1997, *8*, 31–37.

48 KELLEY, S. O., JACKSON, N. M., HILL,
M. G., BARTON, J. K. Long-range elec-
tron transfer through DNA films. *Angew.
Chem. Int. Ed.* **1999**, *38*, 941–945.

49 LENG, F. F., SAVKUR, R., FOKT, I.,
PRZEWLOKA, T., PRIEBE, W., CHAIRES,
J. B. Base specific and regioselective
chemical cross-linking of daunorubicin to DNA. *J. Am. Chem. Soc.*
1996, *118*, 4731–4738.

50 WANG, A. H. J., GAO, Y. G., LIAW, Y.
C., LI, Y. K. Formaldehyde cross-links
daunorubicin and DNA efficiently-
HPLC and X-ray diffraction studies.
Biochemistry **1991**, *30*, 3812–3815.

51 KELLEY, S. O., BOON, E. M., BARTON,
J. K., JACKSON, N. M., HILL, M. G.
Single-base mismatch detection based
on charge transduction through DNA.
Nucleic Acids Res. **1999**, *27*, 4830–4837.

52 BOON, E. M., CERES, D. M.,
DRUMMOND, T. G., HILL, M. G.,
BARTON, J. K. Mutation detection by
electrocatalysis at DNA-modified
electrodes. *Nature Biotechnol.* **2000**, *18*,
1096–1100.

53 BOON, E. M., BARTON, J. K.,
PRADEEPKUMAR, P. I., ISAKSSON, J.,
PETIT, C., CHATTOPADHYAYA, J. An
electrochemical probe of DNA
stacking in an antisense oligonucleotide containing a C3′-endo locked
sugar. Submitted.

54 STEENKEN, S., JOVANOVIC, S. V. How
easily oxidizable is DNA? One-electron
reduction potentials of adenosine and
guanosine radicals in aqueous solu-

tion. *J. Am. Chem. Soc.* **1997**, *119*, 617–618.

55 SUGIYAMA, H., SAITO, I. Theoretical studies of GC-specific photocleavage of DNA via electron transfer: significant lowering of ionization potential and 5′-localization of HOMO of stacked GG bases in B-form DNA. *J. Am. Chem. Soc.* **1996**, *118*, 7063–7068.

56 PRAT, F., HOUK, K. N., FOOTE, C. S. Effect of guanine stacking on the oxidation of 8-Oxoguanine in B-DNA. *J. Am. Chem. Soc.* **1998**, *120*, 845–846.

57 BURROWS, C. J., MULLER, J. G. Oxidative nucleobase modifications leading to strand scission. *Chem. Rev.* **1998**, *98*, 1109–1151.

58 HALL, D. B., HOLMLIN, R. E., BARTON, J. K. Oxidative DNA damage through long-range electron transfer. *Nature* **1996**, *382*, 731–735.

59 NÚÑEZ, M. E., HALL, D. B., BARTON, J. K. Long range oxidative damage to DNA: effects of distance and sequence. *Chem. Biol.* **1999**, *6*, 85–97.

60 HALL, D. B., BARTON, J. K. Sensitivity of DNA-mediated electron transfer to the intervening π-stack: a probe for the integrity of the DNA base stack. *J. Am. Chem. Soc.* **1997**, *119*, 5045–5046.

61 ARKIN, M. R., STEMP, E. D. A., PULVER, S. C., BARTON, J. K. Long-range oxidation of guanine by Ru(III) in duplex DNA. *Chem. Biol.* **1997**, *4*, 389–400.

62 HALL, D. B., KELLEY, S. O., BARTON, J. K. Long-range and short-range oxidative damage to DNA: photoinduced damage to guanines in ethidium-DNA assemblies. *Biochemistry* **1998**, *37*, 15933–15940.

63 NAKATANI, K., DOHNO, C., SAITO, I. Chemistry of sequence-dependent remote guanine oxidation: photoreaction of duplex DNA containing cyanobenzophenone-substituted uridine. *J. Am. Chem. Soc.* **1999**, *121*, 10854–10855.

64 GIESE, B. Long-distance charge transport in DNA: the hopping mechanism. *Acc. Chem. Res.* **2000**, *33*, 631–636.

65 MATSUGO, S., KAWANISHI, S., YAMAMOTO, K., SUGIYAMA, H., MATSUURA, T., SAITO, I.

Bis(hydroperoxy)naphthaldiimide as a photo-fenton reagent: sequence-specific photocleavage of DNA. *Angew. Chem. Int. Ed.* **1991**, *30*, 1351–1353.

66 NÚÑEZ, M. E., NOYES, K. T., GIANOLIO, D. A., McLAUGHLIN, L. W., BARTON, J. K. Long-range guanine oxidation in DNA restriction fragments by a triplex-directed naphthalene diimide intercalator. *Biochemistry* **2000**, *39*, 6190–6199.

67 GASPER, S. M., SCHUSTER, G. B. Intramolecular photoinduced electron transfer to anthraquinones linked to duplex DNA: the effects of gaps and traps on long-range radical cation migration. *J. Am. Chem. Soc.* **1997**, *119*, 12762–12771.

68 HENDERSON, P. T., JONES, D., HAMPIKIAN, G., KAN, Y., SCHUSTER, G. B. Long-distance charge transport in duplex DNA: the phonon-assisted polaron-like hopping mechanism. *Proc. Natl Acad. Sci. USA.* **1999**, *96*, 8353–8358.

69 BIXON, M., JORTNER, J. Energetic control and kinetics of hole migration in DNA. *J. Phys. Chem. B.* **2000**, *104*, 3906–3913.

70 BERLIN, Y. A., BURIN, A. L., RATNER, M. A. Charge Hopping in DNA. *J. Am. Chem. Soc.* **2001**, *123*, 260–268.

71 MEGGERS, E., MICHEL-BEYERLE, M. E., GIESE, B. Sequence dependent long range hole transport in DNA. *J. Am. Chem. Soc.* **1998**. *120*, 12950–12955.

72 LY, D., SANII, L., SCHUSTER, G. B. Mechanism of charge transport in DNA: internally-linked anthraquinone conjugates support phonon-assisted polaron hopping. *J. Am. Chem. Soc.* **1999**, *121*, 9400–9410.

73 HARTWICH, G., CARUANA, D. J., DE LUMLEY-WOODYEAR, T., WU, Y., CAMPBELL, C. N., HELLER, A. Electrochemical study of electron transport through thin DNA films. *J. Am. Chem. Soc.* **1999**, *121*, 10803–10812.

74 WILLIAMS, T. T., ODOM, D. T., BARTON, J. K. Variations in DNA charge transport with nucleotide composition and sequence. *J. Am. Chem. Soc.* **2000**, *122*, 9048–9049.

75 NADEAU, J. G., CROTHERS, D. M. Structural basis for DNA bending. *Proc. Natl Acad. Sci. USA* **1989**, *86*, 2622–2626.

76 DICKERSON, R. E. Sequence-dependent B-DNA conformations in crystals, in *Structure, Motion, Interaction, and Expression of Biological Macromolecules*. Proceedings of Tenth Conversation in Biomolecular Stereodynamics, eds R. Sarma and M.H. Sarma, Academic Press, Schenectady, New York, **1998**, 17–36.

77 WILLIAMS, T. T., BARTON, J. K. The effect of varied ion distributions on long-range DNA charge transport. *J. Am. Chem. Soc.* **2002**, *124*, 1840–1841.

78 BHATTACHARYA, P. K., BARTON, J. K. Influence of intervening mismatches on long-range guanine oxidation in DNA duplexes. *J. Am. Chem. Soc.* **2001**, *123*, 8649–8656.

79 GIESE, B., WESSELY, S. The influence of mismatches on long distance charge transport through DNA. *Angew. Chem. Int. Ed.* **2000**, *39*, 3490–3491.

80 RAJSKI, S. R., BARTON, J. K. How different DNA-binding proteins affect long-range oxidative damage to DNA. *Biochemistry* **2001**, *40*, 5556–5564.

81 RAJSKI, S. R., KUMAR, S., ROBERTS, R. J., BARTON, J. K. Protein-modulated DNA electron transfer. *J. Am. Chem. Soc.* **1999**, *121*, 5615–5616.

82 CHENG, X. D., KUMAR, S., POSFAI, J., PFLUGRATH, J. W., ROBERTS, R. J. Crystal-structure of the HhaI DNA methyltransferase complexed with S-adenosyl-L-methionine. *Cell* **1993**, *74*, 299–307.

83 KIM, Y. C., GEIGER, J. H., HAHN S., SIGLER, P. B. Crystal-structure of a yeast TBP TATA-box complex. *Nature* **1993**, *365*, 512–520.

84 KIM, J. L., NIKOLOV, D. B., BURLEY, S. K. Co-crystal structure of TBP recognizing the minor-groove of a TATA element. *Nature* **1993**, *365*, 520–527.

85 WAGENKNECHT, H.-A, RAJSKI, S. R., PASCALY, M., STEMP, E. D. A., BARTON, J. K. Direct observation of radical intermediates in protein-dependent DNA charge transport. *J. Am. Chem. Soc.* **2001**, *123*, 4400–4407.

86 JOVANOVIC, S. V., STEENKEN, S., SIMIC, M. G. Kinetics and energetics of one-electron transfer reactions involving tryptophan neutral and cation radicals. *J. Phys. Chem.* **1991**, *95*, 684–687.

87 PASCALY, M., YOO, J., BARTON, J. K. DNA mediated charge transport: characterization of a DNA radical localized at an artificial nucleic acid base. Submitted.

88 BOON, E. M., SALAS, J. E., BARTON, J. K. An electrical probe of protein-DNA interactions on DNA-modified surfaces. *Nature Biotechnol.* **2002**, *20*, 282–286.

89 DANDLIKER, P. J., HOLMLIN, R. E., BARTON, J. K. Oxidative thymine dimer repair in the DNA helix. *Science* **1997**, *275*, 1465–1468.

90 SANCAR, A. DNA excision repair. *Annu. Rev. Biochem.* **1996**, *65*, 43–81.

91 JACOBSEN J. R., COCHRAN, A. G., STEPHANS, J. C., KING, D. S., SCHULTZ, P. G. Mechanistic studies of antibody-catalyzed pyrimidine dimer photocleavage. *J. Am. Chem. Soc.* **1995**, *117*, 5453–5461.

92 YOUNG, T., NIEMAN, R., ROSE, S. D. Photo-CIDNP detection of pyrimidine dimer radical cations in anthraquinonesulfonate-sensitized splitting. *Photochem. Photobiol.* **1990**, *52*, 661–668.

93 VICIC, D. A., ODOM, D. T., NÚÑEZ, M. E., GIANOLIO, D. A., McLAUGHLIN, L. W., BARTON, J. K. Oxidative repair of a thymine dimer in DNA from a distance by a covalently linked organic intercalator. *J. Am. Chem. Soc.* **2000**, *122*, 8603–8611.

94 DOTSE, A. K., BOONE, E. K., SCHUSTER, G. B. Remote *cis-syn* thymine [2 + 2] dimers are not repaired by radical cations migrating in duplex DNA. *J. Am. Chem. Soc.* **2000**, *122*, 6825–6833.

95 DANDLIKER, P. J., NÚÑEZ, M. E., BARTON, J. K. Oxidative charge transfer to repair thymine dimers and damage guanine bases in DNA

assemblies containing tethered metallointercalators. *Biochemistry* **1998**, *37*, 6491–6502.

96 LUGER, K., MADER A. W., RICHMOND, R. K., SARGENT, D. F., RICHMOND, T. J. Crystal structure of the nucleosome core particle at 2.8 Å resolution. *Nature* **1997**, *389*, 251–260.

97 NÚÑEZ, M. E., NOYES, K. T., BARTON, J. K. Oxidative charge transport through DNA in nucleosome particles. *Chem. Biol.* **2002**, in press.

98 NÚÑEZ, M. E., HOLMQUIST, G. P., BARTON, J. K. Evidence for DNA charge transport in the nucleus. *Biochemistry* **2001**, *40*, 12465–12471.

8
DNA Interactions of Novel Platinum Anticancer Drugs

Viktor Brabec and Jana Kasparkova

8.1
Introduction

Platinum compounds constitute a discrete class of anticancer agents, which are now widely used in the clinic. The interest in antitumor platinum drugs has its origin in the 1960s, with the serendipitous discovery by Rosenberg of the inhibition of division of bacterial cells by platinum complexes [1]. The first platinum anticancer drug in clinical use was cisplatin (*cis*-diamminedichloroplatinum(II)), which is a very simple and purely inorganic molecule (Fig. 8.1). In spite of its simplicity, it is one of the most potent drugs available for anticancer chemotherapy [2–4]. It is highly effective for the treatment of testicular and ovarian cancer and is used in combination regimens for a variety of other carcinomas, including bladder, small lung tumors and those of head and neck [5]. However, cisplatin toxicity in tumor cells is coupled with several drawbacks (such as nephrotoxicity, neurotoxicity, and emetogenesis) [2]; hence there has been strong interest in the development of improved platinum drugs and consequently in understanding the details of the molecular mechanisms underlying the biological efficacy of the platinum compounds.

There is a large body of experimental evidence that the success of platinum complexes in killing tumor cells mainly results from their ability to form various types of adducts on DNA [6, 7]. Hence, extensive research has been carried out on DNA modifications by platinum antitumor drugs (for reviews, see, for example, Refs [8–10]). This chapter reviews a recent development in the study of DNA modifications by some novel classes of platinum anticancer compounds in cell-free media, which are compared with the effects of cisplatin.

8.2
Modifications by Cisplatin

8.2.1
Adducts and Conformational Distortions

The development of new classes of platinum antitumor drugs is often based on comparisons with the effects of cisplatin. Therefore, in this section the most im-

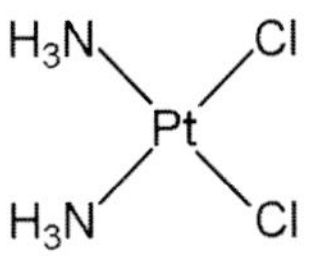

H$_3$N, Cl
  Pt
H$_3$N Cl

Fig. 8.1. Structure of cisplatin, the first platinum antitumor drug used in the clinic.

portant milestones in understanding DNA interactions with cisplatin are briefly addressed. Cisplatin reacts with DNA in the cell nucleus, where the concentration of chlorides is markedly lower than in extracellular fluids. The drug loses its chloride ligands in media containing low concentrations of chloride to form positively charged monoaqua and diaqua species (Fig. 8.2). It has been shown [11] that only these aquated forms bind to DNA.

Cisplatin is administered intravenously. Hence, before it reaches DNA in the nucleus of tumor cells it may interact with various compounds including sulfur-containing molecules (Fig. 8.2). These interactions are generally believed to play a role in mechanisms underlying tumor resistance to antitumor platinum drugs, their inactivation, and side effects [12, 13]. Generally, sulfur-containing biomolecules, such as those containing thiols are known to be highly reactive toward cisplatin so that they inactivate this drug. On the other hand, biomolecules containing thioethers (L-methionine (L-HMet)), including isolated L-HMet molecules, also form platinum–sulfur adducts which have been postulated to be a drug reservoir

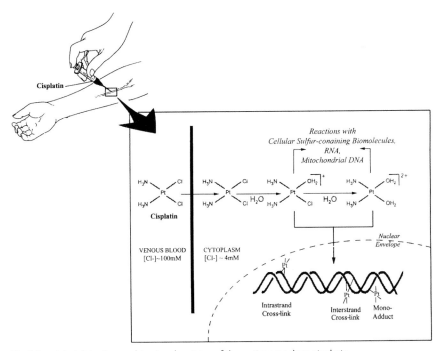

Fig. 8.2. Administration and *in vivo* chemistry of the anticancer drug cisplatin.

for platination of DNA. Thus, it has been proposed that the complexes formed between cisplatin and biomolecules containing thioethers may act as an activated form of platinum compounds with slowly hydrolyzing leaving groups. In other words, it has been suggested [14, 15] that these complexes (or adducts) may function as intermediates that may be transformed into platinum–DNA adducts. The reactions of L-HMet with cisplatin have been examined frequently. The major products of these reactions have been found to contain S,N-chelated L-Met, such as $[Pt(\text{L-Met-}S,N)(NH_3)_2]$ and $[Pt(\text{L-Met-}S,N)_2]$ [16, 17]. In addition, it has been demonstrated [14] that L-HMet increases the rate of reaction of cisplatin with monomeric guanosine 5′-monophosphate (GMP). Thus, these results seem to support the view that methionine residues could mediate the transfer of platinum onto DNA. However, in contrast to these expectations it has been demonstrated (O. Vrana, V. Brabec, unpublished results) that L-HMet decreases the rate of reaction of antitumor cisplatin with base residues in natural, high-molecular-mass DNA (such as plasmid or calf thymus DNA).

It has also been shown convincingly that the formation of the complex of cisplatin with L-HMet makes coordination of base residues in a polynucleotide chain by the platinum atom in cisplatin markedly more difficult than that of bases in monomeric regions of nucleic acids. Hence, the possibility proposed earlier [14, 15] that platinum can be transferred onto DNA via displacement of methionine S-bound to cisplatin by guanine residues in polymeric DNA appears less likely. This also implies that these results do not support the view [12, 13] ·that cisplatin bound to monomeric methionine or to methionine residues in peptides or proteins could potentially act as a drug reservoir available for platination of DNA in the nucleus of tumor cells. In this respect, thioethers seem to behave like thiols that inactivate cisplatin.

Bifunctional cisplatin binds to DNA in a two-step process, first forming monofunctional adducts preferentially at the guanine residues, which subsequently close to intrastrand crosslinks between adjacent purine residues ($\sim$90%) (1,2-GG or 1,2-AG intrastrand crosslinks) [18, 19]. Other minor adducts are 1,3-GXG intrastrand crosslinks (X = A, C, T), interstrand crosslinks ($\sim$6% in linear DNA) preferentially between guanine residues in the 5′-GC/5′-GC sequence [20] and monofunctional lesions. In all adducts, cisplatin is coordinated to the N7 atom of purine residues. The percentage of the 1,2-intrastrand adducts formed by cisplatin is larger than statistically expected, so this crosslink has generally been assumed to be the important adduct for anticancer activity and has therefore been most extensively investigated.

The adducts formed by cisplatin in DNA affect its secondary structure. The formation of major 1,2-GG or 1,2-AG intrastrand crosslinks of cisplatin leads to marked conformational alterations in DNA (Fig. 8.3a) [7, 21]. These adducts induce a roll of 26–50° between the platinated purine residues, displacement of platinum atom from the planes of the purine rings, a directional, rigid bend of the helix axis toward the major groove and a local unwinding of 9–11°. In addition, severe perturbation of hydrogen-bonding within the 5′-coordinated GC base pair, widening and flattening of the minor groove opposite the cisplatin adduct, creation of a hydrophobic notch, global distortion extending over 4–5 base pairs and additional helical parameters characteristic of A-form of DNA have also been reported.

The interstrand crosslink, which is preferentially formed by cisplatin between opposite guanine residues in the 5'-GC/5'-GC sequence [22], also induces several irregularities in DNA (Fig. 8.3b) [20, 23, 24]. The crosslinked guanine residues are not paired with hydrogen bonds to the complementary cytosines, which are located outside the duplex and not stacked with other aromatic rings. All other base residues are paired, but distortion extends over at least four base pairs at the site of the crosslink. In addition, the *cis*-diammineplatinum(II) bridge resides in the minor groove and the double helix is locally reversed to a left-handed, Z-DNA-like form. This adduct induces the helix unwinding by 76–80° relative to B-DNA and also the bending of 20–40° of the helix axis at the crosslinked site toward the minor groove. More detailed descriptions of structures of crosslinks of cisplatin may be found in other recent reviews [7, 20, 21, 25].

The 1,3-GXG intrastrand crosslink formed by cisplatin bends the helix axis toward the major groove by ~30° and locally unwinds DNA (by ~19°) [26, 27]. In addition, DNA is locally denatured and flexible at the site of the adduct.

The monofunctional adducts of cisplatin (preferentially formed at the guanine sites) also distort DNA in a sequence-dependent manner [28, 29]. These distortions disturb stacking interactions in double-helical DNA and make bases around the adduct more accessible. In addition, the monofunctional adduct also unwinds the duplex (unwinding angle is ~6° [30]). On the other hand, no intrinsic bending is induced in DNA by these monofunctional adducts [28].

8.2.2
Effects on DNA Replication and Transcription

DNA replication and transcription are essential processes in rapidly proliferating tumor cells so that their inhibition should result in cytostatic effects. As various adducts formed on DNA are capable of inhibiting DNA replication or transcription by DNA-dependent DNA or RNA polymerases, the effects of platinum adducts on these processes have been thoroughly investigated.

Cisplatin inhibits DNA synthesis by prokaryotic and eukaryotic DNA polymerases both *in vitro* and *in vivo* [31–33]. Inhibition of DNA synthesis occurs mainly at GG sites consistently with the fact that these sites are preferential binding sites of cisplatin. The monofunctional adducts of cisplatin affect replication only negligibly. It has been demonstrated in the studies of replication of DNA modified by cisplatin [33] that the inhibition of DNA elongation is not complete; the polymerases are able to bypass the adducts. The frequency of replication bypass varies for the different polymerases. The ability of DNA polymerases to bypass the lesions of platinum compounds is consistent with the view that these compounds are mutagenic through such a replication bypass. The mutagenic properties of drugs used in the clinic are important factors to consider since mutagenicity may lead to the development of secondary tumors. Interestingly, the 1,2-GG intrastrand crosslinks of cisplatin are considerably less mutagenic than 1,2-AG crosslinks in *E. coli*,

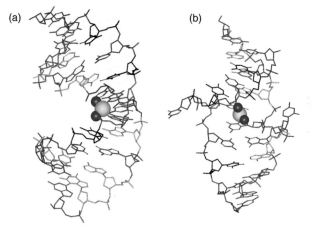

Fig. 8.3. Crystal structures of DNA duplexes containing an 1,2-GG intrastrand crosslink (a) and interstrand crosslink (b) of cisplatin. Images of the duplexes (a) 5'-d(CCTCTG*G*TCTCC)/ 5'-d(GGAGACCAGAGG) [191] and (b) 5'-d(CCTCG*CTCTC)/5'-d(GAGAG*CGAGG) [24], where the asterisks represent the sites involved in the crosslinks of cisplatin. Colors: black, thymine; orange, adenine; green, cytosine; red, guanine; yellow, platinum atom; blue, nitrogen atom.

whereas controversial mutagenicity has been reported for the 1,3-GXG intrastrand crosslink [34, 35]. Although there is no doubt that inhibition of DNA replication is an important part of the mechanism underlying the antitumor effects of cisplatin, it becomes evident that it cannot fully explain the antitumor efficiency of this drug [36] so that this mechanism is more complex [37, 38].

Another essential function of the cellular metabolism affected by lesions formed on DNA by cisplatin is DNA transcription. Transcription by eukaryotic and prokaryotic RNA polymerases (RNA polymerase II and SP6 or T7 RNA polymerases, respectively) of DNA modified by cisplatin *in vitro* has been investigated in detail [22, 39–42]. The RNA polymerases react differently at the various platinum adducts. Bifunctional adducts of cisplatin strongly inhibit transcription of platinated DNA by RNA polymerases. The RNA polymerases entirely bypass the monofunctional adducts of cisplatin [22, 41, 43]. It has been suggested [43] that the platinum adducts which block the RNA polymerases not only constitute a physical barrier to the progress of the enzymes on the template but also specifically alter the properties of transcription complexes as a consequence of the specific conformational changes that they induce in template DNA.

8.2.3
Cellular Resistance, Repair

The limitations of using cisplatin in cancer chemotherapy are also associated with intrinsic and acquired resistance of tumor cells to this drug. Resistance to cisplatin is multifactorial and in general it may consist of mechanisms either limiting the formation of DNA adducts and/or operating downstream of the interaction of cisplatin with DNA to promote cell survival. The formation of DNA adducts by cisplatin can be limited by reduced accumulation of the drug, enhanced drug efflux and cisplatin inactivation by coordination to sulfur-containing biomolecules including metallothioneins (see Section 8.2.1 and Fig. 8.2), whose production may be increased as a consequence of cisplatin treatment. The second group of mechanisms includes enhanced repair of DNA adducts of cisplatin and increased tolerance of the resulting DNA damage.

In human cells, cisplatin intrastrand adducts are removed from DNA mainly by the nucleotide excision repair (NER) mechanism. It has been found using cell-free extracts or a reconstituted NER system that 1,2- and 1,3-intrastrand crosslinks of cisplatin are efficiently repaired [44]. Importantly, this repair of 1,2- but not 1,3-intrastrand crosslinks is blocked upon addition of an HMG-domain protein (HMG = high mobility group) (see Section 8.2.4.1).

An *in vitro* excision repair of a site-specific cisplatin interstrand crosslink has also been studied using mammalian cell-free extracts containing HMG-domain proteins at levels that are not high enough to block excision repair of the 1,2-intrastrand adducts [44]. Excision of the interstrand crosslinks formed by cisplatin has not been detected. DNA interstrand crosslinks pose a special challenge to repair enzymes because they involve both strands of DNA and therefore cannot be repaired using the information in the complementary strand for resynthesis.

There is also evidence that another cellular repair mechanisms, such as recombination or mismatch repair (MMR), can affect antitumor efficiency of cisplatin. The role of recombination in survival of the cells treated with cisplatin has been also demonstrated, but mainly in prokaryotic cells [45]. The sensitivity of many recombination-deficient mutants to cisplatin has been examined and almost all of the strains are very sensitive to cisplatin, in comparison with the corresponding wild-type cells. In addition, mutants deficient in both recombination repair and NER are more sensitive to cisplatin than single mutants. This may imply that the role of recombination in protecting cells from cisplatin toxicity is independent of NER. On the other hand, it cannot be entirely excluded that DNA lesions in prokaryotic cells treated with cisplatin are repaired or tolerated by a mechanism which is different from those effective in eukaryotic cells [46].

Recent observations support the view that MMR mediates the cytotoxicity of cisplatin in tumor cells [47, 48] and that dysfunction of this type of DNA repair may result in the resistance of tumor cells to cisplatin or in drug tolerance [49]. The function of the MMR is to scan newly synthesized DNA and remove mismatches that result from nucleotide incorporation errors made by DNA polymerases. To explain cisplatin tolerance, it is assumed that replication bypass of DNA adducts of cisplatin leads to mutations. During MMR, the strand to be corrected is nicked, a short fragment containing the mismatch is excised, and new DNA fragment is synthesized. The MMR system always replaces the incorrect sequence in the daughter strand, which would leave the cisplatin adduct unrepaired. This activity initiates a futile cycle. During DNA synthesis to replace the excised short fragment, the DNA polymerases would repeatedly incorporate mutations, followed by attempts to remove them. The repeated nicks in DNA formed at each ineffective cycle of repair could trigger a cell death response. Thus, MMR recognition of cisplatin adducts on DNA may trigger a programmed cell death pathway rendering MMR-proficient cells more sensitive to DNA modification by cisplatin than MMR-deficient cells.

8.2.4
Recognition of the Lesions by Cellular Proteins

It is generally believed that antitumor activity of cisplatin is mediated by the recognition of its DNA adducts by cellular proteins. Several classes of these proteins have been identified and mechanisms have been proposed as to how they mediate the antitumor effects of cisplatin. Extensive reviews addressing these questions have been published recently [50–52]. Therefore, in this section only some selected aspects of recognition of DNA adducts formed by cisplatin by damaged DNA-binding proteins are described.

8.2.4.1 HMG-domain proteins
HMGB1 and HMGB2 proteins belong to a group of architectural chromatin proteins that play some kind of structural role in the formation of functional higher order protein–DNA or protein–protein complexes or as signaling molecules in genetically regulated repair pathways. These structure-specific proteins (which recognize DNA with little sequence-specificity and have a high affinity for some non-canonical DNA structures including a kinked DNA) bind selectively to the 1,2-GG or AG adducts of cisplatin, but not to its 1,3-intrastrand crosslinks.

The full-length HMGB1 protein contains two tandem HMG-box domains, A and B, and an acidic C-tail [53]. The domains A and B in HMGB1 protein (HMGB1a and HMGB1b) are linked by a short lysine-rich sequence (containing seven amino acids, A/B linker region). Each "minimal" domain A or B alone specifically recognizes 1,2-GG intrastrand crosslink of cisplatin [54]. The affinity of HMGB1a for DNA containing a 1,2-GG intrastrand crosslink is generally higher than that of HMGB1b and it has been proposed that HMGB1a is the dominating domain in the full-length HMGB1 that binds to the site of the intrastrand crosslink, while HMGB1b facilitates binding providing additional protein–DNA interactions. Interestingly, the binding affinity of HMGB1a and HMGB1b with the duplex containing the site-specific 1,2-GG intrastrand crosslink is modulated by the nature of the bases that flank the platinum lesion [54, 55].

The HMG domain binds to the minor groove of the DNA double helix opposite the platinum 1,2-intrastrand crosslink located in the major groove [56]. Distortions such as prebending, unwinding at the site of platination and preformation of a hydrophobic notch as a consequence of the 1,2-intrastrand crosslink formation are important for the recognition and affinity of the platinum-damaged DNA-binding proteins. Moreover, the A/B linker region in the full-length HMGB1 protein attached to the N-terminus of the domain B markedly enhances binding of the B domain to DNA containing the site-specific 1,2-GG intrastrand crosslink [57].

These studies have been extended to DNA interstrand crosslinks produced by cisplatin or transplatin at the sites where these adducts are formed preferentially [58]. Mammalian HMGB1 protein binds to the interstrand crosslink of cisplatin with a slightly lower affinity, as to the 1,2-GG intrastrand crosslink. Our most recent analysis has revealed that isolated HMGB1a has no affinity to the duplex interstrand crosslinked by cisplatin, whereas the isolated HMGB1b containing the A/B-linker region attached to its N-terminus binds with a noticeable affinity although lower than is that of the "minimal" HMGB1a to 1,2-GG intrastrand crosslink (J. Kasparkova, M. Stros, O. Delalande, J. Kozelka, V. Brabec, unpublished results).

1,3-GXG intrastrand crosslink of cisplatin is not recognized by HMGB1, in spite of the fact that bending and unwinding induced in DNA by this adduct are rather similar to those induced by 1,2-GG intrastrand adduct. This suggests that there are other factors which control the recognition of platinum adducts by DNA-binding proteins [20, 58].

There are a variety of proteins other than HMGB1 and HMGB2 that contain HMG boxes, a conserved 80-amino-acid domain found in a variety of eukaryotic DNA-binding proteins, including a number of transcription factors. Many of these proteins (for instance HMG-D, human and *Drosophila* SSRP1, Cmb1, hUBF, Ixr1, mtTFA, LEF1, SRY, T160, tsHMG, human FACT factor) recognize and bind to DNA modified by cisplatin and several recent reviews describing these binding studies are available [7, 50, 51, 59].

8.2.4.2 Proteins without an HMG domain

The TATA-binding protein (TBP) [60] is essential for transcription initiation in eukaryotes. TBP recognizes and binds to the minor groove of a consensus sequence, TATAAA, known as the TATA box. It binds selectively to and is sequestered by cisplatin-damaged DNA [61, 62]. It has been shown that cisplatin turns the GG sequence into a potential site for TBP and it has been suggested that TBP binds selectively to a specific three-dimensional structure [62]. Another protein lacking an HMG domain that has been shown to bind to cisplatin-modified DNA is Y-box binding protein 1 [63]. A very abundant chromatin protein, the linker histone H1, also binds strongly to DNA modified by cisplatin [64, 65].

A relatively large number of proteins that specifically bind to cisplatin adducts are DNA repair proteins. The repair proteins that recognize cisplatin-modified DNA are mainly those involved in the first step of the repair pathway, i.e. in damage recognition. The repair proteins that have probably attracted the most attention in the studies of recognition of DNA modified by platinum drugs are those that are absent in patients with the NER deficiency characteristic of the disease xeroderma pigmentosum (XP). The minimal factors necessary for removal of damaged nucleotides also include XPA protein and replication protein A (RPA), which are among the major damage-recognition proteins involved in the early stage of NER. XPA (32-kDa protein) and RPA (composed of 70-, 34- and 14-kDa subunits) are able to bind damaged DNA independently, although RPA interaction stimulates XPA binding to damaged DNA. It has been suggested recently [66] that XPA in conjuction with RPA constitutes a regulatory factor that monitors DNA bending and unwinding to verify the damage-specific localization of NER complexes or control their three-dimensional assembly.

The proteins involved in the MMR pathway also display affinity for cisplatin-modified DNA. For instance, human MutSα and its component of MSH2 exhibit an enhanced affinity to DNA containing 1,2-GG intrastrand crosslink [67–69].

DNA photolyase, which is involved in the repair of cyclobutane pyrimidine dimers [70], and Ku autoantigen [71], which takes part in recombination and double-strand break repair, are further examples of proteins lacking an HMG domain that bind to cisplatin-modified DNA.

The human 3-methyladenine DNA glycosylase involved in the base excision repair can also tightly bind to intrastrand crosslinks of cisplatin [72]. Although this protein is unable to release any of these adducts from DNA it has been suggested that it can facilitate their NER.

Another repair protein that has also been thoroughly tested as to whether it specifically recognizes DNA adducts of cisplatin is T4 endonuclease VII, an enzyme with deoxyribonuclease activity [58, 73].

It has been hypothesized that sensitivity of tumor cells might also be associated with the processes involving tumor suppressor protein p53 [74, 75]. The tumor suppressor protein p53 is a nuclear phosphoprotein consisting of 393 amino acids and containing four major functional domains (Fig. 8.4a) [76]. The protein is involved in the control of cell cycle, DNA repair and apoptosis. Hence, p53 is a potent mediator of cellular responses against genotoxic insults [77] and exerts its effect through transcriptional regulation. Upon exposure to genotoxic compounds, p53 protein levels increase due to several post-transcriptional mechanisms.

Interestingly, on average, cells with mutant p53 are more resistant to the effect of cisplatin [78]. Sequence-specific DNA-binding activity of p53 protein is crucial for its tumor suppressor function. Active p53 binds as a tetramer to over 100 different response elements naturally occurring in the human genome (but only approximately 50 show functionality [79]). Free DNA in the segments corresponding to the consensus sequence is already intrinsically bent toward the major groove and the bends are mostly localized at two C(a/t)|(a/t)G tetramers [80, 81]. This localized intrinsic bending has been suggested to contribute in a fundamental way to the stability of the tetrameric p53-DNA complex [80].

DNA interactions of active wild-type human p53 protein with DNA modified by cisplatin has been investigated [82]. DNA adducts reduce the binding affinity of the consensus DNA sequence to p53 whereas the adducts of clinically ineffective *trans* isomer of cisplatin do not. This result has been interpreted to mean that the precise steric fit required for formation and stability of the tetrameric complex of p53 with the consensus sequence in DNA (Fig. 8.4b) cannot be attained as a consequence of additional conformational perturbances induced by cisplatin adducts in the consensus sequence. The results also demonstrate an increase of the binding affinity of p53 to DNA lacking the consensus sequence and modified by cisplatin, but not by transplatin. In addition, major 1,2-GG intrastrand crosslinks of cisplatin are only responsible for this enhanced binding affinity of p53, which is, however,

(a)

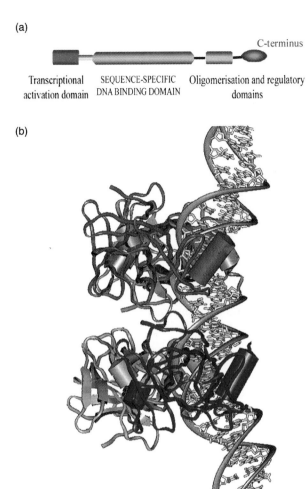

C-terminus

Transcriptional
activation domain

SEQUENCE-SPECIFIC
DNA BINDING DOMAIN

Oligomerisation and regulatory
domains

(b)

Fig. 8.4. (a) Schematic representation of active p53 molecules. The p53 sequence-specific DNA-binding domain (blue) binds to the p53 consensus DNA sequence. At the p53 C-terminus, another sequence-non-specific DNA-binding site occurs which coincides with a negative regulatory segment. The latter segment inhibits DNA-binding functions of the core domain in non-modified full-length p53 protein. (b) Three-dimensional model for the complex of tetrameric p53 bound to DNA bent towards the major groove. The single monomeric p53 subunits are shown in different colors. Adapted from Ref. [80] with permission.

markedly lower than the affinity of p53 to unplatinated DNA containing the consensus sequence.

The distinctive structural features of 1,2-intrastrand crosslinks of cisplatin have been suggested to play a unique role for this adduct in the binding of p53 to DNA lacking the consensus sequence [82]. This intriguing scenario for the mechanism by which cisplatin adducts modulate biological effects of p53 might be some manifestation of the hijacking model (see Section 8.2.5). In addition, comparison of the binding of active, latent and activated p53 to DNA fragments modified by cisplatin has been performed (M. Fojta, H. Pivonkova, M. Brazdova, E. Palecek, S. Pospisilova, B. Vojtesek, J. Kasparkova, V. Brabec, unpublished results). Modifications of DNA lacking consensus sequence by cisplatin enhanced affinity of all three forms of p53. The difference in the binding affinity to platinated and unmodified DNA is even more pronounced in the case of the latent form in comparison with the active form. Consistently with the latter observation, activation of the latent form reduces its preference for platinated DNA. In addition, the p53 core domain appears to be the primary site of sequence-specific binding of non-modified DNA in active p53, while the C-terminus of p53 (Fig. 8.4a) interacts selectively with platinated DNA. These findings suggest that interactions of p53 with platinated DNA may have implications in cell responses to cisplatin treatment.

The cisplatin-damaged DNA-binding proteins apparently occur in nature for purposes other than for specific recognition of platinum adducts in DNA, since platinum compounds do not occur naturally. The affinity of DNA modified by cisplatin to DNA-binding proteins, which may have a fundamental relevance to the antitumor activity of cisplatin and its simple antitumor analogs, is probably a coincidence when some platinum adducts in double-helical DNA adopt a structure that mimics the recognition signal for these proteins.

8.2.5
Mechanism of action of cisplatin

It is generally believed that the key intracellular pharmacological target for platinum compounds is DNA. The adducts formed by cisplatin distort DNA conformation, but it still remains uncertain which of these adducts is the most important in terms of producing anticancer effects. DNA adducts of cisplatin inhibit replication and transcription, but they are also bypassed by DNA or RNA polymerases. In addition, cisplatin adducts are removed from DNA mainly by NER. They are, however, also recognized by a number of proteins which could block DNA adducts of cisplatin from damage recognition needed for repair (Fig. 8.5a). The other hypothesis, based on the observation that a number of proteins recognize cisplatin-modified DNA, is that cisplatin-DNA adducts hijack proteins away from their normal binding sites, thereby disrupting fundamental cellular processes (Fig. 8.5b). Experimental support for these hypothetical aspects of the mechanism underlying the antitumor activity of cisplatin or resistance of some tumors to this drug has been recently thoroughly reviewed [7, 9, 21, 49, 51, 75, 83].

Initially, inhibition of DNA replication was considered to be a process very likely relevant to antitumor efficiency of cisplatin [31]. However, later studies have indicated that cisplatin inhibits tumor cell growth at doses which are considerably lower than those needed to inhibit DNA synthesis [36]. Subsequent observations have revealed that cisplatin can trigger G_2 cell cycle arrest and programmed cell death (apoptosis) [37, 38] exposing a mechanism of cytotoxicity of this drug. However, since apoptosis is a very complex process, a number of possible pathways

(a)

Damaged-DNA
Binding
Protein

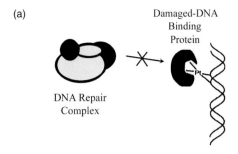

DNA Repair
Complex

(b)

Damaged-DNA
Binding Protein

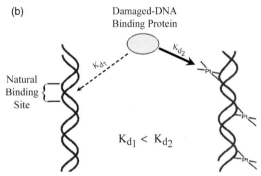

Natural
Binding
Site

$K_{d1} < K_{d2}$

Fig. 8.5. Molecular mechanisms of antitumor activity of cisplatin. (a) The repair shielding model. If the protein recognizing 1,2-GG or AG intrastrand crosslink (for instance HMG-domain protein) binds to these lesions it can inhibit their repair by physically blocking the access to other repair proteins. The crosslinks of cisplatin persist in DNA and potentiate their toxicity. (b) Hijacking model. If the protein recognizing 1,2-GG or AG intrastrand crosslink (for instance transcription factor, p53, TBP) binds to these lesions with a higher affinity than to their natural binding sites ($K_{d1} < K_{d2}$), the proteins could be titrated away from their usual (for instance transcriptional regulatory) function. This may disrupt processes critical for cell survival.

have to be still explored for a complete understanding of the mechanism by which cisplatin triggers apoptosis [84].

8.3
Modifications by Antitumor Analogs of Cisplatin

8.3.1
Carboplatin

The search for an agent less toxic than cisplatin led to the development of carbo-platin (*cis*-diamminecyclobutanedicarboxylatoplatinum(II)) (Fig. 8.6), which has achieved routine clinical use. The leaving group of carboplatin is a cyclobutanedi-carboxylato ligand which undergoes a slower rate of aquation than the chlorides in

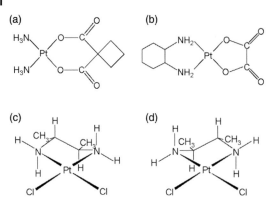

Fig. 8.6. Structures of direct bifunctional analogs of cisplatin:
(a) carboplatin, (b) oxaliplatin, (c) PtCl$_2$(R,R-DAB), and (d)
PtCl$_2$(S,S-DAB).

cisplatin. Carboplatin forms a similar spectrum of DNA adducts as cisplatin with a slightly different sequence preference [85]. The concentrations of carboplatin approximately two orders of magnitude higher are needed to obtain DNA platination levels equivalent to cisplatin. Increased DNA-binding affinity of carboplatin has been observed in the presence of nucleophiles, such as those present in human breast cancer cell cytoplasmatic extracts, thiourea or glutathione [86]. Conformational alterations induced in DNA by carboplatin in cell-free media have been characterized by circular dichroism and differential pulse polarography [87]. The changes of non-denaturational character similar to those observed for cisplatin at the same level of DNA platination have been found. Although less toxic than cisplatin, carboplatin is still only active in the same range of tumors as cisplatin [88].

In spite of some interesting and promising results describing DNA modifications by carboplatin, as yet no substantial advantages over cisplatin have been demonstrated. Carboplatin is more easily administered and is less toxic at standard doses, but exhibits the similar efficacies in most solid tumors and is administered intravenously [89]. From a mechanistic DNA-binding point of view, this is not too surprising since the adducts produced on DNA by cisplatin and carboplatin are similar, although they differ in their relative rates of formation.

8.3.2
Oxaliplatin

Another cisplatin analog that has received approval for use in Asia, Europe, and Latin America is oxaliplatin ((1R,2R-diamminocyclohexane)oxalatoplatinum(II)) (Fig. 8.6b). In terms of clinical activity, oxaliplatin is the first platinum antitumor drug to demonstrate activity in colon and ovarian cancers [89–91]. Importantly, oxaliplatin has been shown to be active in combination with 5-fluorouracil and leucovorin for the treatment of colorectal cancer, a disease in which cisplatin and carboplatin show little activity [92–94]. These important results lend substantial support to the hypothesis that structural modifications of the carrier ligand in cis-platin may alter spectrum of antitumor activity, and so overcome resistance.

In order for its reaction with DNA to occur, the parent compound must become aquated. The hydrolysis of oxaliplatin to form reactive species, 1,2-DACH diaqua platinum(II) ($[Pt(R,R\text{-}DACH)(H_2O)_2]^{2+}$) (DACH = 1,2-diaminocyclohexane), is a slower process than the hydrolysis of cisplatin, but it is facilitated by HCO_3^- and $H_2PO_4^-$ ions [95]. Site- and region-specificity of lesions induced by oxaliplatin in DNA have been determined [96]. The sites of oxaliplatin adducts and their spectrum [97, 98] are nearly identical to the situation when DNA is modified by cisplatin. On the other hand, oxaliplatin is inherently less able than cisplatin to form DNA adducts [99]. Oxaliplatin adducts are repaired with kinetics similar to those of cisplatin adducts. Oxaliplatin, however, is more efficient than cisplatin per equal number of DNA adducts in inhibiting DNA chain elongation. Despite lower DNA reactivity, oxaliplatin exhibits similar or greater cytotoxicity in several human tumor cell lines. Thus, oxaliplatin requires fewer DNA lesions than cisplatin to achieve cell growth inhibition [99].

The conformation of the major DNA adduct formed by oxaliplatin (1,2-GG intrastrand crosslink) has been studied first by using molecular modeling [100]. These studies suggest that the overall conformational alterations induced in DNA by the 1,2-GG intrastrand crosslink of cisplatin and oxaliplatin are similar. The bulky DACH ring of the oxaliplatin adduct fills much of the DNA major groove, making it narrower and less polar at the site of the crosslink. Quite recently, the crystal structure of the 1,2-GG intrastrand crosslink of oxaliplatin in a dodecamer duplex has been reported [101, 102] and confirmed similarity of the overall conformational alterations induced in DNA by the 1,2-GG intrastrand crosslink of cisplatin and oxaliplatin. The crosslink formed by oxaliplatin bends the double helix by $\sim 30°$ toward the major groove. A novel feature of this structure is the presence of a hydrogen bond between pseudoequatorial NH hydrogen atom of the *R,R*-DACH ligand and the O6 atom of the 3'-G of the platinated d(GG) site. This finding has provided structural evidence for the importance of chirality in mediating the interaction between oxaliplatin and double-stranded DNA and a new kind of chiral recognition between an enantiometrically pure metal complex and the DNA double helix.

The subtle differences in overall conformation of 1,2-GG intrastrand crosslinks of cisplatin and oxaliplatin have been suggested [103] to influence further processing of the oxaliplatin crosslink in the cell. Full-length HMGB1 protein and mismatch repair proteins have been found to have a slightly lower affinity to the crosslink of oxaliplatin than to that of cisplatin [47, 103]. In addition, DNA lesions generated by oxaliplatin are also repaired *in vitro* by the mammalian NER pathway with kinetics similar to those of the lesions of cisplatin [104]. Another interesting finding is that eukaryotic DNA polymerases β, γ, ξ, and η bypass 1,2-GG intrastrand crosslink of cisplatin less readily than the same crosslink of oxaliplatin [103, 105–107]. In addition, the misincorporation frequency of DNA polymerase β is slightly greater with 1,2-GG intrastrand crosslink of oxaliplatin than with that of cisplatin on primed single-stranded DNA. Thus, it may be possible that differences in the conformation of 1,2-intrastrand crosslinks of cisplatin and oxaliplatin can be responsible for differences exhibited by these compounds in both the fidelity and efficiency of translesion synthesis by DNA polymerases.

Also interestingly, the MMR protein complex binds to DNA globally modified by cisplatin (see also Section 8.2.4.2), but not to that by oxaliplatin [108]. In addition, *E. coli* mismatch repair protein MutS recognizes the oxaliplatin-modified DNA with 2-fold lower affinity than the cisplatin-modified DNA [109]. Perhaps steric hindrance by the non-polar DACH ligand may prevent the MMR complex or proteins from recognizing the lesion. The latter findings may also be related to the observation that MMR-deficient cells are slightly more resistant to cisplatin but not to oxaliplatin [47, 110].

Besides $[Pt(R,R\text{-}DACH)]^{2+}$ (directly derived from oxaliplatin), other enantiomeric forms of this complex exist. The biological activity of platinum complexes with enantiomeric amine ligands such as $Pt(R,R\text{-}DACH)]^{2+}$ and $[Pt(S,S\text{-}DACH)]^{2+}$ and other enantiomeric pairs has been investigated intensively [111–114]. For instance, the DACH carrier ligand has been shown to significantly affect the ability of platinum–DNA adducts to block essential processes such as replication and transcription [115]. Also importantly, $[PtCl_2(N\text{-}N)]$ complexes with $N\text{-}N = DACH$, 2,3-diaminobutane (DAB) (Fig. 8.6c,d) or 1,2-diaminopropane, having an S configuration at the asymmetric carbon atoms, are markedly more mutagenic toward several strains in *Salmonella typhimurium* than their *R* isomers [116]. Even more importantly, oxaliplatin (having λ-gauche conformation of the chelate ring) exhibits higher activity toward various cancer cell lines than the *S,S*-enantiomer, so that oxaliplatin and not its *S,S*-enantiomer has been approved for clinical use [117]. Hence, although the asymmetry in the amine ligand in these platinum complexes does not involve the coordinated nitrogen atom but rather an adjacent carbon atom, a dependence of the biological activity on the configuration of the amine is observed.

Recently, modifications of DNA by analogs of antitumor cisplatin containing enantiomeric amine ligands, such as $PtCl_2(R,R\text{-}DAB)$ and $PtCl_2(S,S\text{-}DAB)$ (Fig. 8.6c,d), have also been studied [118]. The major differences resulting from the modification by the two enantiomers consist in the thermodynamical destabilization and conformational distortions induced in DNA by the major 1,2-GG intrastrand crosslinks. The crosslink formed by $PtCl_2(S,S\text{-}DAB)$ is more effective in inducing overall destabilization of the duplex and global conformational alterations than that formed by $PtCl_2(R,R\text{-}DAB)$. Bending toward major groove and unwinding angles due to the crosslink of $PtCl_2(S,S\text{-}DAB)$ (24° and 15°, respectively) are smaller than those due to the crosslink of $PtCl_2(R,R\text{-}DAB)$ (35° and 20°, respectively). However, the overall destabilization of the duplex due to the crosslink of $PtCl_2(S,S\text{-}DAB)$ is greater than that due to the crosslink of $PtCl_2(R,R\text{-}DAB)$. In addition, the duplex containing the crosslink of $PtCl_2(R,R\text{-}DAB)$ shows considerably more pronounced distortion of the base pair adjacent to the crosslink on its 3′ side. In contrast, the duplex containing the crosslink of $PtCl_2(S,S\text{-}DAB)$ shows larger distortion of the base pair on the 5′ side of the crosslink.

Interestingly, the intrastrand crosslinks of the $PtCl_2(R,R\text{-}DAB)$ bind to HMGB1a and HMGB1b with a similar affinity as the same crosslink of cisplatin (J. Malina, J. Kasparkova, G. Natile, V. Brabec, unpublished results). In contrast, the crosslink of PtCl$_2$(S,S-DAB) binds to the HMGB1a protein with a noticeably lower affinity, whereas the affinity of HMGB1b to the crosslink of PtCl$_2$(S,S-DAB) is only slightly lower than that to the crosslinks of PtCl$_2$(R,R-DAB) or cisplatin. As the intrastrand crosslink of the PtCl$_2$(S,S-DAB) thermodynamically destabilizes DNA more than the crosslink of the PtCl$_2$(R,R-DAB) [118], this result supports the previously established correlation [119] that the increasing thermodynamical destabilization due to the 1,2-GG intrastrand crosslink of cisplatin lowers its affinity to HMG-domain proteins.

The 1,2-GG intrastrand crosslinks of both enantiomers are efficiently removed from DNA by NER in the *in vitro* assay using mammalian or rodent cell-free extracts capable of removing the 1,2-intrastrand crosslinks of cisplatin (J Malina, J Kasparkova, G Natile, V Brabec, unpublished results). Consistent with the different affinity of the intrastrand crosslinks of both enantiomers to HMG-domain proteins, addition of HMGB1a only blocks the repair of the crosslinks of PtCl$_2$(R,R-DAB), whereas the repair of the crosslinks of PtCl$_2$(S,S-DAB) remains unaffected. Similar results have been also obtained with DNA modified by PtCl$_2$(DACH) enantiomers.

It has been suggested that during replication, translesion synthesis past platinum adduct catalyzed by DNA polymerases may be error prone and result in mutations. No data which would allow comparison of translesion synthesis past 1,2-GG intrastrand crosslink of Pt(R,R-DACH) and Pt(S,S-DACH) are available. On the other hand, it has been shown [103, 120] that translesion synthesis past 1,2-GG intrastrand crosslink of cisplatin or oxaliplatin is markedly or entirely inhibited if this crosslink is bound to HMGB1 protein. Hence, it is reasonable to expect that HMGB1 proteins can modulate the efficiency of DNA polymerases to bypass the 1,2-GG intrastrand crosslink also formed by Pt-DACH compounds and consequently affect their mutagenicity. Thus, considerably lower mutagenic effects of Pt(R,R-DACH) or Pt(R,R-DAB) in comparison with S,S-enantiomers [116] could be due to markedly tighter binding of HMG-domain proteins to the major 1,2-GG intrastrand crosslink of R,R-enantiomers than to the crosslink of S,S-enantiomers; this tighter binding would inhibit translesion synthesis past the crosslink more efficiently and consequently would reduce its mutagenic effects.

Thus, these results suggest that very fine structural modifications, such as those promoted by enantiomeric ligands in bifunctional platinum(II) compounds, can modulate the "downstream effects" such as specific protein recognition by DNA-processing enzymes and other cellular components. More specifically, as a consequence of enantiomorphism of carrier amine ligands of cisplatin analogs their major DNA crosslinks not only exhibit different conformational features, but also are processed differently by cellular components. It has also been suggested that these differences are associated with a different antitumor and other biological activities of the two enantiomers. As such, the results expand our knowledge of how stereochemistry of the carrier amine moiety in antitumor platinum(II) compounds influences some crucial processes underlying their toxicity toward cancer cells and provide a rational basis for the design of new platinum antitumor drugs and chemotherapeutic strategies.

Additional carrier ligands with aliphatic cyclic components have been also tested. One compound, which continues in phase II clinical trials and for which DNA reactions have been described, is lobaplatin (1,2-diaminomethylcyclobutane-platinum(II) lactate). This platinum drug is active in ovarian tumors in patients previously treated with cisplatin and/or carboplatin [121, 122]. Like cisplatin and oxaliplatin, lobaplatin also forms predominantly 1,2-GG and 1,2-AG intrastrand crosslinks in DNA [98]. Lobaplatin forms in cell-free media about four times less crosslinks than cisplatin. However, this difference is smaller in cells, suggesting enhancement of adduct formation by certain cellular mechanisms and/or compounds.

8.3.3
Other Analogs

One way to affect biological activity of cisplatin is to target this platinum compound to DNA differently by the attachment of the platinum moiety to a suitable carrier [123, 124]. Compared with untargeted analogs, such compounds may exhibit a different DNA-binding mode including sequence selectivity [125, 126]. Importantly, this change of the DNA-binding selectivity has already been shown to result in good *in vitro* and *in vivo* activity in tumors resistant to conventional agents. Thus, the objective of the research focused on this type of cisplatin analog is to design multifunctional molecules that bind DNA in predictable ways and that may have novel pharmacological activities.

8.3.3.1 **Bidentate analogs**
One class of platinum complexes targeted to DNA involves cisplatin linked to an intercalator. An example is (1,2-diaminoethane)dichloroplatinum(II) linked to acridine orange by trimethylene or hexamethylene chains (Fig. 8.7a) [127]. In the reaction with DNA, the platinum residues crosslink two base residues, while acridine orange is intercalated between the base pairs at a distance of one to two base pairs from the platinum-binding site.

The poor oral bioavailability properties of cisplatin or carboplatin were the impetus for a search for oral activity in platinum drugs. One compound selected for clinical evaluation from this class is also a bifunctional analog of cisplatin AMD473 [*cis*-amminedichloro(2-methylpyridine)platinum(II)] (Fig. 8.7b). AMD473 is less reactive toward the sulfur-containing molecules, such as methionine and thiourea [128]. The DNA-binding properties of AMD473 differ from those of cisplatin. On naked DNA, several adducts unique to AMD473 have been observed [128]. Within cells, AMD473 forms interstrand crosslinks much more slowly than cisplatin. An interesting finding is that antibodies elicited against DNA modified by AMD473 do not recognize DNA modified by cisplatin.

The idea of developing an orally available platinum drug has led to the design of antitumor platinum(IV) analogs [129, 130]. In addition, being inert to substitution,

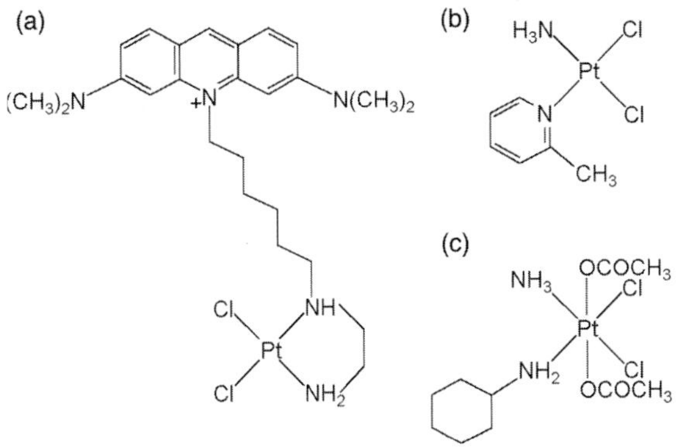

(a)

(CH$_3$)$_2$N N(CH$_3$)$_2$

(b)

H$_3$N Cl
Pt
N Cl
CH$_3$

Cl NH
Pt
Cl NH$_2$

(c)

NH$_3$ OCOCH$_3$
Cl
Pt
NH$_2$ Cl
OCOCH$_3$

Fig. 8.7. Structures of targeted bifunctional analogs of cisplatin: (a) (1,2-diamino-ethane)dichloroplatinum(II) linked to acridine orange by hexamethylene chain, (b) *cis*-amminedichloro(2-methylpyridine) platinum(II) (AMD473), and (c) *bis*-acetatoamminedichloro (cyclohexylamine)platinum(IV) (JM216).

platinum(IV) complexes theoretically have the advantage of demonstrating fewer side effects than their platinum(II) counterparts. The oxidation state of the platinum atom in platinum coordination compounds determines the steric configuration of the molecule. Platinum(II) structures are planar molecules, while platinum(IV) derivatives assume an octahedral shape.

Octahedral platinum(IV) complexes undergo ligand substitution reactions that are slow relative to those of their platinum(II) analogs. They have been considered the compounds which are unable to react directly with DNA. The antitumor activity of platinum(IV) compounds has been suggested to require *in vivo* reduction to the kinetically more labile, and therefore reactive, platinum(II) derivatives [131–133]. Thus, platinum(IV) complexes are frequently designated as pro-drugs that have to be first, after their administration, activated by a reaction with reducing agents present in body liquids. On the other hand, mechanisms other than reduction to active platinum(II) species may be important for cytostatic efficiency of platinum(IV) drugs [134]. Several platinum(IV) analogs of cisplatin covalently bind simple nucleic acid bases or their monomeric derivatives [135, 136] or an isolated DNA [87, 137] directly without addition of any reducing agent. In contrast to cisplatin, however, the rate of the reaction of platinum(IV) complexes with DNA is markedly lower [87]. In addition, at the same level of the modification in cell-free media, thermal stability, renaturation and conformational alterations in DNA modified by the platinum(II) and platinum(IV) analogs are different [87].

One of the tetravalent analogs of cisplatin is also JM216 [*bis*-acetatoamminedi-chloro(cyclohexylamine)platinum(IV)] (Fig. 8.7c), the first orally administerable platinum complex that has entered clinical trials. The DNA-binding properties of JM216 on naked DNA or within tumor cells are similar to those of cisplatin, although there are some differences in the nature of the adducts [138]. JM216 forms interstrand crosslinks. As regards major intrastrand crosslinks, some differ-

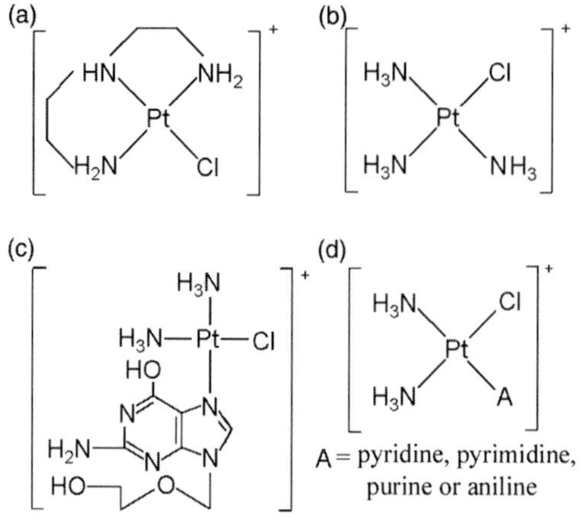

Fig. 8.8. Structures of platinum(II) triamine cations: (a) [PtCl(dien)]$^+$, (b) [PtCl(NH$_3$)$_3$]$^+$, (c) cis-[PtCl(NH$_3$)$_2$(N7-ACV)]$^+$, and (d) antitumor trisubstituted platinum(II) compounds containing pyridine, pyrimidine, purine, or aniline ligand.

ences between these adducts formed by JM216 and cisplatin may be deduced from the analysis by using antibodies elicited against DNA modified by cisplatin.

8.3.3.2 Monodentate analogs

Exploring new structural classes of platinum antitumor drugs has also resulted in the discovery of new platinum(II) complexes including those of formula cis-[PtCl(NH$_3$)$_2$(A)]$^+$ (where A is a heterocyclic amine ligand) (Fig. 8.8d). Thus, these formally monofunctional complexes are analogs of cisplatin only containing one leaving chloride group, similar to closely related and simpler, but inactive, platinum–triamine complexes, such as [PtCl(dien)]Cl or [PtCl(NH$_3$)$_3$]Cl (Fig. 8.8a,b). The adducts formed between double-helical DNA and [PtCl(dien)]Cl comprise mainly monofunctional adducts at N7 atoms of guanine residues [139]. These monofunctional adducts distort DNA and reduce its thermal stability in a sequence-dependent manner [28, 29, 140]. The most pronounced effects are observed if the platinated guanine residue is flanked by single pyrimidines on both 3′ and 5′ sides. Also importantly, the conformational distortion is always more pronounced in the flanking base pairs containing the base on the 5′ site of the monofunctional adduct. These distortions disturb stacking interactions in double-helical DNA and make bases around the adduct more accessible. In addition, the monofunctional adduct also unwinds the duplex [30] – the [PtCl(dien)]Cl–DNA structure exhibits an unwinding angle of 6°, but no bending [28].

The DNA-binding mode of the new antitumor monodentate platinum compounds has been also intensively investigated. An interesting example of this class of platinum compounds is the trisubstituted platinum(II) compounds in which A = pyridine, pyrimidine, purine, piperidine or aniline (Fig. 8.8d) [141, 142]. These compounds have demonstrated activity against a number of murine tumors and human tumor cell lines. These trisubstituted platinum(II) compounds form on DNA monofunctional adducts which have been suggested to be responsible for the antitumor activity of these agents. The monofunctional adducts of these compounds are capable of blocking DNA replication *in vitro* almost as efficiently as the major adducts of cisplatin, but are not recognized by cellular damaged DNA recognition proteins. The principal sites at which the replication is blocked are single guanines. Interestingly, the compound in which A = 4-methyl pyridine distorts double-helical DNA in the base pairs flanking the adduct on its 5′ side due to significant contacts of the 4-methyl pyridine ring with the backbone of the DNA on the 5′ side of the adduct, but there is no indication for the intercalation of the pyridine ring [143, 144]. Taken together, the platinum-triamines have been characterized as a new class of platinum anticancer agents which modify DNA differently from cisplatin. These differences have been proposed to be associated with different biological effects of these monofunctional compounds in comparison with cisplatin.

In the search for new, therapeutically more effective platinum drugs, platinum(II) compounds containing, as a part of the coordination sphere, certain selected antiviral nucleosides have been synthesized as well [145, 146]. Several compounds of this type exhibit similar or enhanced antiviral activities *in vitro* and in many instances are less toxic to normal cells than either component. A novel compound, *cis*-$[PtCl(NH_3)_2(N7\text{-}ACV)]^+$ has been synthesized [146] which contains antiviral acyclovir (ACV) in the coordination sphere of cisplatin (Fig. 8.8c). It exhibits activity against various herpes viruses and has been found to be as effective as cisplatin when equitoxic doses are administered *in vivo* to P388 leukemia-bearing mice. It has also been found to be active against a cisplatin-resistant subline of the P388 leukemia. *cis*-$[PtCl(NH_3)_2(N7\text{-}ACV)]^+$ binds to DNA forming monofunctional adducts preferentially at guanine sites. These adducts are capable of terminating DNA and RNA synthesis *in vitro* [147] and affect some structural and other physical properties of DNA in a way that is similar to those produced by the major adduct of cisplatin. It has been suggested that the ACV ligand itself interacts with DNA in a non-covalent manner, producing certain structural features similar to those of the major adduct of cisplatin. These could be recognized and further processed by some components of tumor cells in a fundamentally different way from other monofunctional platinum adducts such as $[PtCl(dien)]Cl$.

Whatever the detailed mechanism of biological activity of this new class of monodentate antitumor platinum(II) agents, these compounds are interesting from a mechanistic point of view and, therefore, worthy of additional testing.

8.4
Modification by Antitumor Analogs of Clinically Ineffective Transplatin

Since the discovery of the antitumor activity of cisplatin, none of the structural analogs of cisplatin which have advanced to clinical trials are likely to display novel clinical properties in comparison to the parent drug. Therefore, the search con-

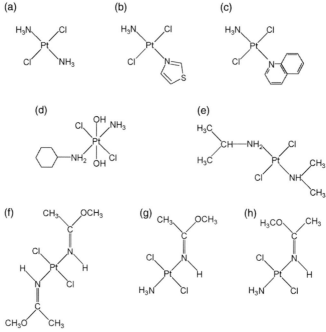

Fig. 8.9. Structures of transplatin and its antitumor analogs: (a) transplatin, (b) *trans*-[PtCl₂(NH₃)(thiazole)], (c) *trans*-[PtCl₂(NH₃)(quinoline)], (d) *trans*-amine(cyclohexylamine) dichlorodihydroxoplatinum(IV) (JM335), (e) *trans*-[PtCl₂(dimethylamine)(isopropylamine)], (f) *trans*-[PtCl₂(E-iminoether)₂], (g) *trans*-[PtCl₂(NH₃)(E-iminoether)], and (h) *trans*-[PtCl₂(NH₃)(Z-iminoether)].

tinues for an improved platinum antitumor agent, motivated by the desire to design a less toxic compound that is non-cross-resistant with cisplatin or carboplatin. In this search the hypothesis that platinum drugs that bind to DNA in a fundamentally different manner to cisplatin will have altered pharmacological properties has been tested. This concept has already led to the synthesis of several new unconventional platinum antitumor compounds that violate the original structure–activity relationships. The clinical inactivity of transplatin (Fig. 8.9a) is considered a paradigm for the classical structure–activity relationships of platinum drugs [8], but to this end several new analogs of transplatin which exhibit a different spectrum of cytostatic activity, including activity in tumor cells resistant to cisplatin, have been identified.

The antitumor *trans*-platinum complexes whose DNA-binding mode will be reviewed in this section include three distinct series (Fig. 8.9): (1) analogs containing iminoether groups of general formula *trans*-[PtCl₂(E-iminoether)₂]; (2) analogs containing planar amine ligand of general structure *trans*-[PtCl₂(NH₃)(L)], where L = planar amine such as quinoline or thiazole; (3) analogs with asymmetric aliphatic ligands and *trans*-[PtCl₂(NH₃)(L)], where L = cyclohexylamine (CHA). As the DNA-binding mode of these new antitumor transplatin analogs is compared with that of the parent inactive compound, we will summarize briefly the fundamental features of DNA modified by transplatin first.

8.4.1
Modifications by Transplatin

Structure–activity studies often employ inactive compounds such as transplatin (Fig. 8.9a). It has been widely used to investigate the mechanism of action of cisplatin. In this approach, differences between active and inactive compounds that may be responsible for the pharmacological effect are searched for. Hence, the clinical inactivity of the *trans* isomer of cisplatin is considered a paradigm for the classical structure–activity relationships of platinum drugs.

Transplatin-DNA adducts are mainly interstrand crosslinks ($\sim$12% in linear DNA) preferentially formed between guanine and complementary cytosine residues and a relatively large proportion ($\sim$50%) of adducts remain monofunctional [41, 148]. Intrastrand crosslinks between non-adjacent guanine residues or between guanine and either adenine or cytosine residues have been also deduced from the results of the analyses of isolated DNA incubated with transplatin at a relatively high extent (one or more Pt atoms per 100 nucleotides). However, it has been suggested [25, 149] that in cells transplatin forms only a small amount of interstrand crosslinks because of the slow closure of the monofunctional adducts coupled to their trapping by intracellular sulfur nucleophiles.

Transplatin also forms in DNA various types of adducts which affect DNA conformation less severely than the crosslinks of cisplatin. Monofunctional lesions, which are the major lesions, affect DNA conformation very little, like the adducts of [PtCl(dien)]Cl (see Section 8.3.3.2). The conformational alterations induced by the interstrand crosslinks are, however, much less severe than those induced by the crosslinks of cisplatin. The duplex is only slightly distorted on both sides of the crosslink but all bases are still paired and hydrogen bonded. The interstrand crosslink of transplatin unwinds the double helix by $\sim$12° and induces a slight, flexible bending of $\sim$20° of its axis toward minor groove [150–152]. The 1,3-intrastrand crosslink of transplatin, which is a less likely lesion in cells [149], unwinds the double helix by 45°, bends the DNA by 26°, and locally denatures, imparting a flexibility to the DNA which acts like a hinge joint without producing a directed bend [26, 153].

8.4.2
Analogs Containing Iminoether Groups

One class of the *trans*-platinum complexes that exhibit antitumor activity comprises analogs of transplatin in which NH_3 is substituted by iminoether ligands. The *trans*-[$PtCl_2$(E-iminoether)$_2$] complex (Fig. 8.9f) is not only more cytotoxic than its *cis* congener, but is also endowed with significant antitumor activity [154, 155] (iminoether = $HN=C(OCH_3)CH_3$; it can have either E or Z configuration depending on the relative position of OCH_3 and N-bonded Pt with respect to the C=N double bond, *cis* in the *Z* isomer and *trans* in the *E* isomer). These results strongly imply a new mechanism of action for *trans*-[PtCl$_2$(*E*-iminoether)$_2$]. By analogy with the diamminedichloroplatinum(II) isomers, the inhibition of DNA synthesis by *trans*-[PtCl$_2$(*E*-iminoether)$_2$] implies a role for DNA binding in the mechanism of action. The presence of the iminoether group may result in altered hydrogen bonding and steric effects affecting the kinetics of DNA binding, the structures, and/or the stability of the adducts formed and resulting local conformational alterations in DNA. Bifunctional *trans*-[PtCl$_2$(*E*-iminoether)$_2$] preferentially forms stable monofunctional adducts at guanine residues in double-helical DNA even when DNA is incubated with the platinum complex for a relatively long time (48 h at 37°C in 10 mM NaClO$_4$) [156]. The random modification of natural DNA in cell-free media results in the non-denaturational alterations in the conformation of DNA similar to the modification by cisplatin, but differently from the modification by transplatin, which produces denaturational distortions in DNA [87]. The most striking feature of the lesions of *trans*-[PtCl$_2$(*E*-iminoether)$_2$] is that they prematurely terminate RNA synthesis at sites and with an efficiency similar to those of major DNA adducts of cisplatin [157]. This is a very interesting finding since the prevalent lesions formed on DNA by *trans*-[PtCl$_2$(*E*-iminoether)$_2$] are monofunctional adducts at guanine residues and monofunctional DNA adducts of other simpler platinum(II) complexes (such as those of [PtCl(dien)]Cl, cisplatin, transplatin) do not terminate RNA synthesis [22, 41]. In addition, it is generally accepted that monofunctional DNA adducts of cisplatin or transplatin are not relevant to the cytostatic effects of these metal complexes.

Recently, the short duplex containing the single, central monofunctional adduct of *trans*-[PtCl$_2$(*E*-iminoether)$_2$] at the guanine residue has been analyzed by NMR spectroscopy [158]. This analysis has yielded the model from which it is possible to roughly estimate the bending induced by the monofunctional adduct of *trans*-[PtCl$_2$(*E*-iminoether)$_2$]. The bending angle estimated in this way was ~45° toward the minor groove, i.e. in a direction opposite to that of the major 1,2-GG intrastrand crosslink of cisplatin. This result has been confirmed by studying the bending induced by single, site-specific monofunctional adduct of *trans*-[PtCl$_2$(*E*-iminoether)$_2$] in the oligodeoxyribonucleotide duplexes (19–22 bp) using electrophoretic retardation as a quantitative measure of the extent of planar curvature (O. Novakova, J. Kasparkova, J. Malina, G. Natile, V. Brabec, unpublished results). The bending angle estimated by this phasing assay was ~20° toward the minor groove. Hence, the distortion induced in DNA by the monofunctional adduct of *trans*-[PtCl$_2$(*E*-iminoether)$_2$] is apparently distinctly different from that produced in DNA by major 1,2-GG intrastrand crosslink of antitumor cisplatin. Then, it is also not surprising that HMG-domain proteins have no affinity to the monofunctional adduct of *trans*-[PtCl$_2$(*E*-iminoether)$_2$]. On the other hand, this adduct is removed from DNA by NER with an efficiency similar to that of 1,2-GG intrastrand crosslink of cisplatin. However, on the basis of the analogy with the mechanism of antitumor activity of cisplatin, the NER of *trans*-[PtCl$_2$(*E*-iminoether)$_2$] monofunctional adducts should be blocked, allowing the damage to persist.

As HMG-domain proteins do not bind to *trans*-[PtCl$_2$(*E*-iminoether)$_2$] lesion, it is reasonable to expect that a pathway different from that including non-covalent binding of damaged DNA-binding proteins is effective in blocking NER of the *trans*-[PtCl$_2$(*E*-iminoether)$_2$] lesions. Importantly, monofunctional adducts formed by *trans*-[PtCl$_2$(*E*-iminoether)$_2$] on DNA can crosslink various types of proteins (including histone H1) and this crosslinking markedly enhances the efficiency of monofunctional adducts of *trans*-[PtCl$_2$(*E*-iminoether)$_2$] to terminate DNA and RNA synthesis *in vitro* (O. Novakova, J. Kasparkova, J. Malina, G. Natile, V. Brabec, unpublished results). In addition, the crosslinking of proteins to DNA monofunctional adducts of *trans*-[PtCl$_2$(*E*-iminoether)$_2$] blocks NER of these lesions, suggesting a different mechanism for "repair shielding" of genotoxic *trans*-[PtCl$_2$(*E*-iminoether)$_2$] adducts. Unique properties of monofunctional DNA adducts of *trans*-[PtCl$_2$(*E*-iminoether)$_2$] and mainly their ability to crosslink proteins might be of fundamental importance in explaining the anticancer activity of this class of *trans*-platinum(II) complexes.

Recently, in order to contribute further to our understanding of the DNA-binding mode of *trans*-platinum complexes with iminoether ligands, DNA modifications by the complexes only containing one iminoether group have been examined [159]. These complexes were *trans*-[PtCl$_2$(NH$_3$)(*Z*-iminoether)] and *trans*-[PtCl$_2$(NH$_3$)(*E*-iminoether)] (mixed *Z* and mixed *E*, respectively) (Fig. 8.9g,h). In a panel of human tumor cell lines, both mixed *Z* and mixed *E* show a cytotoxic potency higher than that of transplatin and mixed *Z* is more active and less toxic than mixed *E*. In the reaction with DNA in a cell-free medium, mixed *Z* forms monofunctional adducts that do not evolve into intrastrand crosslinks but close slowly into interstrand crosslinks between complementary guanine and cytosine residues. These interstrand crosslinks behave as hinge joints, increasing the flexibility of the DNA double helix. These data demonstrate that several features of DNA-binding mode of the antitumor-active mixed *Z* are very similar to those of transplatin, which suggests that the clinical inactivity of transplatin does not depend only on its specific DNA-binding mode.

8.4.3
Analogs Containing Planar Amine Ligand

The *trans*-platinum compounds of this class that have been studied can be grouped into three distinct series of the general structure *trans*-[PtCl$_2$(L)(L′): (1) L = L′ = pyridine or thiazole; (2) L = quinoline and L′ = substituted sulfoxide R′R″SO, R′ = CH$_3$, CH$_2$Ph, Ph; and (3) L = quinoline, thiazole, or pyridine and L′ = NH$_3$ [160]. There are two major chemical differences between NH$_3$ and a planar ligand. Steric hindrance by the appropriately positioned H atoms of the ring system reduces chemical reactivity and surface and stacking interactions become more pronounced over the hydrogen bonding associated with NH$_3$ groups. In this section the results of several biophysical, biochemical, and molecular modeling studies of DNA modifications by antitumor *trans*-[PtCl$_2$(NH$_3$)(quinoline)] or *trans*-[PtCl$_2$(NH$_3$)(thiazole)] (Fig. 8.9b,c) are described as examples [161, 162].

trans-[PtCl$_2$(NH$_3$)(quinoline)] or *trans*-[PtCl$_2$(NH$_3$)(thiazole)] bind monofunctionally to DNA with a rate similar to that of transplatin. The overall rate of the rearrangement to bifunctional adducts is also similar to that observed in the case of DNA modification by transplatin, i.e. it is relatively slow (after 48 h about 34% of adducts remain monofunctional). In contrast to transplatin, however, its analogs containing the planar quinoline or thiazole ligand form considerably more interstrand crosslinks after 48 h ($\sim$30%) with a much shorter half-time ($\sim$5 h) ($\sim$12% for transplatin, $t_{1/2} > 11$ h [20, 41]). In addition, the planar ligand in all or in a significant fraction of DNA adducts of these transplatin analogs, in which platinum is coordinated by base residues, is well positioned to interact with the duplex.

The adducts of the transplatin analogs containing the planar quinoline or thiazole ligand terminate *in vitro* RNA synthesis preferentially at guanine residues. Interestingly, DNA modified by the *trans*-platinum compounds containing the planar ligand is recognized by the antibodies that specifically recognize DNA modified by cisplatin, which suggests that these transplatin analogs behave in some respects like cisplatin. Models for both monofunctional adducts and bifunctional interstrand crosslinks have been proposed. Computer-generated models show that the combination of monofunctional covalent binding and a stacking interaction between the planar ligand and the DNA bases can produce a kink in the duplex which is strongly suggestive of the directed bend produced by the major cisplatin-DNA adduct (1,2-GG intrastrand crosslink).

Further investigations have been focused on the analysis of short duplexes containing the single site-specific adduct of *trans*-[PtCl$_2$(NH$_3$)(thiazole)] (Fig. 8.9b) (J. Kasparkova, N. Farrell, V. Brabec, unpublished results). The monofunctional adduct creates a local conformational distortion, which is localized to five base pairs around the adduct and includes a stable curvature (34° toward major groove) and unwinding (13°). Hence, this distortion is very similar to that produced in DNA by major 1,2-GG intrastrand crosslink of antitumor cisplatin. In addition, these monofunctional adducts are recognized by HMGB1-domain proteins with an affinity similar to that of the 1,2-GG intrastrand crosslink of cisplatin. The effective removal of these adducts from DNA by NER has been also observed, but it is inhibited if the NER system is supplemented by HMGB1a or full-length HMGB1 proteins. Structural properties, recognition by HMG-domain proteins and NER of monofunctional adducts of [PtCl(dien)]Cl or *trans*-[PtCl$_2$(NH$_3$)(thiazole)] on the one hand and 1,2-GG intrastrand crosslink of cisplatin on the other are compared in Tab. 8.1.

Additional work has shown [162], using Maxam–Gilbert footprinting, that *trans*-[PtCl$_2$(NH$_3$)(quinoline)] and *trans*-[PtCl$_2$(NH$_3$)(thiazole)] preferentially form DNA interstrand crosslinks between guanine residues at the 5′-GC/5′-GC sites. Thus, DNA interstrand crosslinking by transplatin analogs containing the planar ligand is formally equivalent to that by antitumor cisplatin, but different from clinically ineffective transplatin which preferentially forms these adducts between complementary guanine and cytosine residues [41].

These results have shown for the first time that the simple chemical modification of structure of an inactive platinum compound alters its DNA-binding mode

Tab. 8.1. Comparison of the structural properties, recognition by HMG-domain proteins and NER of monofunctional adducts of [PtCl(dien)]Cl and trans-[PtCl$_2$(NH$_3$)(thiazole)] with 1,2-GG intrastrand crosslink of cisplatin (see Figs 8.1, 8.8a, and 8.9b for structures of the platinum complexes)

Properties	Monofunctional adduct of [PtCl(dien)]Cl	Monofunctional adduct of trans-[PtCl$_2$(NH$_3$)(thiazole)][d]	1,2-GG intrastrand crosslink of cisplatin
Bending	No[b]	34° toward major groove	32–34° toward major groove[f,g]
Unwinding	6°[c]	12°	13°[d,h]
Recognition by:			
HMGB1a[a]	No[d]	Yes, $K_{D(app)} \sim 38.5$ nM[e]	Yes, $K_{D(app)} \sim 30.8$ nM[a]
HMGB1b[a]	No[d]	Yes, $K_{D(app)} \sim 2.1$ μM[e]	Yes, $K_{D(app)} \sim 1.9$ μM[a]
Mammalian NER	No[d]	Yes	Yes[a,i]

[a] The results reported here were obtained with the same samples of the domain A and domain B of HMGB1 protein under identical conditions [178].
[b] Brabec et al., 1992 [28].
[c] Keck and Lippard, 1992 [30].
[d] Kasparkova, J., Farrell, N., Brabec, V., unpublished results.
[e] Apparent dissociation constants, $K_{D(app)}$, were estimated in the same way as described [186].
[f] Rice et al., 1988 [187].
[g] Bellon et al., 1990 [26].
[h] Bellon et al., 1991 [188].
[i] Zamble et al., 1996 [44].

into that of an active drug and that the processing of the monofunctional DNA adducts of these trans-platinum analogs in tumor cells may be similar to that of the major bifunctional adducts of "classical" cisplatin.

8.4.4
Other Analogs

Recently, the DNA-binding mode of a new cytotoxic trans-platinum compound containing different aliphatic amines (trans-[PtCl$_2$(dimethylamine)(isopropylamine)]) (Fig. 8.9e) has been evaluated in cell-free media [163]. trans-[PtCl$_2$(dimethylamine)(isopropylamine)] readily forms DNA interstrand crosslinks. The number of these crosslinks is considerably higher than the number formed under the same conditions by transplatin or cisplatin. In addition, the compound exhibits a preferential binding affinity to alternating purine–pyrimidine sequences in DNA, forms the adduct that inhibits the B–Z transition in DNA, and blocks DNA synthesis in vitro more efficiently than the adducts of cisplatin. These particular DNA-binding properties have been proposed to be related to the efficiency of trans-[PtCl$_2$(dimethylamine)(isopropylamine)] in inducing apoptosis and some selective killing in an H-ras overexpressing cell line [163–165].

The search for a clinically useful orally available antitumor platinum drug also led to the design of several platinum(IV) compounds with *trans* geometry of leaving ligands [129]. An example is JM335 [*trans*-amine(cyclohexylamine) dichlorodihydroxoplatinum(IV)] (Fig. 8.9d). Mechanistic studies with JM335 in carcinoma cell lines [138] have revealed DNA-binding properties different from those of cisplatin. JM335 induces single-strand breaks in DNA (formation of this DNA lesion is cell line-dependent) and interstrand crosslinks. Interestingly, DNA extracted from cells treated with JM335 is not recognized by an antibody elicited against DNA modified by cisplatin. In addition, the kinetics of apoptosis is more rapid for JM335 than for its *cis* isomer. However, JM335 is generally less cytotoxic than cisplatin, probably due to its inactivation by thiols, so that it has not been selected for clinical trials. Nevertheless it remains one of the more interesting and active *trans*-platinum complexes.

8.5
Modifications by Polynuclear Platinum Antitumor Drugs

Polynuclear platinum compounds comprising di- or trinuclear platinum centers linked by variable length diamine chains constitute another new class of antitumor drugs with chemical and biological properties different from cisplatin. The first drug of this class which already entered phase II clinical trials in cancer patients is the trinuclear compound [{*trans*-PtCl(NH$_3$)$_2$}$_2$ μ-*trans*-Pt(NH$_3$)$_2${H$_2$N(CH$_2$)$_6$ · NH$_2$}$_2$]$^{4+}$ (designated as BBR3464) (Fig. 8.10b). Notable chemical features of this

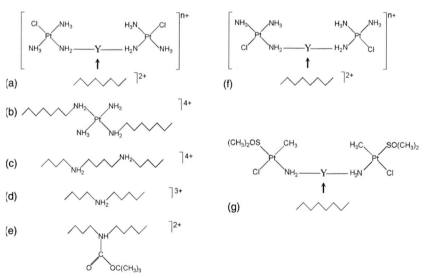

Fig. 8.10. Structures of dinuclear (a, c–g) and trinuclear (b) platinum(II) compounds. (a) 1,1/t,t, (b) 1,0,1/t,t,t (BBR3464), (c) 1,1/t,t-spermine, (d) 1,1/t,t-spermidine, (e) 1,1/t,t BOC-spermidine, (f) 1,1/c,c, and (g) ORGANObisPt.

compound that differ from those of cisplatin or other antitumor mononuclear counterparts are the 4+ charge and the bifunctional binding character, where the binding sites are separated by large distance. Initial clinical trials have revealed activity in pancreatic, lung, and melanoma cancers, suggesting the complementary clinical anticancer activity of BBR3464 in comparison to cisplatin. Preclinical studies have shown cytotoxicity of BBR3464 at 10-fold lower concentration than cisplatin and collateral sensitivity in cisplatin-resistant cell lines [166–169]. Importantly, BBR3464 also displays consistently high antitumor activity in human tumor xenografts characterized as mutant p53. This important feature suggests that the new agent may find utility in the over 60% of cancer cases where mutant p53 status has been indicated. DNA damage by chemotherapeutic agents is in many cases mediated through the p53 pathway. Consistently, cytotoxicity displayed in mutant cell lines would suggest an ability to bypass this pathway.

BBR3464 has been designed on the basis of systematic studies on various dinuclear compounds. The history of the development of this class of platinum drugs has been recently thoroughly reviewed [170, 171] so that here only the major features of DNA-binding modes of dinuclear and trinuclear compounds capable of bifunctional binding will be described. DNA interactions in cell-free media of the compounds, which are the agents most closely related to trinuclear BBR3464, will be summarized first. Then the focus will be on studies of DNA modifications by BBR3464.

8.5.1
Dinuclear Compounds

Exploring new structural classes of platinum antitumor drugs resulted in the discovery of dinuclear bis(platinum) complexes with equivalent coordination spheres with the single chloride leaving group on each platinum linked by a variable-length alkanediamine chain, represented by the general formula $[\{PtCl(NH_3)_2\}_2 \cdot (H_2N(CH_2)_nNH_2)]^{2+}$ ($n = 2$–6). The leaving chloride ligands are either *cis* $(1,1/c,c)$ or *trans* $(1,1/t,t)$ (Fig. 8.10a,f). These dinuclear platinum complexes exhibit antitumor activity *in vitro* and *in vivo* comparable with that of cisplatin, but importantly they retain activity in acquired cisplatin-resistant cell lines [171]. However, the $1,1/c,c$ complexes have been shown to be less efficient in overcoming cisplatin resistance than their $1,1/t,t$ counterparts. This situation represents a fundamental difference between mononuclear and dinuclear platinum chemistry and biology – in the mononuclear case cisplatin is active, while its direct isomer transplatin is antitumor-inactive. These differences in biological activity between dinuclear and mononuclear platinum complexes on the one hand and the differences between the dinuclear compounds themselves on the other have provided the impetus for the studies of molecular mechanisms underlying these differences.

Like cisplatin or carboplatin, the dinuclear platinum complexes bind to DNA and inhibit DNA replication and transcription, which indicates that DNA modification by dinuclear platinum complexes plays an important role in the mechanism of their biological action. Initial studies have already shown some significant differences in DNA modification in cell-free media by individual dinuclear platinum complexes and cisplatin. $1,1/t,t$ isomers bind to DNA more readily than $1,1/c,c$ complexes. Both isomers unwind globally modified DNA by 10–12° (i.e. similar to cisplatin). In contrast to cisplatin, both dinuclear isomers induce the *B–Z* transition in poly(dG-dC).poly(dG-dC) [172] and preferentially form interstrand cross-links in DNA, the $1,1/c,c$ isomer being more efficient [173, 174]. $1,1/t,t$ complexes have also been shown to form minor 1,2-GG intrastrand crosslinks, producing a flexible, non-directional bend in DNA [175]. Intrastrand crosslinks have not been observed in double-helical DNA if it is modified by $1,1/c,c$. A more recent study [176] has investigated the reasons underlying the observed inability of $1,1/c,c$ complexes to form a 1,2-GG intrastrand crosslink with double-helical DNA. ^{1}H NMR spectroscopy of samples of very short single-stranded di- or tetranucleotides containing the GG sequence modified by $1,1/c,c$ has provided evidence for re-stricted rotation around the 3' G in single-stranded 1,2-GG intrastrand crosslinks of $1,1/c,c$. This steric hindrance, not present in 1,2-GG intrastrand crosslinks of $1,1/t,t$ complexes, has been suggested to be responsible for the inability of $1,1/c,c$ complexes to form 1,2-GG intrastrand crosslinks with sterically more demanding double-helical DNA.

Interesting results have been obtained when modifications of natural DNA by the dinuclear platinum(II) organometallic complex [{PtCH$_3$Cl((CH$_3$)$_2$SO)}$_2$(μ-N-N)] (N-N = H$_2$N(CH$_2$)$_6$NH$_2$) (ORGANObisPt, Fig. 8.10g) were investigated [177]. The complex binds irreversibly to DNA, but its DNA-binding mode is different from that of the formally equivalent $1,1/c,c$ (Fig. 8.10f). Interestingly, ORGANObisPt binds to DNA considerably faster than $1,1/c,c$ and cisplatin. In addition, when ORGANObisPt binds to DNA it exhibits a strong base sequence specificity to guanine residues. ORGANObisPt forms on double-helical DNA mainly monofunctional adducts and also a small amount of DNA interstrand crosslinks ($\sim$2%), i.e. a radically smaller amount in comparison with the complex $1,1/c,c$. Importantly, these interstrand crosslinks of ORGANObisPt are capable of terminating RNA synthesis *in vitro* while its major monofunctional adducts are not. In addition, the adducts of ORGANObisPt affect conformation of DNA, but in a different way than its dinuclear analog $1,1/c,c$ or cisplatin. Thus, these results demonstrate that some structural features of ORGANObisPt, such as the charge or nature of the *trans* and *cis* activating groups relative to the labile chloride, are responsible for the altered DNA-binding mode and biological activity in comparison with the $1,1/c,c$ compound.

More recent studies have been mainly focused on structural details of major interstrand crosslinks formed by $1,1/c,c$ and $1,1/t,t$ in DNA [174, 178]. The interstrand crosslinks of $1,1/c,c$ are preferentially formed between guanine residues. Besides 1,2 interstrand crosslinks (between guanine residues in neighboring base pairs), 1,3 or 1,4 interstrand crosslinks are also possible. In the latter two long-range adducts, the sites involved in the crosslinks are separated by one or two base pairs. 1,2, 1,3, and 1,4 interstrand crosslinks are formed at a similar rate and are preferentially oriented in the 5' → 5' direction. In addition, the DNA adducts of these complexes inhibit DNA transcription *in vitro*.

(a) 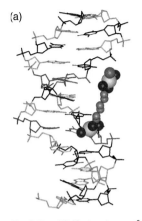   (b)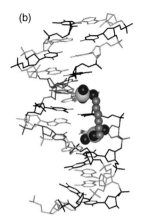

Fig. 8.11. NMR structures of DNA duplexes containing a monofunctional (a) or 1,4-GG interstrand adduct (b) of dinuclear 1,1/*t,t*. Images of the self-complementary duplexes 5′-d(ATATG*TACATAT)/5′-d(ATATGTACATAT) (a) and 5′-d(ATATG*TACATAT)/5′- d(ATATG*TACATAT) (b), where the asterisk represents the sites involved in the adducts of 1,1/*t,t* crosslink [179]. Colors: black, thymine; orange, adenine; green, cytosine; red, guanine; yellow, platinum atom; blue, nitrogen atom; gray, carbon atom; green, chlorine atom.

The properties of the interstrand crosslinks formed in DNA by 1,1/*t,t* complexes have been investigated more thoroughly [178]. Preferential DNA-binding sites in these lesions are G residues in the base pairs separated by at least one other base pair. The crosslinks between G residues in neighboring base pairs are formed with a pronouncedly slower rate than long-range 1,3 or 1,4 crosslinks. Importantly, the length of the diamine bridge linking the two platinum units does not appear to be a substantial factor affecting DNA interstrand crosslinking by the bifunctional dinuclear platinum compounds. As in the case of the interstrand crosslinks of 1,1/*c,c* [174], these lesions of 1,1/*t,t* complexes are also preferentially formed in the 5′ → 5′ direction so that this feature of the interstrand crosslinking might be common for this class of dinuclear platinum compound. The reasons for this preference in the orientation of DNA interstrand crosslinks of dinuclear compounds are unknown.

The conformational distortions induced in DNA by the major 1,3- or 1,4-interstrand crosslinks of 1,1/*t,t* have been further evaluated more in detail (Fig. 8.11) [178, 179]. Interestingly, this lesion results in only a very small directional bending of helix axis (~10°), duplex unwinding (9°), and the conformational flexibility of the crosslink, a feature expected to contribute to the distinct differences in biological profile between dinuclear and mononuclear platinum antitumor drugs.

HMG-domain proteins play a role in sensitizing cells to cisplatin (see Sections 8.2.4.1 and 8.2.5). An important structural motif recognized by HMG-domain proteins on DNA modified by cisplatin is a stable, directional bend of the helix axis. The major interstrand crosslinks and even minor 1,2-d(GpG) intrastrand crosslinks of 1,1/*t,t* compounds bend the helix axis much less efficiently than the crosslinks of cisplatin [175, 178]. Therefore, it is not surprising that very weak or no recognition of DNA adducts of $1,1/t,t$ complexes by HMGB1 protein has been observed [178]; this is consistent with the assumption that an important structural motif recognized by HMG-domain proteins is bent or kinked duplex axis. The affinity of HMG-domain proteins to the duplex containing the 1,2-GG intrastrand crosslink of cisplatin is sequence-dependent and is reduced with increasing destabilization of the duplex due to the crosslink. Thus, a lack of an affinity of the minor 1,2-GG intrastrand crosslink of $1,1/t,t$ to HMG-domain proteins is also consistent with the observation that this lesion reduces the thermal and thermodynamical stability of DNA markedly more than the same lesion of cisplatin [180]. Thus, it is clear that the major DNA adducts of antitumor $1,1/t,t$ compounds may present a block to DNA or RNA polymerase [173, 181] but are not a substrate for recognition by HMG-domain proteins. From these considerations it can be concluded that the mechanism of antitumor activity of bifunctional dinuclear platinum complexes does not involve recognition by HMG-domain proteins as a crucial step, in contrast to the proposals for cisplatin and its direct analogs (see Section 8.2.5).

One possible role for binding of HMG-domain proteins to DNA modified by cisplatin is that these proteins shield damaged DNA from intracellular NER (see Section 8.2.5). No recognition of the 1,2 intrastrand crosslink of $1,1/t,t$ by HMGB1 proteins has been noticed [178], but effective removal of these adducts from DNA by NER has been observed [66]. These results suggest that the processing of the intrastrand crosslinks of $1,1/t,t$ in tumor cells sensitive to this drug may not be relevant to its antitumor effect. Major adducts formed in DNA by $1,1/t,t$ compounds are, however, interstrand crosslinks, which are, in general, believed to be repaired much less easily than intrastrand adducts (see also Section 8.2.3 and reference [20]). Thus, 1,3- or 1,4-interstrand crosslinks of $1,1/t,t$ are not removed in the *in vitro* assay using mammalian and rodent cell-free extracts capable of removing the 1,2-intrastrand crosslinks of cisplatin (J. Kasparkova, N. Farrell, V. Brabec, unpublished results). Hence, the major DNA adducts of bifunctional dinuclear platinum compounds do not have to be shielded by damaged DNA-recognition proteins, such as those containing HMG domains, to prevent their repair.

As mentioned above, addition of the central tetraamine(platinum) unit as in BBR3464 (Fig. 8.10b) greatly enhances the cytotoxicity and antitumor activity in comparison to the prototype dinuclear compound linked by a simple diamine as in $1,1/t,t$ or $1,1/c,c$. In examining structure–activity relationships within this class of compounds a logical step was a substitution of the central tetraamine platinum unit by other H-bonding groups that would retain antitumor activity. Designed linear polyamine-linked dinuclear platinum complexes display the same biological profile as BBR3464 and represent a further subclass of dinuclear compounds with promising preclinical activity [170].

The DNA-binding profiles of three bifunctional dinuclear platinum(II) polyamine-linked compounds, $[\{trans\text{-}PtCl(NH_3)_2\}_2\{\mu\text{-}spermine\text{-}N^1,N^{12}\}]^{4+}$, $[\{trans\text{-}PtCl(NH_3)_2\}_2\{\mu\text{-}spermidine\text{-}N^1,N^8\}]^{3+}$, and $[\{trans\text{-}PtCl(NH_3)_2\}_2\{\mu\text{-}BOC\text{-}spermidine\}]^{2+}$ ($1,1/t,t$-spermine, $1,1/t,t$-spermidine, and $1,1/t,t$ BOC-spermidine, respectively in Fig. 8.10c–e), have been evaluated [182]. All of the compounds bind preferentially in a bifunctional manner. The kinetics of binding of these compounds corresponds to their relative high charge (2+ to 4+). The preference for the formation of interstrand crosslinks, however, does not follow a charge-based pattern. The preferred sites of binding of the dinuclear polyamine-linked compounds are similar to their trinuclear BBR3464 counterpart (see Section 8.5.2), and charge differences do not contribute solely to the variances between the compounds.

8.5.2
Trinuclear Compound

The identification of bifunctional trinuclear BBR3464 (Fig. 8.10b) as the first polynuclear platinum compound used in the clinic was also the impetus for its DNA-binding studies [183]. The high charge on BBR3464 facilitates rapid binding to DNA with a $t_{1/2}$ of $\sim$40 min, significantly faster than the neutral cisplatin. The melting temperature of DNA adducted by BBR3464 increases at low ionic strength but decreases in high salt for the same level of the modification. This unusual behavior is in contrast to that of cisplatin. BBR3464 produces an unwinding angle of 14° in negatively supercoiled pSP73 plasmid DNA, indicative of bifunctional DNA binding. Quantitation of interstrand DNA crosslinking in linearized plasmid DNA has indicated approximately 20% of the DNA to be interstrand crosslinked. While this is significantly higher than the value for cisplatin, it is, interestingly, lower than that for dinuclear platinum compounds such as 1,1/t,t or 1,1/c,c (Tab. 8.2). Either the presence of charge in the linker backbone or the increased distance between platinating moieties may contribute to this relatively decreased ability of BBR3464 to induce DNA interstrand crosslinking. Moreover, BBR3464 rapidly forms long-range delocalized lesions on DNA with sequence selectivity and strong sequence preference for single dG or d(GG) sites. By choosing an appropriate sequence on the basis of sequence-specificity studies, molecular modeling on 1,4-GG interstrand and 1,5-GG intrastrand crosslinks has further confirmed the similarity in energy between the two forms of crosslink [183].

Finally, immunochemical analysis, which has shown that antibodies raised to cisplatin-adducted DNA do not recognize DNA modified by BBR3464, has confirmed the unique nature of the DNA adducts formed by BBR3464. In contrast, DNA modified by BBR3464 inhibits the binding of antibodies raised to transplatin-adducted DNA. Thus, the bifunctional binding of BBR3464 contains few similarities to that of cisplatin, but may have a subset of adducts recognized as similar to transplatin.

We examined how the structures of the various types of the crosslinks of BBR3464 affect conformational properties of DNA and how these adducts are recognized by HMGB1 protein and removed from DNA during *in vitro* NER reactions. The first analyses [184] focused on the 1,2-GG and other long-range intrastrand crosslinks between guanine residues and revealed that these lesions create a local conformational distortion, but that none of these crosslinks results in a stable curvature. In addition, we have observed no recognition of these crosslinks by HMGB1 proteins, but we have observed effective removal of these adducts from DNA by NER. Thus, as in the case of the intrastrand crosslinks of dinuclear platinum compounds (see Section 8.5.1), the processing of the intrastrand crosslinks of

Tab. 8.2. Summary of the structural properties and DNA-binding characteristics of dinuclear platinum compounds and trinuclear BBR3464 (see Figs 8.1 and 8.10 for structures).

Properties	Cisplatin	1,1/t,t (n = 6)	1,1/t,t BOC-spermidine	1,1/t,t-spermidine	1,1/t,t-spermine	1,0,1/t,t,t (n = 6,6)
Reactive Pt–Pt distance (Å)	N/A	12.4	14.9	14.9	21.1	25.3
Linker charge	N/A	0	0	1+	2+	2+
H bonding in linker	N/A	No	No	Yes	Yes	Yes
Total charge	0	2+	2+	3+	4+	4+
DNA binding ($t_{1/2}$, min)	~240[b]	200–300[h]	9.2[j]	8.2[j]	2.8[j]	40[e]
Unwinding angle/adduct (degrees)	13[c]	10–14[c]	13[j]	12[j]	15[j]	14[e]
Decrease of EtBr fluorescence[a]	Medium[d,e]	Strong[e]	Strong[j]	Strong[j]	Strong[j]	Strong[e]
% interstrand crosslink/adduct at $r_b = 2 \times 10^{-4}$	6[f]	70–90[e]	74[j]	40[j]	57[j]	20[e]
Affinity of the major adducts to HMG proteins	Strong[g]	Weak[i]	ND	ND	ND	No[k]

[a] Ethidium bromide (EtBr) is used to characterize perturbations induced in DNA by bifunctional adducts of platinum compounds [183, 189, 190]. Delocalization of these distortions due to long-range crosslinks reduces EtBr intercalation into DNA and consequently its fluorescence more than monofunctional adducts or crosslinks between neighboring base pairs.
[b] Bancroft *et al.*, 1990 [11].
[c] Keck and Lippard, 1992 [30].
[d] Kasparkova *et al.*, 1999 [174].
[e] Brabec *et al.*, 1999 [183].
[f] Brabec and Leng, 1993 [41].
[g] Ohndorf *et al.*, 1999 [56].
[h] Zaludova *et al.*, 1997 [173].
[i] Kasparkova *et al.*, 2000 [178].
[j] McGregor *et al.*, 2002 [182].
[k] Kasparkova *et al.*, 2002, unpublished results.

BBR3464 in tumor cells sensitive to this new drug may not be relevant to its anti-tumor effect.

The analysis of long-range interstrand crosslinks formed by BBR3464 in DNA (J. Zehnulova, J. Kasparkova, N. Farrell, V. Brabec, unpublished results) has demonstrated that these lesions distort DNA only weakly and that these lesions are not recognized by DNA-damaged binding proteins, including those containing HMG domains. On the other hand, the interstrand crosslinks are not removed from DNA in *in vitro* NER reactions and are not recognized by XPA and RPA proteins, i.e. the major damage-recognition proteins involved in the early stage of NER (see Section 8.2.4.2). Thus, as in the case of the dinuclear complex 1,1/t,t, long-range interstrand crosslinks rather than intrastrand adducts appear to be more likely candidates for the genotoxic lesion relevant to the antitumor effects of BBR3464. An intriguing feature of interstrand crosslinking efficiency of BBR3464 is its ability to form crosslinks in the $3' \rightarrow 3'$ direction (J. Zehnulova, J. Kasparkova, N. Farrell, V. Brabec, unpublished results), confirming an ability of this agent to form more types of adducts than the dinuclear counterparts. Thus, these results are consistent with the view that the presence of charge and hydrogen-bonding capacity within the central linker of BBR3464 in the form of a tetraamineplatinum(II) moiety could result in significant pre-association prior to covalent binding, which may affect the extent and direction of interstrand crosslinks.

An important feature of the antitumor efficacy of BBR3464 is also the hypersensitivity of human tumors with mutant p53 to this drug. It has been suggested that apoptosis induced in tumor cells by BBR3464 is not mediated by p53 and that "bypassing" of the p53 pathway may have its origin in the novel DNA-binding mode of this trinuclear platinum agent [170]. In connection with this suggestion the results of our recent studies on the effect of DNA adducts of BBR3464 on the binding of active p53 protein to DNA (J. Kasparkova, S. Pospisilova, N. Farrell, V. Brabec, unpublished results) appear particularly interesting. The adducts of BBR3464 inhibit binding of p53 to the consensus sequence with an efficiency similar to that exhibited by DNA adducts of cisplatin (see Section 8.2.4.2). On the other hand, no adduct of BBR3464 increases the binding affinity of p53 to DNA lacking the consensus sequence. This is in fundamental contrast to cisplatin, whose major 1,2-intrastrand crosslinks increase binding affinity of p53 to DNA lacking consensus sequence [82]. On average, cells with mutant p53 are more resistant to the effects of cisplatin [78], whereas BBR3464 maintains activity in several tumors with mutant p53 with both acquired and inherent resistance to cisplatin [185]. As the DNA-binding activity of p53 protein is crucial to its tumor suppressor function, the plausible explanation for different sensitivities of human tumors with mutant p53 to BBR3464 and cisplatin may be the different efficiency of the adducts of these drugs to increase the binding affinity of DNA lacking the consensus sequence to p53. Confirmation of this hypothesis remains an exciting and potentially significant area of research aimed at further understanding mechanisms underlying the unique antitumor properties of polynuclear platinum agents.

In summary, BBR3464 and antitumor dinuclear bifunctional platinum agents form on DNA adducts which are clearly not those found for cisplatin (Tab. 8.2). Hence, these compounds exhibit a unique profile of DNA binding, strengthening the original hypothesis that modification of DNA binding in manners distinct from that of cisplatin will also lead to a distinct and unique profile of antitumor activity.

8.6
Concluding Remarks

Platinum antitumor drugs represent important agents for the treatment of several different tumors. It appears that changing the chemical structure of platinum compounds may substantially modulate their DNA-binding mode, subsequent processing of DNA damage, and consequently the mechanism of biological efficacy of these compounds. These structural modifications may also affect not only the spectrum of biological activity of the platinum agents and the development of drug resistance, but also their toxicity profile. Hence, information summarized in this article has implications for the future development of platinum-based clinical agents that new, clinically relevant agents not based on cisplatin analogs can also be found. A further understanding of how new platinum compounds modify DNA and how these modifications are further processed in cells may lead to further insight of "downstream" effects initiated through differential protein recognition and repair that may produce unique antitumor effects. On the other hand, there is still a gap between these molecular events involving DNA interactions and understanding why platinum anticancer drugs are more poisonous to cancer cells than to normal cells. The studies so far performed in this area have implicated multiple systems, including several classes of DNA repair, replication, transcription, cell cycle, and cell death responses involved in the processes associated with cellular sensitivity to the platinum drugs. It is also likely that many other determinants remain to be identified.

Complete knowledge of how modifications of DNA by antitumor platinum compounds and other metal-based drugs affect the components of these pathways should provide a basis for understanding the mechanism of action of platinum antitumor drugs and, thereby, a rational basis for the design of new anticancer agents and chemotherapeutic strategies.

Acknowledgments

This work was supported by the Grant Agency of the CR (grant no. 305/02/1552/A), the Internal Grant Agency of the Ministry of Health of the CR (grant no. NL6058-3/2000) and the Grant Agency of the Academy of Sciences of CR (grant no. A5004101). The work was also supported in part by an International Research Scholar's award from the Howard Hughes Medical Institute (USA) and the Wellcome Trust (UK). Note that the reference list is not an exhaustive review of all literature associated with the subject of this article. In many cases, the most recent papers or reviews are cited to enable the reader to trace back to earlier contributions.

References

1 ROSENBERG, B., VAN CAMP, L., KRIGAS, T. Inhibition of division in Escherichia coli by electrolysis products from a platinum electrode. *Nature* **1965**, *205*, 698–699.

2 WONG, E., GIANDOMENICO, C. M. Current status of platinum-based antitumor drugs. *Chem. Rev.* **1999**, *99*, 2451–2466.

3 O'DWYER, P. J., STEVENSON, J. P., JOHNSON, S. W. Clinical status of cisplatin, carboplatin, and other platinum-based antitumor drugs, in *Cisplatin. Chemistry and Biochemistry of a Leading Anticancer Drug*, ed. B. Lippert, VHCA, Wiley-VCH, Zürich, Weinheim, 1999, 31–72.

4 GIACCONE, G. Clinical perspectives on platinum resistance. *Drugs* 2000, *59*, 9–17.

5 WEISS, R. B., CHRISTIAN, M. C. New cisplatin analogs in development. A review. *Drugs* 1993, *46*, 360–377.

6 JOHNSON, N. P., BUTOUR, J.-L., VILLANI, G. *et al.* Metal antitumor compounds: The mechanism of action of platinum complexes. *Prog. Clin. Biochem. Med.* 1989, *10*, 1–24.

7 JAMIESON, E. R., LIPPARD, S. J. Structure, recognition, and processing of cisplatin-DNA adducts. *Chem. Rev.* 1999, *99*, 2467–2498.

8 REEDIJK, J. Improved understanding in platinum antitumour chemistry. *Chem. Commun.* 1996, 801–806.

9 GUO, Z. J., SADLER, P. J. Metals in medicine. *Angew. Chem. Int. Ed.* 1999, *38*, 1513–1531.

10 COHEN, S. M., LIPPARD, S. J. Cisplatin: From DNA damage to cancer chemotherapy, in *Progress in Nucleic Acid Research and Molecular Biology*, ed. K. Moldave, Academic Press, San Diego, 2001, 93–130.

11 BANCROFT, D. P., LEPRE, C. A., LIPPARD, S. J. Pt-195 NMR kinetic and mechanistic studies of *cis*-diamminedichloroplatinum and *trans*-diamminedichloroplatinum(II) binding to DNA. *J. Am. Chem. Soc.* 1990, *112*, 6860–6871.

12 REEDIJK, J. Why does cisplatin reach guanine-N7 with competing S-donor ligands avaiable in the cell? *Chem. Rev.* 1999, *99*, 2499–2510.

13 REEDIJK, J., TEUBEN, J. M. Platinum-sulfur interactions involved in antitumor drugs, rescue agents, and biomolecules, in *Cisplatin. Chemistry and Biochemistry of a Leading Anticancer Drug*, ed. B. Lippert, VHCA, Wiley-VCH, Zürich, Weinheim, 1999, 339–362.

14 BARNHAM, K. J., DJURAN, M. I., MURDOCH, P. D. S., RANFORD, J. D., SADLER, P. J. L-Methionine increases the rate of reaction of 5'-guanosine monophosphate with anticancer drug cisplatin: mixed-ligand adducts and reversible methionine binding. *J. Chem. Soc. Dalton Trans.* 1995, 3721–3726.

15 VAN BOOM, S. S. G. E., CHEN, B. W., TEUBEN, J. M., REEDIJK, J. Platinum-thioether bonds can be reverted by guanine-N7 bonds in Pt(dien)$^{2+}$ model adducts. *Inorg. Chem.* 1999, *38*, 1450–1455.

16 NORMAN, R. E., RANFORD, J. D., SADLER, P. J. Studies of platinum(II) methionine complexes – metabolites of cisplatin. *Inorg. Chem.* 1992, *31*, 877–888.

17 MURDOCH, P. D. S., RANDORF, J. D., SADLER, P. J., BERNERS-PRICE, S. J. Cis-trans isomerization of [Pt(L-Methionine)$_2$]. Metabolite of the anticancer drug cisplatin. *Inorg. Chem.* 1993, *32*, 2249–2255.

18 FICHTINGER-SCHEPMAN, A. M. J., VAN DER VEER, J. L., DEN HARTOG, J. H. J., LOHMAN, P. H. M., REEDIJK, J. Adducts of the antitumor drug cis-diamminedichloroplatinum(II) with DNA: Formation, identification, and quantitation. *Biochemistry* 1985, *24*, 707–713.

19 EASTMAN, A. The formation, isolation and characterization of DNA adducts produced by anticancer platinum complexes. *Pharmacol. Ther.* 1987, *34*, 155–166.

20 BRABEC, V. Chemistry and structural biology of 1,2-interstrand adducts of cisplatin, in *Platinum-Based Drugs in Cancer Therapy*, eds L. R. Kelland and N. P. Farrell, Humana Press, Totowa, NJ, 2000, 37–61.

21 GELASCO, A., LIPPARD, S. J. Anticancer activity of cisplatin and related complexes, in *Metallopharmaceuticals I. DNA Interactions*, eds M. J. Clarke and P. J. Sadler, Springer, Berlin, 1999, 1–43.

22 LEMAIRE, M. A., SCHWARTZ, A., RAHMOUNI, A. R., LENG, M. Interstrand cross-links are preferentially formed at the d(GC) sites in the reaction between *cis*-diamminedichloroplatinum(II) and DNA. *Proc. Natl Acad. Sci. USA* 1991, *88*, 1982–1985.

23 HUANG, H. F., ZHU, L. M., REID, B. R., DROBNY, G. P., HOPKINS, P. B. Solution structure of a cisplatin-induced DNA interstrand cross-

link. *Science* **1995**, *270*, 1842–1845.

24 COSTE, F., MALINGE, J. M., SERRE, L. *et al.* Crystal structure of a double-stranded DNA containing a cisplatin interstrand cross-link at 1.63 A resolution: hydration at the platinated site. *Nucleic Acids Res.* **1999**, *27*, 1837–1846.

25 LENG, M., SCHWARTZ, A., GIRAUD-PANIS, M. J. Transplatin-modified oligonucleotides as potential antitumor drugs, in *Platinum-Based Drugs in Cancer Therapy*, eds L. R. Kelland and N. P. Farrell, Humana Press, Totowa, NJ, 2000, 63–85.

26 BELLON, S. F., LIPPARD, S. J. Bending studies of DNA site-specifically modified by cisplatin, trans-diamminedichloroplatinum(II) and cis-Pt(NH₃)₂(N3-cytosine)Cl⁺. *Biophys. Chem.* **1990**, *35*, 179–188.

27 TEUBEN, J. M., BAUER, C., WANG, A. H. J., REEDIJK, J. Solution structure of a DNA duplex containing a cis-diammineplatinum(II) 1,3-d(GTG) intrastrand cross-link, a major adduct in cells treated with the anticancer drug carboplatin. *Biochemistry* **1999**, *38*, 12305–12312.

28 BRABEC, V., REEDIJK, J., LENG, M. Sequence-dependent distortions induced in DNA by monofunctional platinum(II) binding. *Biochemistry* **1992**, *31*, 12397–12402.

29 BRABEC, V., BOUDNY, V., BALCAROVA, Z. Monofunctional adducts of platinum(II) produce in DNA a sequence-dependent local denaturation. *Biochemistry* **1994**, *32*, 1316–1322.

30 KECK, M. V., LIPPARD, S. J. Unwinding of supercoiled DNA by platinum ethidium and related complexes. *J. Am. Chem. Soc.* **1992**, *114*, 3386–3390.

31 PINTO, A. L., LIPPARD, S. J. Sequence-dependent termination of in vitro DNA synthesis by cis- and trans-diamminedichloroplatinum(II). *Proc. Natl Acad. Sci. USA* **1985**, *82*, 4616–4620.

32 VILLANI, G., HUBSCHER, U., BUTOUR, J.-L. Sites of termination of in vitro

DNA synthesis on cis-diammine-dichloroplatinum(II) treated single-stranded DNA: a comparison between E. coli DNA polymerase I and eucaryotic DNA polymerases alpha. *Nucleic Acids Res.* **1988**, *16*, 4407–4418.

33 COMESS, K. M., BURSTYN, J. N., ESSIGMANN, J. M., LIPPARD, S. J. Replication inhibition and translesion synthesis on templates containing site-specifically placed cis-diammine-dichloroplatinum(II) DNA adducts. *Biochemistry* **1992**, *31*, 3975–3990.

34 BURNOUF, D., DAUNE, M., FUCHS, R. P. Spectrum of cisplatin-induced mutations in Escherichia coli. *Proc. Natl Acad. Sci. USA* **1987**, *84*, 3758–3762.

35 YAREMA, K. J., LIPPARD, S. J., ESSIGMANN, J. M. Mutagenic and genotoxic effects of DNA adducts formed by the anticancer drug cis-diamminedichloroplatinum(II). *Nucleic Acids Res.* **1995**, *23*, 4066–4072.

36 SORENSON, C. M., EASTMAN, A. Mechanism of cis-diammine-dichloroplatinum(II)-induced cytotoxicity – Role of G2 arrest and DNA double-strand breaks. *Cancer Res.* **1988**, *48*, 4484–4488.

37 CHU, G. Cellular responses to cisplatin. The roles of DNA-binding proteins and DNA repair. *J. Biol. Chem.* **1994**, *269*, 787–790.

38 ALLDAY, M. J., INMAN, G. J., CRAWFORD, D. H., FARRELL, P. J. DNA damage in human B cells can induce apoptosis, proceeding from G1/S when p53 is transactivation competent and G2/M when it is transactivation defective. *EMBO J.* **1995**, *14*, 4994–5005.

39 CORDA, Y., JOB, C., ANIN, M. F., LENG, M., JOB, D. Transcription by eucaryotic and procaryotic RNA polymerases of DNA modified at a d(GG) or a d(AG) site by the antitumor drug cis-diamminedichloroplatinum(II). *Biochemistry* **1991**, *30*, 222–230.

40 CORDA, Y., ANIN, M. F., LENG, M., JOB, D. RNA polymerases react differently at d(ApG) and d(GpG) adducts in DNA modified by cis-

diamminedichloroplatinum(II). *Biochemistry* **1992**, *31*, 1904–1908.

41 BRABEC, V., LENG, M. DNA interstrand cross-links of trans-diamminedichloroplatinum(II) are preferentially formed between guanine and complementary cytosine residues. *Proc. Natl Acad. Sci. USA* **1993**, *90*, 5345–5349.

42 CULLINANE, C., MAZUR, S. J., ESSIGMANN, J. M., PHILLIPS, D. R., BOHR, V. A. Inhibition of RNA polymerase II transcription in human cell extracts by cisplatin DNA damage. *Biochemistry* **1999**, *38*, 6204–6212.

43 CORDA, Y., JOB, C., ANIN, M.-F., LENG, M., JOB, D. Spectrum of DNA-platinum adduct recognition by prokaryotic and eukaryotic DNA-dependent RNA polymerases. *Biochemistry* **1993**, *32*, 8582–8588.

44 ZAMBLE, D. B., MU, D., REARDON, J. T., SANCAR, A., LIPPARD, S. J. Repair of cisplatin-DNA adducts by the mammalian excision nuclease. *Biochemistry* **1996**, *35*, 10004–10013.

45 ZDRAVESKI, Z. Z., MELLO, J. A., MARINUS, M. G., ESSIGMANN, J. M. Multiple pathways of recombination define cellular responses to cisplatin. *Chem. Biol.* **2000**, *7*, 39–50.

46 GERMANIER, M., DEFAIS, M., JOHNSON, N. P., VILLANI, G. Repair of platinum – DNA lesions in E. coli by a pathway which does not recognize DNA damage caused by MNNG or UV light. *Mutat. Res.* **1984**, *145*, 35–41.

47 FINK, D., NEBEL, S., AEBI, S. *et al.* The role of DNA mismatch repair in platinum drug resistance. *Cancer Res.* **1996**, *56*, 4881–4886.

48 AEBI, S., FINK, D., GORDON, R., KIM, H. K., ZHENG, H., FINK, J. L., HOWELL, S. B. Resistance to cytotoxic drugs in DNA mismatch repair-deficient cells. *Clin. Cancer Res.* **1997**, *3*, 1763–1767.

49 JOHNSON, S. W., FERRY, K. V., HAMILTON, T. C. Recent insights into platinum drug resistance in cancer. *Drug Res. Updates* **1998**, *1*, 243–254.

50 ZLATANOVA, J., YANEVA, J., LEUBA, S. H. Proteins that specifically recognize cisplatin-damaged DNA: a clue to anticancer activity of cisplatin. *FASEB J.* **1998**, *12*, 791–799.

51 ZAMBLE, D. B., LIPPARD, S. J. The response of cellular proteins to cisplatin-damaged DNA, in *Cisplatin. Chemistry and Biochemistry of a Leading Anticancer Drug*, ed. B. Lippert, VHCA, Wiley-VCH: Zürich, Weinheim, 1999, 73–110.

52 KARTALOU, M., ESSIGMANN, J. M. Recognition of cisplatin adducts by cellular proteins. *Mutat. Res.* **2001**, *478*, 1–21.

53 THOMAS, J. O. HMG 1 and 2: architectural DNA-binding proteins. *Biochem. Soc. Trans.* **2001**, *29*, 395–401.

54 DUNHAM, S. U., LIPPARD, S. J. DNA sequence context and protein composition modulate HMG-domain protein recognition of ciplatin-modified DNA. *Biochemistry* **1997**, *36*, 11428–11436.

55 COHEN, S. M., MIKATA, Y., HE, Q., LIPPARD, S. J. HMG-Domain protein recognition of cisplatin 1,2-intrastrand d(GpG) cross-links in purine-rich sequence contexts. *Biochemistry* **2000**, *39*, 11771–11776.

56 OHNDORF, U. M., ROULD, M. A., HE, Q., PABO, C. O., LIPPARD, S. J. Basis for recognition of cisplatin-modified DNA by high-mobility-group proteins. *Nature* **1999**, *399*, 708–712.

57 STROS, M. Two mutations of basic residues within the N-terminus of HMG-1 B domain with different effects on DNA supercoiling and binding to bent DNA. *Biochemistry* **2001**, *40*, 4769–4779.

58 KASPARKOVA, J., BRABEC, V. Recognition of DNA interstrand cross-links of cis-diamminedichloroplatinum(II) and its trans isomer by DNA-binding proteins. *Biochemistry* **1995**, *34*, 12379–12387.

59 YARNELL, A. T., OH, S., REINBERG, D., LIPPARD, S. J. Interaction of FACT, SSRP1, and the high mobility group (HMG) domain of SSRP1 with DNA damaged by the anticancer drug cisplatin. *J. Biol. Chem.* **2001**, *276*, 25736–25741.

60 ORPHANIDES, G., LAGRANGE, T., REINBERG, D. The general transcription factors of RNA polymerase II. *Genes Dev.* **1996**, *10*, 2657–2683.

61 VICHI, P., COIN, F., RENAUD, J. P. *et al.* Cisplatin- and UV-damaged DNA lure the basal transcription factor TFIID/TBP. *EMBO J.* **1997**, *16*, 7444–7456.

62 COIN, F., FRIT, P., VIOLLET, B., SALLES, B., EGLY, J. M. TATA binding protein discriminates between different lesions on DNA, resulting in a transcription decrease. *Mol. Cell. Biol.* **1998**, *18*, 3907–3914.

63 ISE, T., NAGATANI, G., IMAMURA, T. *et al.* Transcription factor Y-box binding protein 1 binds preferentially to cisplatin-modified DNA and interacts with proliferating cell nuclear antigen. *Cancer Res.* **1999**, *59*, 342–346.

64 YANEVA, J., LEUBA, S. H., VAN HOLDE, K., ZLATANOVA, J. The major chromatin protein histone H1 binds preferentially to cis-platinum-damaged DNA. *Proc. Natl Acad. Sci. USA* **1997**, *94*, 13448–13451.

65 PANEVA, E. G., SPASSOVSKA, N. C., GRANCHAROV, K. C., ZLATANOVA, J. S., YANEVA, J. N. Interaction of histone H1 with cis-platinum modified DNA. *Z. Naturforsch.* **1998**, *53c*, 135–138.

66 MISSURA, M., BUTERIN, T., HINDGES, R. *et al.* Double-check probing of DNA bending and unwinding by XPA-RPA: an architectural function in DNA repair. *EMBO J.* **2001**, *20*, 3554–3564.

67 DUCKETT, D. R., DRUMMOND, J. T., MURCHIE, A. I. H. *et al.* Human MutSα recognizes damaged DNA base pairs containing O^6-methylguanine, O^4-methylthymine, or the cisplatin-d(GpG) adduct. *Proc. Natl Acad. Sci. USA* **1996**, *93*, 6443–6447.

68 MELLO, J. A., ACHARYA, S., FISHEL, R., ESSIGMANN, J. M. The mismatch-repair protein hMSH2 binds selectively to DNA adducts of the anticancer drug cisplatin. *Chem. Biol.* **1996**, *3*, 579–589.

69 YAMADA, M., O'REGAN, E., BROWN, R., KARRAN, P. Selective recognition of a cisplatin-DNA adduct by human mismatch repair proteins. *Nucleic Acids Res.* **1997**, *25*, 491–495.

70 ÖZER, Z., REARDON, J. T., HSU, D. S., MALHOTRA, K., SANCAR, A. The other function of DNA photolyase: Stimulation of excision repair of chemical damage to DNA. *Biochemistry* **1995**, *34*, 15886–15889.

71 TURCHI, J. J., HENKELS, K. M., ZHOU, Y. Cisplatin-DNA adducts inhibit translocation of the Ku subunits of DNA-PK. *Nucleic Acids Res.* **2000**, *28*, 4634–4641.

72 KARTALOU, M., SAMSON, L. D., ESSIGMANN, J. M. Cisplatin adducts inhibit 1,N-6-ethenoadenine repair by interacting with the human 3-methyladenine DNA glycosylase. *Biochemistry* **2000**, *39*, 8032–8038.

73 MURCHIE, A. I. H., LILLEY, D. M. J. T4 endonuclease VII cleaves DNA containing a cisplatin adduct. *J. Mol. Biol.* **1993**, *233*, 77–82.

74 RIVA, C. M. Restoration of wild-type p53 activity enhances the sensitivity of pleural metastasis to cisplatin through an apoptotic mechanism. *Anticancer Res.* **2000**, *20*, 4463–4471.

75 JORDAN, P., CARMO-FONSECA, M. Molecular mechanisms involved in cisplatin cytotoxicity. *Cell. Mol. Life Sci.* **2000**, *57*, 1229–1235.

76 MAY, P., MAY, E. Twenty years of p53 research: structural and functional aspects of the p53 protein. *Oncogene* **1999**, *18*, 7621–7636.

77 JANUS, F., ALBRECHTSEN, N., DORNREITER, I., WIESMÜLLER, L., GROSSE, F., DEPPERT, W. The dual role model for p53 in maintaining genomic integrity. *Cell. Mol. Life Sci.* **1999**, *55*, 12–27.

78 O'CONNOR, P. M., JACKMAN, J., BAE, I. *et al.* Characterization of the p53 tumor suppressor pathway in cell lines of the National Cancer Institute anticancer drug screen and correlations with the growth-inhibitory potency of 123 anticancer agents. *Cancer Res.* **1997**, *57*, 4285–4300.

79 TOKINO, T., THIAGALINGAM, S., ELDEIRY, W. S., WALDMAN, T., KINZLER, K. W., VOGELSTEIN, B. p53 tagged sites from human genomic DNA. *Hum. Mol. Genet.* **1994**, *3*, 1537–1542.

80 NAGAICH, A. K., ZHURKIN, V. B., DURELL, S. R., JERNIGAN, R. L., APPELLA, E., HARRINGTON, R. E. p53-induced DNA bending and twisting: p53 tetramer binds on the outer side of a DNA loop and increases DNA twisting. *Proc. Natl Acad. Sci. USA* **1999**, *96*, 1875–1880.

81 LEBRUN, A., LAVERY, R., WEINSTEIN, H. Modeling multi-component protein-DNA complexes: the role of bending and dimerization in the complex of p53 dimers with DNA. *Protein Eng.* **2001**, *14*, 233–243.

82 KASPARKOVA, J., POSPISILOVA, S., BRABEC, V. Different recognition of DNA modified by antitumor cisplatin and its clinically ineffective trans isomer by tumor suppressor protein p53. *J. Biol. Chem.* **2001**, *276*, 16064–16069.

83 KELLAND, L. R. Preclinical perspectives on platinum resistance. *Drugs* **2000**, *59*, 1–8.

84 GONZALEZ, V. M., FUERTES, M. A., ALONSO, C., PEREZ, J. M. Is cisplatin-induced cell death always produced by apoptosis? *Mol. Pharmacol.* **2001**, *59*, 657–663.

85 BLOMMAERT, F. A., VAN DIJK-KNIJNENBURG, H. C. M., DIJT, F. J. *et al.* Formation of DNA adducts by the anticancer drug carboplatin: Different nucleotide sequence preferences in vitro and in cells. *Biochemistry* **1995**, *34*, 8474–8480.

86 NATARAJAN, G., MALATHI, R., HOLLER, E. Increased DNA-binding activity of cis-1,1-cyclobutanedicarboxy latodiammineplatinum(II) (Carboplatin) in the presence of nucleophiles and human breast cancer MCF-7 cell cytoplasmic extracts: Activation theory revisited. *Biochem. Pharmacol.* **1999**, *58*, 1625–1629.

87 VRANA, O., BRABEC, V., KLEINWÄCHTER, V. Polarographic studies on the conformation of some platinum complexes: relations to anti-tumour activity. *Anti-Cancer Drug Des.* **1986**, *1*, 95–109.

88 HIGHLEY, M. S., CALVERT, A. H. Clinical experience with cisplatin and carboplatin, in *Platinum-Based Drugs in Cancer Therapy*, eds L. R. Kelland and N. P. Farrell, Humana Press, Totowa, NJ, 2000, 171–194.

89 LOKICH, J. What is the "best" platinum: Cisplatin, carboplatin, or oxaliplatin? *Cancer Invest.* **2001**, *19*, 756–760.

90 CHOLLET, P., BENSMAINE, M., BRIENZA, S. *et al.* Single agent activity of oxaliplatin in heavily pretreated advanced epithelial ovarian cancer. *Ann. Oncol.* **1996**, *7*, 1065–1070.

91 WISEMAN, L. R., ADKINS, J. C., PLOSKER, G. L., GOA, K. L. Oxaliplatin – A review of its use in the management of metastatic colorectal cancer. *Drugs Aging* **1999**, *14*, 459–475.

92 RIXE, O., ORTUZAR, W., ALVAREZ, M. *et al.* Oxaliplatin, tetraplatin, cisplatin, and carboplatin: Spectrum of activity in drug-resistant cell lines and in the cell lines of the National Cancer Institute's Anticancer Drug Screen panel. *Biochem. Pharmacol.* **1996**, *52*, 1855–1865.

93 LEVI, F. A., ZIDANI, R., VANNETZEL, J. M. *et al.* Chronomodulated versus fixed-infusion-rate delivery of ambulatory chemotherapy with oxaliplatin, fluorouracil, and folinic acid (leucovorin) in patients with colorectal cancer metastases: a randomized multi-institutional trial. *J. Natl Cancer Inst.* **1994**, *86*, 1608–1617.

94 LEVI, F., METZGER, G., MASSARI, C., MILANO, G. Oxaliplatin – Pharmacokinetics and chronopharmacological aspects. *Clin. Pharmacokinetics* **2000**, *38*, 1–21.

95 MAULDIN, S. K., PLESCIA, M., RICHARD, F. A., WYRICK, S. D., VOYKSNER, R. D., CHANEY, S. G. Displacement of the bidentate malonate ligand from (d,l-*trans*-1,2-diaminecyclohexane) malonato-platinum(II) by physiologically important compounds in vitro. *Biochem. Pharmacol.* **1998**, *37*, 3321–3333.

96 WOYNAROWSKI, J. M., CHAPMAN, W. G., NAPIER, C., HERZIG, M. C. S., JUNIEWICZ, P. Sequence- and region-specificity of oxaliplatin adducts in naked and cellular DNA. *Mol. Pharmacol.* **1998**, *54*, 770–777.

97 JENNERWEIN, M. M., EASTMAN, A., KHOKHAR, A. Characterization of adducts produced in DNA by isomeric 1,2-diaminocyclohexaneplatinum(II) complexes. *Chem.-Biol. Interactions* 1989, *70*, 39–50.

98 SARIS, C. P., VAN DE VAART, P. J. M., RIETBROEK, R. C., BLOMMAERT, F. A. In vitro formation of DNA adducts by cisplatin, lobaplatin and oxaliplatin in calf thymus DNA in solution and in cultured human cells. *Carcinogenesis* 1996, *17*, 2763–2769.

99 WOYNAROWSKI, J. M., FAIVRE, S., HERZIG, M. C. S. *et al.* Oxaliplatin-induced damage of cellular DNA. *Mol. Pharmacol.* 2000, *58*, 920–927.

100 SCHEEFF, E. D., BRIGGS, J. M., HOWELL, S. B. Molecular modeling of the intrastrand guanine-guanine DNA adducts produced by cisplatin and oxaliplatin. *Mol. Pharmacol.* 1999, *56*, 633–643.

101 SPINGLER, B., WHITTINGTON, D. A., LIPPARD, S. J. 2.4 A crystal structure of an oxaliplatin 1,2-d(GpG) intrastrand cross-link in a DNA dodecamer duplex. *Inorg. Chem.* 2001, *40*, 5596–5602.

102 SPINGLER, B., WHITTINGTON, D. A., LIPPARD, S. J. 1,2-d(GpG) intrastrand cross-link formed by oxaliplatin with duplex DNA: a crystallographic study. *J. Inorg. Biochem.* 2001, *86*, 440–440.

103 VAISMAN, A., LIM, S. E., PATRICK, S. M. *et al.* Effect of DNA polymerases and high mobility group protein 1 on the carrier ligand specificity for translesion synthesis past platinum-DNA adducts. *Biochemistry* 1999, *38*, 11026–11039.

104 REARDON, J. T., VAISMAN, A., CHANEY, S. G., SANCAR, A. Efficient nucleotide excision repair of cisplatin, oxaliplatin, and bis-aceto-ammine-dichloro-cyclohexylamine-platinum(IV) (JM216) platinum intrastrand DNA diadducts. *Cancer Res.* 1999, *59*, 3968–3971.

105 VAISMAN, A., CHANEY, S. G. The efficiency and fidelity of translesion synthesis past cisplatin and oxaliplatin GpG adducts by human DNA polymerase beta. *J. Biol. Chem.* 2000, *275*, 13017–13025.

106 VAISMAN, A., MASUTANI, C., HANAOKA, F., CHANEY, S. G. Efficient translesion replication past oxaliplatin and cisplatin GpG adducts by human DNA polymerase eta. *Biochemistry* 2000, *39*, 4575–4580.

107 VAISMAN, A., WARREN, M. W., CHANEY, S. G. The effect of DNA structure on the catalytic efficiency and fidelity of human DNA polymerase *β* on templates with platinum-DNA adducts. *J. Biol. Chem.* 2001, *276*, 18999–19005.

108 NEBEL, S., FINK, D., AEBI, S., NEHME, A., CHRISTEN, R. D., HOWELL, S. B. Role of the DNA mismatch repair proteins in the recognition of platinum DNA adducts. *Proc. Am. Assoc. Cancer. Res.* 1997, *38*, A2402.

109 ZDRAVESKI, Z. Z., MELLO, J. A., FARINELLI, C. K., ESSIGMANN, J. M., MARINUS, M. G. MutS preferentially recognizes cisplatin- over oxaliplatin-modified DNA. *J. Biol. Chem.* 2002, *277*, 1255–1260.

110 FINK, D., ZHENG, H., NEBEL, S. *et al.* In vitro and in vivo resistance to cisplatin in cells that have lost DNA mismatch repair. *Cancer Res.* 1997, *57*, 1841–1845.

111 KIDANI, Y., INAGAKI, K., IIGO, M., HOSHI, A., KURETANI, K. Antitumor activity of 1,2-diamminocyclohexane-platinum complexes against Sarcoma 180 ascites form. *J. Med. Chem.* 1978, *21*, 1315–1318.

112 COLUCCIA, M., CORREALE, M., GIORDANO, D. *et al.* Mutagenic activity of some platinum complexes with monodentate and bidentate amines. *Inorg. Chim. Acta* 1986, *123*, 225–229.

113 PASINI, A., ZUNINO, F. New cisplatin analogues – on the way to better antitumor agents. *Angew. Chem. Int. Ed.* 1987, *26*, 615–624.

114 VICKERY, K., BONIN, A. M., FENTON, R. R. *et al.* Preparation, characterization, cytotoxicity, and mutagenicity of a pair of enantiomeric platinum(II) complexes with the potential to bind enantioselectively to DNA. *J. Med. Chem.* 1993, *36*, 3663–3668.

115 PAGE, J. D., HUSAIN, I., SANCAR, A., CHANEY, S. G. Effect of the diamino-

cyclohexane carrier ligand on platinum adduct formation, repair, and lethality. *Biochemistry* **1990**, *29*, 1016–1024.

116 FANIZZI, F. P., INTINI, F. P., MARESCA, L. *et al.* Biological activity of platinum complexes containing chiral centers on the nitrogen or carbon atoms of a chelate diamine ring. *Inorg. Chim. Acta* **1987**, *137*, 45–51.

117 MISSET, J. L. Oxaliplatin in practice. *Br. J. Cancer* **1998**, *77*, S4, 4–7.

118 MALINA, J., HOFR, C., MARESCA, L., NATILE, G., BRABEC, V. DNA interactions of antitumor cisplatin analogs containing enantiomeric amine ligands. *Biophys. J.* **2000**, *78*, 2008–2021.

119 PILCH, D. S., DUNHAM, S. U., JAMIESON, E. R., LIPPARD, S. J., BRESLAUER, K. J. DNA sequence context modulates the impact of a cisplatin 1,2-d(GpG) intrastrand cross-link an the conformational and thermodynamic properties of duplex DNA. *J. Mol. Biol.* **2000**, *296*, 803–812.

120 HOFFMANN, J. S., LOCKER, D., VILIANI, G., LENG, M. HMG1 protein inhibits the translesion synthesis of the major DNA cisplatin adduct by cell extracts. *J. Mol. Biol.* **1997**, *270*, 539–543.

121 GIETEMA, J. A., DE VRIES, E. G. E., SLEIJFER, D. T. *et al.* A phase-I study of 1,2-diamminomethyl-cyclobutane-platinum(II)-lactate (D-19466 lobaplatin) administered daily for 5 days. *Br. J. Cancer* **1993**, *67*, 396–401.

122 GIETEMA, J. A., VELDHUIS, G. J., GUCHELAAR, H. J. *et al.* Phase-II and pharmacokinetic study of lobaplatin in patients with relapsed ovarian cancer. *Br. J, Cancer* **1995**, *71*, 1302–1307.

123 SUNDQUIST, W. I., LIPPARD, S. J. The coordination chemistry of platinum anticancer drugs and related compounds with DNA. *Coord. Chem. Rev.* **1990**, *100*, 293–322.

124 LENG, M., BRABEC, V. DNA adducts of cisplatin, transplatin and platinum-intercalating drugs, in *DNA Adducts: Identification and Biological Significance*, eds K. Hemminki, A. Dipple, D. E. G. Shuker, F. F.

Kadlubar, D. Segerbäck, H. Bartsch, International Agency for Research on Cancer, Lyon, 1994, 339–348.

125 BROGGINI, M., ERBA, E., PONTI, M. *et al.* Selective DNA interaction of the novel distamycin derivative FCE 24517. *Cancer Res.* **1991**, *51*, 199–204.

126 CHURCH, K. M., WURDEMAN, R. L., ZHANG, Y., CHEN, F. X., GOLD, B. N-(2-chloroethyl)-N-nitrosoureas covalently bound to nonionic and monocationic lexitropsin dipeptides. Synthesis, DNA affinity binding characteristics, and reactions with P-32-end-labeled DNA. *Biochemistry* **1990**, *29*, 6827–6838.

127 BOWLER, B. E., HOLLIS, L. S., LIPPARD, S. J. Synthesis and DNA binding and photonicking properties of acridine orange linked by a polymethylene tether to (1,2-diaminoethane)dichloroplatinum(II). *J. Am. Chem. Soc.* **1984**, *106*, 6102–6104.

128 HOLFORD, J., RAYNAUD, F., MURRER, B. A. *et al.* Chemical, biochemical and pharmacological activity of the novel sterically hindered platinum co-ordination complex, cis-[amminedichloro(2-methylpyridine)] platinum(II) (AMD473). *Anti-Cancer Drug Des.* **1998**, *13*, 1–18.

129 KELLAND, L. R. New platinum drugs. The pathway to oral therapy, in *Platinum-Based Drugs in Cancer Therapy*, eds L. R. Kelland, N. P. Farrell, Humana Press, Totowa, NJ, 2000, 299–319.

130 JURASKOVA, V., BRABEC, V. Evaluation of cytotoxic and antitumour effects of tetravalent analog of carboplatin. *Neoplasma* **1989**, *36*, 297–303.

131 ROTONDO, E., FIMIANI, V., CAVALLARO, A., AINIS, T. Does the antitumoral activity of platinum(IV) derivatives result from their in vivo reduction? *Tumori* **1983**, *69*, 31–36.

132 BLATTER, E. E., VOLLANO, J. F., KRISHNAN, B. S., DABROWIAK, J. C. Interaction of the antitumor agents cis,cis,trans-Pt^{IV}(NH$_3$)$_2$Cl$_2$(OH)$_2$ and cis,cis,trans-Pt^{IV}[(CH$_3$)$_2$CHNH$_2$]$_2$ · Cl$_2$(OH)$_2$ and their reduction products with PM2 DNA. *Biochemistry* **1984**, *23*, 4817–4820.

133 PENDYALA, L., ARAKALI, A. V., SANSONE, P., COWENS, J. W., CREAVEN, P. J. DNA binding of iproplatin and its divalent metabolite cis-dichloro-bis-isopropylamine platinum(II). *Cancer Chemother. Pharmacol.* **1990**, *27*, 248–250.

134 KELLAND, L. R., MURRER, B. A., ABEL, G., GIANDOMENICO, C. M., MISTRY, P., HARRAP, K. R. Ammine/Amine platinum(IV) dicarboxylates: A novel class of platinum complex exhibiting selective cytotoxicity to intrinsically cisplatin-resistant human ovarian carcinoma cell lines. *Cancer Res.* **1992**, *52*, 822–828.

135 CHOI, H. K., HUANG, S. K.-S., BAU, R. Octahedral complex of anticancer Pt(IV)(cyclohexyldiamine) agents with 9-methylguanine. *Biochem. Biophys. Res. Commun.* **1988**, *156*, 1125–1129.

136 ROAT, R. M., REEDIJK, J. Reaction of mer-trichloro(diethylenetriamine)-platinum(IV) chloride, (mer-[Pt(dien)Cl$_3$]Cl), with purine nucleosides and nucleotides results in formation of platinum(II) as well as platinum(IV) complexes. *J. Inorg. Biochem.* **1993**, *52*, 263–274.

137 NOVAKOVA, O., VRANA, O., KISELEVA, V. I., BRABEC, V. DNA interactions of antitumor platinum(IV) complexes. *Eur. J. Biochem.* **1995**, *228*, 616–624.

138 MELLISH, K. J., BARNARD, C. F. J., MURRER, B. A., KELLAND, L. R. DNA-binding properties of novel cis- and trans platinum-based anticancer agents in 2 human ovarian carcinoma cell lines. *Int. J. Cancer* **1995**, *62*, 717–723.

139 JOHNSON, N. P., MACQUET, J.-P., WIEBERS, J. L., MONSARRAT, B. Structures of adducts formed between [Pt(dien)Cl]Cl and DNA in vitro. *Nucleic Acids Res.* **1982**, *10*, 5255–5271.

140 VAN GARDEREN, C. J., VAN DEN ELST, H., VAN BOOM, J. H., REEDIJK, J., VAN HOUTE, L. P. A. A double-stranded DNA fragment shows a significant decrease in double-helix stability after binding of monofunctional platinum amine compounds. *J. Am. Chem. Soc.* **1989**, *111*, 4123–4125.

141 HOLLIS, L. S., AMUNDSEN, A. R., STERN, E. W. Chemical and biological properties of a new series of cis-diammineplatinum(II) antitumor agents containing three nitrogen donors: cis-[Pt(NH$_3$)$_2$(N-donor)Cl]$^+$. *J. Med. Chem.* **1989**, *32*, 128–136.

142 HOLLIS, L. S., SUNDQUIST, W. I., BURSTYN, J. N. *et al.* Mechanistic studies of a novel class of trisub-stituted platinum(II) antitumor agents. *Cancer Res.* **1991**, *51*, 1866–1875.

143 BAUER, C., PELEG-SHULMAN, T., GIBSON, D., WANG, A. H.-J. Monofunctional platinum amine complexes destabilize DNA significantly. *Eur. J. Biochem.* **1998**, *256*, 253–260.

144 PELEG-SHULMAN, T., KATZHENDLER, J., GIBSON, D. Effects of monofunctional platinum binding on the thermal stability and conformation of a self-complementary 22-mer. *J. Inorg. Biochem.* **2000**, *81*, 313–323.

145 TAYLOR, R. C., WARD, S. G. The antiviral activity of some selected inorganic and organometallic complexes. Possible new chemo-therapeutic strategies, in *Lectures in Bioinorganic Chemistry*, eds M. Nicolini and L. Sindellari, Raven Press, New York, 1991, 63–90.

146 COLUCCIA, M., BOCCARELLI, A., CERMELLI, C., PORTOLANI, M., NATILE, G. Platinum(II)-acyclovir complexes: synthesis, antiviral and antitumour activity. *Metal-Based Drugs* **1995**, *2*, 249–256.

147 BALCAROVA, Z., KASPARKOVA, J., ZAKOVSKA, A., NOVAKOVA, O., SIVO, M. F., NATILE, G., BRABEC, V. DNA interactions of a novel platinum drug, cis-[PtCl(NH$_3$)$_2$(N7-acyclovir)]$^+$. *Mol. Pharmacol.* **1998**, *53*, 846–855.

148 EASTMAN, A., JENNERWEIN, M. M., NAGEL, D. L. Characterization of bifunctional adducts produced in DNA by trans-diamminedichloro-platinum(II). *Chem.-Biol. Interactions* **1988**, *67*, 71–80.

149 BOUDVILLAIN, M., DALBIES, R., AUSSOURD, C., LENG, M. Intrastrand cross-links are not formed in the reaction between transplatin and native DNA: Relation with the clinical inefficiency of transplatin. *Nucleic Acids Res.* **1995**, *23*, 2381–2388.

150 BRABEC, V., SIP, M., LENG, M. DNA conformational distortion produced by site-specific interstrand cross-link of trans-diamminedichloroplatinum(II). *Biochemistry* **1993**, *32*, 11676–11681.

151 MALINGE, J.-M., LENG, M. Interstrand cross-links in cisplatin or transplatin-modified DNA, in *Cisplatin. Chemistry and Biochemistry of a Leading Anticancer Drug*, ed. B. Lippert, Verlag Helvetica Chimica Acta, Wiley-VCH, Zürich, Weinheim, 1999, 159–180.

152 PAQUET, F., BOUDVILLAIN, M., LANCELOT, G., LENG, M. NMR solution structure of a DNA dodecamer containing a transplatin interstrand GN7-CN3 cross-link. *Nucleic Acids Res.* **1999**, *27*, 4261–4268.

153 ANIN, M. F., LENG, M. Distortions induced in double-stranded oligonucleotides by the binding of cis-diamminedichloroplatinum(II) or trans-diamminedichloroplatinum(II) to the d(GTG) sequence. *Nucleic Acids Res.* **1990**, *18*, 4395–4400.

154 COLUCCIA, M., NASSII, F., LOSETO, F. *et al.* A trans-platinum complex showing higher antitumor activity than the cis congeners. *J. Med. Chem.* **1993**, *36*, 510–512.

155 COLUCCIA, M., BOCCARELLI, A., MARIGGIO, M. A., CARDELLICCHIO, N., CAPUTO, P., INTINI, F. P., NATILE, G. Platinum(II) complexes containing iminoethers: A trans platinum antitumour agent. *Chem.-Biol. Interactions* **1995**, *98*, 251–266.

156 BRABEC, V., VRÁNA, O., NOVÁKOVÁ, O. *et al.* DNA adducts of antitumor trans-[PtCl$_2$(E-imino ether)$_2$]. *Nucleic Acids Res.* **1996**, *24*, 336–341.

157 ZALUDOVA, R., ZAKOVSKA, A., KASPARKOVA, J. *et al.* DNA modifications by antitumor trans-[PtCl$_2$(E-iminoether)$_2$]. *Mol. Pharmacol.* **1997**, *52*, 354–361.

158 ANDERSEN, B., MARGIOTTA, N., COLUCCIA, M., NATILE, G., SLETTEN, E. Antitumor trans platinum DNA adducts: NMR and HPLC study of the interaction between a *trans*-Pt iminoether complex and the deoxy decamer d(CCTCGCTCTC). d(GAGAGCGAGG). *Metal-Based Drugs* **2000**, *7*, 23–32.

159 LENG, M., LOCKER, D., GIRAUD-PANIS, M. J. *et al.* Replacement of an NH$_3$ by an iminoether in transplatin makes an antitumor drug from an inactive compound. *Mol. Pharmacol.* **2000**, *58*, 1525–1535.

160 FARRELL, N. Current status of structure-activity relationships of platinum anticancer drugs: Activation of the trans geometry, in *Metal Ions in Biological Systems*, eds A. Sigel and H. Sigel, Marcel Dekker, New York, Basel, Hong Kong, 1996, 603–639.

161 ZAKOVSKA, A., NOVAKOVA, O., BALCAROVA, Z., BIERBACH, U., FARRELL, N., BRABEC, V. DNA interactions of antitumor trans-[PtCl$_2$(NH$_3$)(quinoline)]. *Eur. J. Biochem.* **1998**, *254*, 547–557.

162 BRABEC, V., NEPLECHOVA, K., KASPARKOVA, J., FARRELL, N. Steric control of DNA interstrand cross-link sites of trans platinum complexes: specificity can be dictated by planar nonleaving groups. *J. Biol. Inorg. Chem.* **2000**, *5*, 364–368.

163 PEREZ, J. M., MONTERO, E. I., GONZALEZ, A. M. *et al.* X-ray structure of cytotoxic trans-[PtCl$_2$(dimethyl-amine) (isopropylamine)]: Interstrand cross-link efficiency, DNA sequence specificity, and inhibition of the B-Z transition. *J. Med. Chem.* **2000**, *43*, 2411–2418.

164 PEREZ, J. M., MONTERO, E. I., GONZALEZ, A. M., ALVAREZ-VALDES, A., ALONSO, C., NAVARRO-RANNINGER, C. Apoptosis induction and inhibition of H-ras overexpression by novel trans-[PtCl$_2$(isopropylamine)(amine')] complexes. *J. Inorg. Biochem.* **1999**, *77*, 37–42.

165 MONTERO, E. I., DIAZ, S., GONZALEZ-VADILLO, A. M., PEREZ, J. M., ALONSO, C., NAVARRO-RANNINGER, C. Preparation and characterization of novel trans-[PtCl$_2$(amine) (isopropylamine)] compounds: Cytotoxic activity and apoptosis

induction in ras-transformed cells. *J. Med. Chem.* **1999**, *42*, 4264–4268.

166 CALVERT, P. M., HIGHLEY, M. S., HUGHES, A. N. *et al.* A phase I study of a novel, trinuclear, platinum analogue, BBR3464, in patients with advanced solid tumors. *Clin. Cancer Res.* **1999**, *5*, *Suppl. S*, 333.

167 SESSA, C., CAPRI, G., GIANNI, L. *et al.* Clinical and pharmacological phase I study with accelerated titration design of a daily times five scedule of BBR3464, a novel cationictriplatinum complex. *Ann. Oncol.* **2000**, *11*, 977–983.

168 CALVERT, A. H., THOMAS, H., COLOMBO, N. *et al.* Phase II clinical study of BBGR 3464, a novel, bifunctional analogue, in patients with advanced ovarian cancer. *Eur. Conf. Clin. Oncol.* **2001**, *October 21*.

169 SCAGLIOTTI, G., CRINÓ, L., DE MARINIS, F. *et al.* Phase II trial of BBR-3464, a novel, bifunctional platinum analogue, in advanced but favourable out-come, non-small cell lung cancer patients. *Eur. Conf. Clin. Oncol.* **2001**, *October 2001*.

170 FARRELL, N. Polynuclear charged platinum compounds as a new class of anticancer agents: Toward a new paradigm, in *Platinum-based Drugs in Cancer Therapy*, eds L. R. Kelland and N. P. Farrell, Humana Press, Totowa/NJ, 2000, 321–338.

171 FARRELL, N., QU, Y., BIERBACH, U., VALSECCHI, M., MENTA, E. Structure–activity relationship within di- and trinuclear platinum phase I clinical agents, in *Cisplatin. Chemistry and Biochemistry of a Leading Anticancer Drug*, ed. B. Lippert, VHCA, Wiley-VCH, Zurich, Weinheim, 1999, 479–496.

172 FARRELL, N., APPLETON, T. G., QU, Y. *et al.* Effects of geometric isomerism and ligand substitution in bifunctional dinuclear platinum complexes on binding properties and conformational changes in DNA. *Biochemistry* **1995**, *34*, 15480–15486.

173 ZALUDOVA, R., ZAKOVSKA, A., KASPARKOVA, J. *et al.* DNA interactions of bifunctional dinuclear platinum(II)

antitumor agents. *Eur. J. Biochem.* **1997**, *246*, 508–517.

174 KASPARKOVA, J., NOVAKOVA, O., VRANA, O., FARRELL, N., BRABEC, V. Effect of geometric isomerism in dinuclear platinum antitumor complexes on DNA interstrand cross-linking. *Biochemistry* **1999**, *38*, 10997–11005.

175 KASPARKOVA, J., MELLISH, K. J., QU, Y., BRABEC, V., FARRELL, N. Site-specific d(GpG) intrastrand cross-links formed by dinuclear platinum complexes. Bending and NMR studies. *Biochemistry* **1996**, *35*, 16705–16713.

176 MELLISH, K. J., QU, Y., SCARSDALE, N., FARRELL, N. Effect of geometric isomerism in dinuclear platinum antitumour complexes on the rate of formation and structure of intrastrand adducts with oligonucleotides. *Nucleic Acids Res.* **1997**, *25*, 1265–1271.

177 MARINI, V., KASPARKOVA, J., NOVAKOVA, O., MONSU-SCOLARO, L., ROMEO, R., BRABEC, V. Biophysical analysis of natural, double-stranded DNA modified by a dinuclear platinum(II) organometallic compound in a cell-free medium. *J. Biol. Inorg. Chem.* **2002**, *7*, 725–734.

178 KASPARKOVA, J., FARRELL, N., BRABEC, V. Sequence specificity, conformation, and recognition by HMG1 protein of major DNA interstrand cross-links of antitumor dinuclear platinum complexes. *J. Biol. Chem.* **2000**, *275*, 15789–15798.

179 COX, J. W., BERNERS-PRICE, S., DAVIES, M. S., QU, Y., FARRELL, N. Kinetic analysis of the stepwise formation of a long-range DNA interstrand cross-link by a dinuclear platinum antitumor complex: Evidence for aquated intermediates and formation of both kinetically and thermodynamically controlled conformers. *J. Am. Chem. Soc.* **2001**, *123*, 1316–1326.

180 HOFR, C., FARRELL, N., BRABEC, V. Thermodynamic properties of duplex DNA containing a site-specific d(GpG) intrastrand crosslink formed by an antitumor dinuclear platinum complex. *Nucleic Acids Res.* **2001**, *29*, 2034–2040.

181 Zou, Y., Vanhouten, B., Farrell, N. Sequence specificity of DNA-DNA interstrand cross-link formation by cisplatin and dinuclear platinum complexes. *Biochemistry* **1994**, *33*, 5404–5410.

182 McGregor, T. D., Hegmans, A., Kasparkova, J. *et al.* A comparison of DNA binding profiles of dinuclear platinum compounds with polyamine linkers and the trinuclear platinum phase II clinical agent BBR3464. *J. Biol. Inorg. Chem.* **2002**, *7*, 397–404.

183 Brabec, V., Kasparkova, J., Vrana, O. *et al.* DNA modifications by a novel bifunctional trinuclear platinum Phase I anticancer agent. *Biochemistry* **1999**, *38*, 6781–6790.

184 Zehnulova, J., Kasparkova, J., Farrell, N., Brabec, V. Conformation, recognition by high mobility group domain proteins, and nucleotide excision repair of DNA intrastrand cross-links of novel antitumor trinuclear platinum complex BBR3464. *J. Biol. Chem.* **2001**, *276*, 22191–22199.

185 Pratesi, P., Righetti, S. C., Supino, R. *et al.* High antitumor activity of a novel multinuclear platinum complex against cisplatin-resistant p53 mutant human tumors. *Br. J. Cancer* **1999**, *80*, 1912–1919.

186 He, Q., Ohndorf, U.-A., Lippard, S. J. Intercalating residues determine the mode of HMG1 domains A and B binding to cisplatin-modified DNA. *Biochemistry* **2000**, *39*, 14426–14435.

187 Rice, J. A., Crothers, D. M., Pinto, A. L., Lippard, S. J. The major adduct of the antitumor drug cis-diamminedichloroplatinum(II) with DNA bends the duplex by 40o toward the major groove. *Proc. Natl Acad. Sci. USA* **1988**, *85*, 4158–4161.

188 Bellon, S. F., Coleman, J. H., Lippard, S. J. DNA unwinding produced by site-specific intrastrand cross-links of the antitumor drug cis-diamminedichloroplatinum(II). *Biochemistry* **1991**, *30*, 8026–8035.

189 Butour, J. L., Macquet, J. P. Differentiation of DNA–platinum complexes by fluorescence. The use of an intercalating dye as a probe. *Eur. J. Biochem.* **1977**, *78*, 455–463.

190 Butour, J. L., Alvinerie, P., Souchard, J. P., Colson, P., Houssier, C., Johnson, N. P. Effect of the amine nonleaving group on the structure and stability of DNA complexes with cis-$[Pt(R-NH_2)_2(NO_3)_2]$. *Eur. J. Biochem.* **1991**, *202*, 975–980.

191 Takahara, P. M., Rosenzweig, A. C., Frederick, C. A., Lippard, S. J. Crystal structure of double-stranded DNA containing the major adduct of the anticancer drug cisplatin. *Nature* **1995**, *377*, 649–652.

9
Electrochemical Detection of DNA with Small Molecules

Shigeori Takenaka

9.1
Introduction

Small molecules that can bind to nucleic acids (DNA and RNA) often show a characteristic change in their absorption spectra upon complexation with nucleic acids [1, 2]. Studies on the interaction of small molecules with DNA have been based on these phenomena. Since most of these DNA-binding small molecules are also electrochemically active (reducible or oxidative, i.e. redox active), and their interaction with DNA can be studied electrochemically. Some examples of electrochemically active DNA-binding molecules are shown in Fig. 9.1. They are classified into the following groups: (1) intercalating molecules that can insert adjacent base pairs, and (2) groove binders such as organic dyes and metal complexes, which can bind to the groove of DNA. Nucleic bases also undergo redox reaction under certain "extreme" conditions and therefore an electrochemical response is obtained by proper selection of electrodes and pH of the electrolyte.

Here, the electrochemistry of nucleic acids is reviewed first and an electrochemical change accompanying the binding of an electrochemical ligand with DNA under conditions in which the DNA is inactive electrochemically will then be described. The ligands exhibiting a preference for double-stranded DNA can be used as hybridization indicators in DNA detection and can be applied to the electrochemical detection of target genes coupled with a DNA probe (a DNA fragment which can form a duplex with the desired DNA sequence) immobilized on the electrode. This system is called a DNA sensor or genosensor. Integration of a multiple DNA probe–immobilized electrode on a small surface is also discussed from the viewpoint of DNA microarray and chip technology. together with their applications.

9.2
Electrochemistry of Nucleic Acids

Since electrochemical analysis of DNA and RNA was reported for the first time by Berg in 1957 [3], many papers have been reported, mainly by Berg's and Palecek's

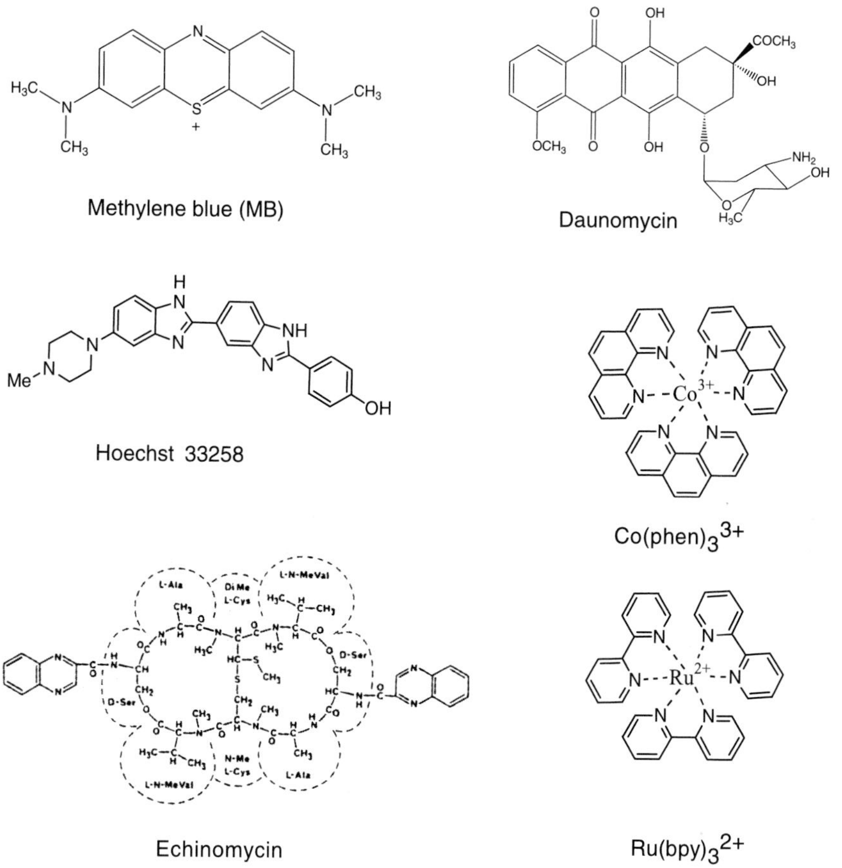

Methylene blue (MB)

Daunomycin

Hoechst 33258

Co(phen)$_3$$^{3+}$

Echinomycin

Ru(bpy)$_3$$^{2+}$

Fig. 9.1. Chemical structures of six electrochemically active DNA-binding ligands: methylene blue (MB), daunomycin, Hoechst 33258, echinomycin, Co(phen)$_3$$^{3+}$, and Ru(bpy)$_3$$^{2+}$.

groups [4–6]. DNA and RNA have electrochemically reactive parts on nucleic bases, some of which are illustrated in Fig. 9.2. The imino parts at N1–C6 of adenine, at N7–C8 of guanine, and at N3–C4 of cytosine are reducible by mercury electrodes, as these electrodes have a hydrogen excess potential and hence have a wider potential window in the reducing region. The N1–C6 of adenine and N3–C4 of cytosine are protected from redox reaction by engaging in hydrogen bonding of DNA duplex and therefore protonation at N1 of adenine and N3 of cytosine is necessary for their reduction. On the other hand, C8 of guanine and C2 of adenine are oxidized by graphite electrodes [8].

Differential pulse (d.p.) voltammetry of denatured DNA at pH 7.0 gives rise to two anodic peaks at +0.89 and +1.17 V resulting from guanine and adenine bases of DNA adsorbed on the surface of the electrode, respectively [9, 10]. In 1981, Brabec observed distinct electrochemical oxidation peaks for intact (double-

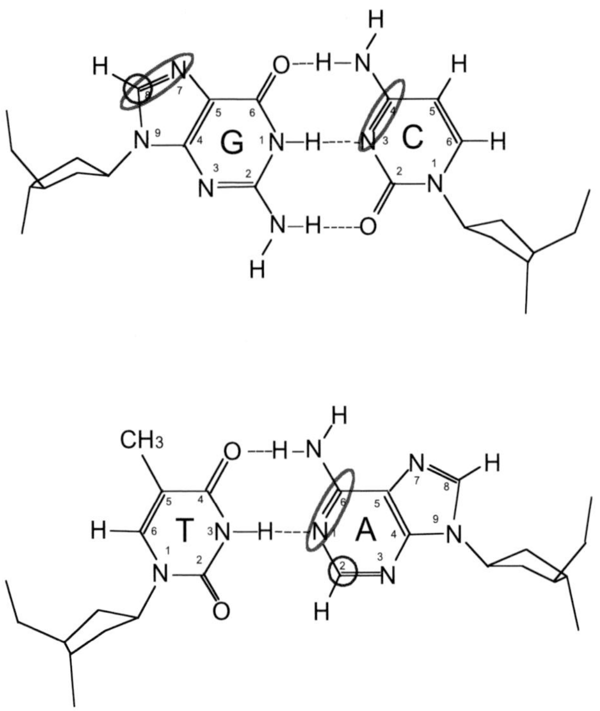

Fig. 9.2. Watson–Crick base pairs and their electroactive sites. Upper and lower sites of base pairs are located in major and minor groove, respectively. Reduction site: N7–C8 of guanine, N3–C4 of cytosine, and N1–C6 of adenine. Oxidation site: C8 of guanine and C2 of adenine.

stranded, ds) and denatured (single-stranded, ss) DNA [11]. The oxidation peak currents of native DNA were markedly smaller than those of denatured one. Since the guanine and adenine bases undergoing oxidation reaction are not involved in the hydrogen bonding of DNA duplex, these differences were ascribed to a difference in the mode of adsorption of nucleic bases of ssDNA and dsDNA on the surface of the graphite electrode; more flexible ssDNA should be adsorbed more and was susceptible to the oxidation reaction. Pang *et al.* reported the electrochemical oxidation of calf thymus DNA on a gold microelectrode using cyclic voltammetry [12]. An oxidation reaction of intact calf thymus DNA took place at +1.09 V, whereas that for denatured DNA took place at +1.16 V, having shifted toward the more anodic potential. The oxidation active bases do not take part in the hydrogen bonding of DNA duplex but they are masked deep in the groove of DNA duplex, making them inaccessible to oxidation reaction on the electrode surface. In all of the cases, a well-defined peak was observed only in the first scan because the oxidation product of the nucleic bases seems to remain adsorbed on the electrode surface, presumably making the electrode unresponsive.

Electrochemical reduction of DNA has been studied at neutral pH by direct current (d.c.) polarography [4, 6, 7]. In this case, ssDNA gave an electrochemical signal, whereas dsDNA was electrochemically inactive. This difference appears to be associated with different accessibility of bases in the two types of DNA, since the reduction sites are involved in hydrogen bonding with each other and away from the electrode. These results can be used for the discrimination of ssDNA from dsDNA by d.c. polarography [4, 5, 7]. Palecek and co-workers showed that the reduction peak of dsDNA is observed at potentials different from that of ssDNA [13]. The current intensity for ssDNA was only 1% that for dsDNA. The detection limit of this technique was 20 mg mL^{-1} of ssDNA. Since DNA can be adsorbed on the surface of mercury electrodes, DNA can be detected electrochemically after its concentration onto the surface. This technique is called adsorptive stripping (AdS) analysis. The combination of AdS with a.c. or cyclic voltammetry improved the sensitivity of detection to 1 ppm with calf thymus DNA [14]. Palecek then developed adsorptive transfer stripping voltammetry (AdtSV), in which the electrode is immersed in a solution of nucleic acid (5–10 mL) for a short period of time, washed, and transferred to an electrolyte solution without DNA [15]. A cathodic scan of this electrode gave reduction signals due to adenine and guanine.

9.3
DNA Labeling Through a Covalent Bond

Redox reactions of nucleic acids take place only at "extreme" potentials and are often accompanied by irreversible reactions. Palecek and co-workers found for the first time that osmium tetroxide-pyridine can bind covalently to the thymine base of ssDNA as depicted in Fig. 9.3 [5, 13]. This complex undergoes a reversible redox reaction at −0.2 V or −0.7 V on carbon and mercury electrodes, respectively. This was the first electrochemical marker ever known for DNA. Since then, electrochemical studies of nucleic bases modified by small molecules such as benzo[a]-pyrene–DNA adducts have been reported [16].

9.4
Electrochemistry of Metal Complexes Bound to DNA

Since DNA is a polyanion of phosphate esters, it can interact with various metal ions. Until now, the interaction of several metal ions with DNA has been reported

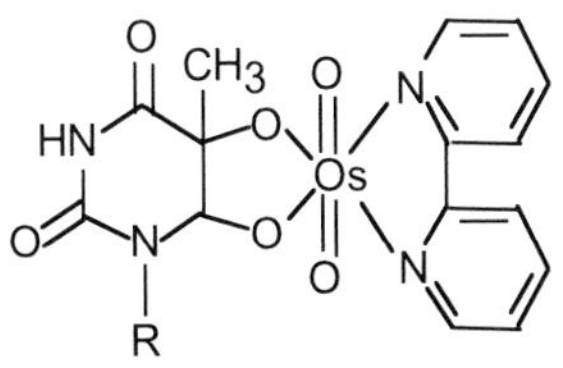

Fig. 9.3. Chemical structure of thymine modified by OsO$_4$ and bipyridyl.

together with their biological response [17]. For example, copper ion interacts with the phosphate anions or nucleic bases at low or high metal concentrations, respectively [18, 19]. Cis-diaminodichloroplatinum(II), an anticancer drug, is known as a crosslinking reagent for neighboring guanine bases [20, 21]. Metal complexes such as tris(1,10-phenanthrolinato)ruthenium chelate, $Ru(phen)_3{}^{2+}$, are known to bind to DNA reversibly. Especially, the optical isomers, Δ and Λ, of $Ru(phen)_3{}^{2+}$ can discriminate left- and right-handed DNA helices [22] and this discrimination was achieved electrochemically [23]. When bound to DNA, absorption and fluorescence spectra of $Ru(phen)_3{}^{2+}$ also changed significantly. However, monitoring of the interaction of tris(2,2'-bipyridyl)cobalt(III) chelate, $Co(bpy)_3{}^{3+}$, with DNA is difficult because of a small absorption change. In cases like this where spectrophotometric monitoring is difficult, an electrochemical method is particularly useful to study the DNA interaction with small molecules.

The metal complexes that can bind to DNA reported to date are summarized in Tab. 9.1. Bard and co-workers studied electrochemically the interaction of metal complexes with DNA in detail [24, 25]. Here, tris(1,10-phenanthrolinato)cobalt(III) chelate, $Co(phen)_3{}^{3+}$, is taken as an example. A cyclic voltammogram of $Co(phen)_3{}^{3+}$ in the absence and presence of calf thymus DNA with a glassy carbon electrode is shown in Fig. 9.4 [25]. A cathodic peak (E_{pc}) at 0.107 V and anodic peak (E_{pa}) at 0.173 V due to the reduction and oxidation reaction, respectively, are observed. The difference between both peak potentials is $\Delta E_p = 60$ mV, close enough to the theoretical value of 59 mV, indicating a diffusion-controlled reversible one electron transfer is taking place. Both peak potentials shifted toward the more anodic side of 0.120 and 0.182 V, concomitant with a decrease in peak currents in the presence of an excess of calf thymus DNA. The latter phenomenon can be explained by an increase in the apparent size of the molecule of $Co(phen)_3{}^{3+}$ bound to DNA duplex to result in a smaller diffusion rate. The peak currents in the ab-

Tab. 9.1. Electrochemically active metal complexes, which can bind to DNA.

Compound	References	Compound	References
$Co(phen)_3{}^{3+}$	24–28	$Os(phen)_2(dppz)^{2+}$	37, 41
$Fe(phen)_3{}^{2+}$	25, 29	$Os(phen)_3{}^{2+}$	41–44
$Co(bpy)_3{}^{3+}$	25, 30–34, 41	$(tpy)(bpy)ROH_2{}^{2+}$	41
$Fe(bpy)_3{}^{2+}$	25, 29	$(tpy)(phen)RuOH_2{}^{2+}$	41
$Ru(phen)_3{}^{2+}$	23, 28, 35, 36	$Mn(TMPyP)^{3+}$	45
$Ru(bpy)_3{}^{2+}$	37, 38	$Fe(TMPyP)^{3+}$	45
$Cu(phen)_2{}^{2+}$	29	TMAP	46
$Ru(NH_3)_6{}^{3+}$	33, 39	Hemine	47
$Os(bpy)_3{}^{2+}$	27, 40, 41	$Ir(bpy)(phen)(phi)^{3+}$	39

phen, 1,10-phenanthrolinato; bpy, 2,2'-bipyrididyl; dppz, dipyrido-phenazine; tpy, 2,2',2''-terpyridine; TMPyP, 5,10,15,20-tetra-(N-methyl-4-pyridyl)-porphyrin; TMAP, meso-tetra(p-N-trimethylanilinium)porphyrin; phi, 9,10-phenanthrenequinone diimine.

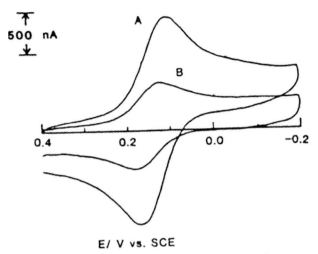

Fig. 9.4. Cyclic voltammogram of 1.0×10^{-4} M Co(phen)$_3{}^{3+}$ in the absence (A) and presence (B) of 5.0 mM calf thymus DNA in 5 mM Tris–HCl, 50 mM NaCl, pH 7.1. Sweep rate 100 mV s^{-1}.

sence and presence of DNA duplex were proportional to the square root of scan rate, $v^{1/2}$, indicating that a diffusion-controlled electrochemical reaction is taking place. From these plots diffusion coefficients $D_f = 5.0 \ (\pm 0.6) \times 10^{-6}$ cm^2 s^{-1} and $D_b = 3.2 \ (\pm 1.2) \times 10^{-7}$ cm^2 s^{-1} were obtained for Co(phen)$_3{}^{3+}$ alone and that bound to DNA, respectively.

It is manifest that the diffusion rate of Co(phen)$_3{}^{3+}$ decreased to about 1/20th upon binding to the DNA duplex. Bard and co-workers verified that the current decrease is not due to a decrease in the surface area of the electrode upon binding of the complex by showing that the CV of a mixture of Co(phen)$_3{}^{3+}$ and octa-cyanomolybdate(IV), Mo(CN)$_8{}^{4-}$, was the same, irrespective of the absence and presence of a DNA duplex [25]. The peak potential and current of Mo(CN)$_8{}^{4-}$ do not change in the presence of DNA duplex, whereas those of Co(phen)$_3{}^{3+}$ change in the presence of DNA duplex.

On the other hand, the current shift in the presence of DNA reflects the strength of the interaction of reduced or oxidized species of Co(phen)$_3{}^{3+}$ with DNA. The thermodynamic cycle in the DNA binding process of these species is shown in Scheme 9.1.

This one electron redox process is expressed by the following modified Nernst equation:

$$E_b{}^{o'} - E_f{}^{o'} = 0.059 \log(K_{2+}/K_{3+}) \tag{1}$$

where K_{2+} and K_{3+} represent the binding constants for the reduced (M^{2+}) and oxidized (M^{3+}) species of Co(phen)$_3{}^{3+}$, respectively. In other words, the difference

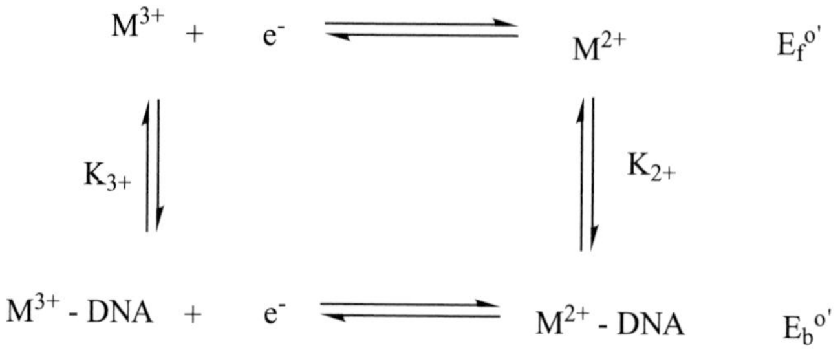

$$M^{3+} + e^- \rightleftharpoons M^{2+} \qquad E_f^{o'}$$

$$\Big\updownarrow K_{3+} \qquad\qquad\qquad \Big\updownarrow K_{2+}$$

$$M^{3+}\text{-DNA} + e^- \rightleftharpoons M^{2+}\text{-DNA} \qquad E_b^{o'}$$

Scheme 9.1

between the redox potentials of the bound and free species represents the binding ratio of the oxidized to reduced species. In the case of $Co(phen)_3^{3+}$, the current shift in the presence of DNA was +17 mV and $K_{2+}/K_{3+} = 1.74$, implying that the binding of $Co(phen)_3^{2+}$ is nearly twice greater than that of $Co(phen)_3^{3+}$. Bard and co-workers concluded that the hydrophobic interaction contributes more than the electrostatic one in the binding of $Co(phen)_3^{3+}$ with DNA.

Next, the determination of binding constants by electrochemical treatments will be described. As DNA is added, the current decreases to a constant value, which represents that for the bound form. Since the concentration of the ligand bound to DNA, C_b, can be estimated from this value, one can obtain the binding constant, K, from the following Scatchard-type equation:

$$C_b = b - \{b^2 - (2K^2 C_t[DNA])/s\}^{1/2}/2K$$
$$b = 1 + KC_t + K[DNA]/2s \qquad (2)$$

where s is the site size in base pairs excluded by the bound ligand complex and C_t is the total concentration of ligand. If it is assumed that the electron transfer occurs in a reversible, diffusion-controlled manner, two limiting cases may be described for the current in CV from Eq. (2). When the bound and free forms of the ligand cannot interconvert to each other on the time scale of electrochemical measurements (static, S), the square root of the diffusion coefficient ($D^{1/2}$) is simply the sum of the free and bound forms of the ligand weighed by the individual mole fractions and the equation is rearranged into

$$i_{pc} = B(D_f^{1/2} C_f + D_b^{1/2} C_b) \qquad (3)$$

If the complexes interconvert rapidly on the experimental time scale, $D^{1/2}$ is the mole-fraction-weighed time average of free and bound values (mobile, M) and the equation is rearranged into

$$i_{pc} = BC_t(D_f C_f/C_t + D_b C_b/C_t)^{1/2} \qquad (4)$$

where C_f is the concentration of the free ligand. For Nernstian reaction in CV at 25°C, $B = 2.69 \times 10^5 \, n^{3/2} \, A \, v^{1/2}$, where n is the number of electrons transferred per metal complex and A is the surface area of the electrode. Using this procedure, the binding constant of Co(phen)$_3$$^{3+}$ with calf thymus DNA was determined as $K = 1.6\ (\pm 0.2) \times 10^4\ M^{-1}$ (S, $s = 6$ bp) or $2.6\ (\pm 0.4) \times 10^4\ M^{-1}$ (M, $s = 5$ bp) [25].

9.5
Electrochemistry of DNA-binding Small Molecules

Most of the DNA-binding small molecules reported to date (Tab. 9.2) are more or less electrochemically active, exhibiting redox activity at cathodic potentials. Natural products carrying an anthraquinone skeleton were the first to be studied as this kind of molecules and daunomycin is the ligand used most often for electro-

Tab. 9.2. Electrochemically active small molecules, which can bind to DNA.

Compound	References	Compound	References
Daunorubicin	48, 49	Ethidium homodimer	51
Duborimycin	48	Ethidium monoazide	51
RP33921	48	Chlortetracycline	51
RP21080	48	Tetracycline	51
Daunomycin	39, 50–57	Deoxycline	51
Adriamycin	50, 57, 58	Minocycline	51
4'-Deoxyadriamycin	57	Hoechst 33258	51, 63, 79
Iremycin	50	Hoechst 33342	51
Carminomycin	50	7-Aminoactinomycin D	51
N,N'-Dibenzyl	50	Chromomycin A	51
daunomycin		Olivomycin	51
Steffimycin	50	Vinblastine	51
3'-Hydroxydaunomycin	50	Propidium iodide	51
Musetlamycin	50	Quinacrine mustard	51
Violamycin BI	50	DAPI	51
component		Rifampicin	51
MA144-S1	50	Stains-all	51
Nogalamycin	50, 57	Phenothiazine	54
Aclacinomycin	50	Chlorpromazine	64
Cinerubin B	50	Promethazine	64
Maecellomycin	50	Mitoxantrone	65
Neutral red	50	Khellin	66
Crystal violet	59	Benzyl viologene	67
Methylene blue (MB)	39, 59–61	Echinomycin	4, 68
9-Aminoacridine	28, 51	Chloroquine	28
Acridine orange	51	Epirubicin	69
Aclarubicin	51	Bis-acridine carrying	70–73
Doxorubicin	28, 51	viologen unit(s)	
Pirarubicin	51	FND-1	74, 75
Ethidium bromide	46, 51, 62	FND-2	76

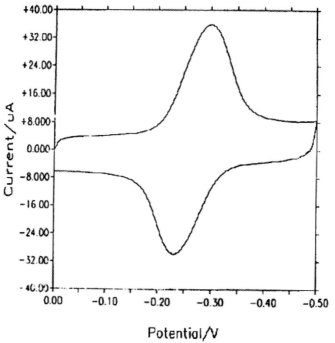

Fig. 9.5. Cyclic voltammetry of 1.0 μM methylene blue (MB) in 50 mM phosphate (pH 7) on a DNA-modified electrode (area = 0.7 cm^2). Scan rate 100 mV s^{-1}.

chemical studies. The molecules carrying an acridine or ethidium skeleton also are electrochemically active, but they often undergo irreversible reactions. Methylene blue (MB) is known to undergo a reversible redox reaction and was studied in detail by Barton and co-workers [60, 88]. Figure 9.5 shows a cyclic voltammogram of MB on a DNA immobilized gold electrode in 50 mM phosphate (pH 7). Peak potentials due to the redox reaction of MB were observed at −0.25 V and −0.30 (versus SCE) with ΔE_p = 0.05 V. The reduced form of MB is known as leuco-MB. The binding constant and its difference between the reduced and oxidized forms of such a kind of molecules can be estimated according to the treatment analogous to that for DNA-binding metal complexes. The redox reaction occurs on the same site of these molecules as that of their binding (interacting site).

On the other hand, Takenaka and co-workers synthesized a bis-acridine derivative carrying a viologen linker as a reversible electrochemically active moiety at two acridine parts [70] (Fig. 9.6). Because of the bis-intercalating character, this molecule was expected to exhibit higher affinity for the DNA duplex. Electrochemical studies of the molecule revealed that the reduced form of the molecule possesses higher affinity than the oxidized one, suggesting that the electrostatic interaction is the major contributor in the DNA binding. It was also suggested that its viologen parts take part actively in the DNA interaction process. This notion that the viol-

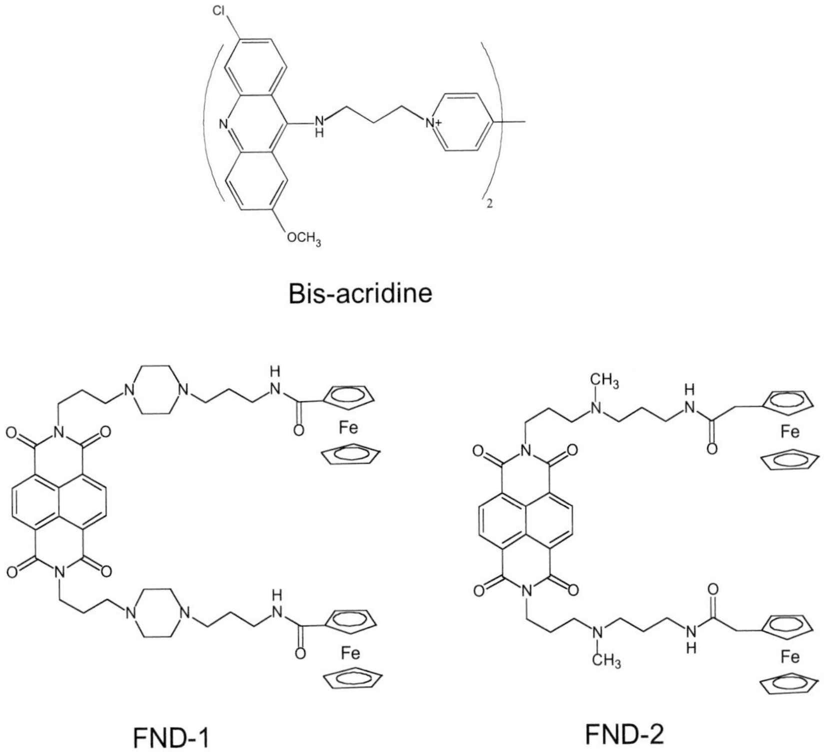

Bis-acridine

FND-1   FND-2

Fig. 9.6. Chemical structures of bis-acridine and threading intercalators FND-1 and FND-2 carrying electrochemical signal parts.

ogen interacts with the DNA groove was supported by recent electrochemical studies of dibenzylviologen [67].

Recently, Palecek and co-workers reported an electrochemical study of echino-mycin as bis-intercalator in a hope of improving affinity for the DNA duplex (see below) [4, 68]. The quinoxaline parts of echnomycin are electrochemically active. Takenaka and co-workers also succeeded in synthesizing mixed ligands carrying multiple viologen units [72, 73] and the acridinyl viologen connected through a different length of alkyl chain [71]. These bis-intercalators underwent a reversible redox reaction and showed slight preference for AT-rich DNA duplexes.

9.6
DNA Sensor Based on an Electrochemically Active DNA-binding Molecule as a Hybridization Indicator

As described above, the diffusion of small molecules is retarded upon binding to DNA in solution to result in a decrease in the redox current. In addition, the peak

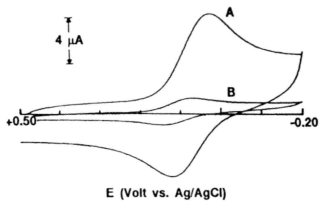

E (Volt vs. Ag/AgCl)

Fig. 9.7. Cyclic voltammogram of 1.6×10^4 M^{-1} Co(phen)$_3{}^{3+}$ at a glassy carbon electrode alone or modified with [poly(dG-dC)]$_2$.

potential also shifted upon binding to DNA, suggesting that the microenvironment around the molecule has changed. In contrast, where DNA is immobilized on the electrode, the molecule is concentrated on the surface of the electrode and the current intensity increases accordingly. Such an example is shown in Fig. 9.7 in a cyclic voltammogram for a glassy carbon electrode carrying [poly(dG-dC)]$_2$ in an electrolyte containing Co(phen)$_3{}^{3+}$ [31]. The target DNA can be detected by using such molecules coupled with a DNA-immobilized electrode. The DNA sensor or genosensor shown in Fig. 9.8 is based on this principle [77, 78].

First, a DNA probe, which is a denatured DNA fragment capable of forming a duplex with the complementary gene sequence, is immobilized on the surface of an electrode. A double-strand formation (hybridization) reaction of sample DNA with the DNA probe is conducted on the surface of the electrode. Second, the electrode is dipped in an electrolyte containing a hybridization indicator and then electrochemical measurements are made. The amount of molecule bound on the electrode and the amount of DNA duplex formed are estimated from the signal intensity. In this sense, the molecule is regarded as a transducer to covert the DNA duplex formation to an electrochemical signal and can be called a hybridization indicator. In this technology, the selectivity of the indicator for dsDNA is very important to determine the sensitivity of a DNA sensor; if the indicator binds to ssDNA, the DNA probe-immobilized electrode gives rise to an electrochemical signal even in the absence of target DNA, which results in a background noise.

So far, several groups have developed an electrochemical genosensor. Ishimori and co-worker used Hoechst 33258 (Fig. 9.1) as a hybridization indicator [63]. Since this ligand binds to the groove of the DNA duplex, it was expected to have a low affinity for ssDNA. Wang and co-workers achieved highly sensitive detection of target DNA using Co(phen)$_3{}^{3+}$ coupled with a DNA probe-immobilized electrode by a potentiometric stripping analysis (PSA) technique [78]. Marrazza and co-workers detected PCR products containing a mismatched sequence using dau-

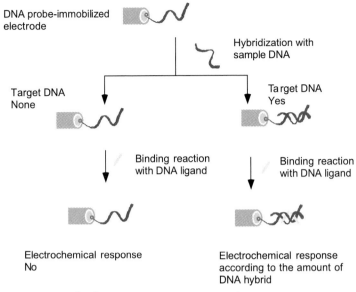

Fig. 9.8. Principle of a DNA sensor or genosensor using a ligand specific for double-stranded DNA coupled with a DNA probe immobilized on the electrode.

nomycin by the PSA technique [56]. However, small molecules such as metal complexes, intercalators, and groove binders can bind to ssDNA as well due to electrostatic and hydrophobic or stacking interactions.

Bis-acridine carrying a viologen unit was synthesized to improve the affinity for dsDNA [70]. Echinomycin is an example; its cyclic voltammogram with a mercury electrode carrying single or double-stranded DNA in an electrolyte is shown in Fig. 9.9 [4, 68]. A peak current due to the quinoxaline rings of echinomycin was observed for the dsDNA-immobilized mercury electrode at a cathodic potential. Both

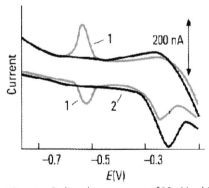

Fig. 9.9. Cyclic voltammograms of 10 μM echinomycin before immersion of native (2) and denatured (1) DNA at a hanging drop mercury electrode.

bis-intercalators undergo redox reactions at cathodic potentials, where the reduction of dissolved oxygen overlaps, hampering sensitive detection of target DNA.

Takenaka and co-workers synthesized ferrocenyl naphthalene diimide [74–77] FND-1 and -2 (Fig. 9.6) as a hybridization indicator with a strong preference for dsDNA and reversible redox response at an anodic potential. When naphthalene diimide binds to the DNA duplex, both of its substituents project out in the major and minor grooves simultaneously. These substituents serve as an anchor to prevent the complex from dissociation, and as a result a very stable complex of naphthalene diimide with DNA duplex is formed. In contrast, the complex of naphthalene diimide with ssDNA is expected to be less stable, as no such a stabilizing factor is present. When the binding process of naphthalene diimide to a DNA duplex is envisaged further, it is easily seen that one of the substituents should penetrate adjacent base pairs. These types of ligand are thus called threading intercalators.

On the basis of these considerations, threading intercalators were expected to be highly selective indicators for dsDNA. FND-1 was designed in such a way that ferrocene moieties are connected to the terminus of a substituent on naphthalene diimide. FND-1 should intercalate between adjacent base pairs with the ferrocene moieties projecting in the major and minor grooves of DNA duplex, with the naphthalene diimide and ferrocene parts acting as a threading intercalator and reversible redox reaction parts at anodic potentials, respectively. Although the discrimination of double-stranded from single-stranded DNA by FND-1 is not complete, because it still binds to ssDNA through electrostatic and stacking interactions, the selectivity for the two types of DNA has been improved significantly.

Thiolated oligonucleotide dT_{20} was immobilized on a gold electrode through a thiol–gold linkage as a DNA probe and the resulting electrode was allowed to hybridize with complementary dA_{20} or non-complementary dT_{20} [75]. After dipping in an electrolyte containing FND-1, the electrode was transferred to another electrolyte without FND-1 and a cyclic voltammogram was measured. As shown in Fig. 9.10, a peak current was obtained only for the complementary oligonucleotide with the bare electrode, whereas the signal intensity for the non-complementary DNA was barely above background. Differential pulse voltammetry was applied to this measurement to suppress background due to the capacitance component and Takenaka and co-workers succeeded in detecting plasmid DNA carrying a target gene at the femtomole level [75]. They also developed FND-2 undergoing a redox reaction at a more negative potential [76]. It was also attempted from an environmental viewpoint, using DNA-immobilized electrodes to detect DNA-binding molecules present in nature [80].

9.7
Mismatched DNA Detection by Hybridization Indicator

There is a diversity by one nucleobase in individual genes. This phenomenon, called single nucleotide polymorphism (SNP), is the cause of the susceptibility of

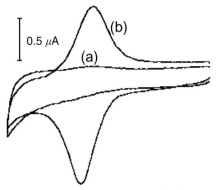

Fig. 9.10. Cyclic voltammograms of a dT_{20}-immobilized gold electrode after hybridization with complementary dA_{20} (b) and non-complementary dT_{20} (a) in an electrolyte containing FND-1.

individuals to drugs and diseases. It is thus important to know about SNPs in advance in order to diagnose and prevent various syndromes. When a DNA probe carrying a certain SNP site is allowed to hybridize with sample DNA, a fully matched or mismatched DNA duplex is formed. Takenaka and co-workers developed a mismatched base detection system based on FND coupled with a DNA probe-immobilized gold electrode, the principle of which is illustrated in Fig. 9.11 [82, 83]. The DNA probe is immobilized on the electrode by the sulfur–gold linkage. This method of immobilization is often used because of the ease of controlling the amount of DNA probe immobilized on the electrode.

Electrochemical measurements are made before and after hybridization with denatured sample DNA. First, the d.p. voltammetry of the DNA probe-immobilized electrode is measured in an electrolyte containing FND, and the peak current obtained (i_o) is used to estimate the amount of DNA immobilized on the electrode. Since the binding of FND with ssDNA is proportional to the number of phosphate anions, the peak current should also be proportional to the amount of DNA probe on the surface of the electrode (the diffusion current of FND-1 is also included in this current). After hybridization with sample DNA, d.p. voltammetry is measured and the peak current i is obtained. The current shift, $\Delta i = (i/i_o - 1) \times 100$ is used for the evaluation of the amount of dsDNA per ssDNA immobilized on the electrode. Under conditions where fully matched and mismatched sample DNA can form a double strand, the Δi value for the mismatched DNA duplex is smaller than that for the fully matched DNA duplex due to the fact that the binding of FND around the mismatched base area is impaired. It was found from other experiments that one mismatched base results in loss of two to three FND molecules otherwise bound there.

One of the advantages of this method lies in the fact that it is not necessary to set the hybridization temperature rigorously, in contrast to the conventional hybrid-

(a)

(b)

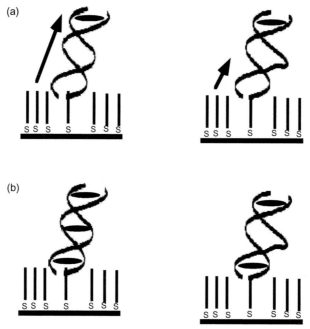

Fig. 9.11. Electrochemical detection of a single base mismatch based on charge transduction through DNA (a) and by reducing the amount of FND-1 bound to mismatched DNA duplex (b).

ization methods where a proper setting of the temperature is essential. Barton and co-workers established a mismatched DNA detection technique using an electron transfer reaction through a π-stack of nucleic bases of DNA duplex [85] (Fig. 9.11). They prepared a gold electrode carrying DNA duplex to high-density (DNA film) [61, 86]. Methylene blue (MB) binds to only the surface of this film and therefore electron transfer reaction occurs between MB and the electrode. Where a mismatched base(s) exists, the efficiency of electron transfer is reduced and hence the existence of a mismatched base on the DNA duplex can be estimated from a change of the amount of their electric charge.

Thus far, discrimination of mismatched bases has been based on the difference in the stability of mismatched and fully matched DNA duplex and hence stringent control of temperatures for hybridization and subsequent washing has been critical. However, it is often difficult to find a common optimum condition for fully matched and mismatched DNA duplexes to discriminate the two. In this sense, these two techniques described above are especially useful in mismatched base detection, as they do not rely on the stability of the DNA duplex.

Another mismatched base detection was achieved by Wang's group using peptide nucleic acid (PNA), in which a "duplex" of PNA with DNA is formed more easily than the DNA/DNA duplex [87].

9.8
DNA-detecting System using Hybridization Indicator as a Mediator

The redox reaction of ferricyanide–ferrocyanide, an anionic marker, is blocked on a DNA film prepared by Barton's group because of electrostatic repulsion between the anionic marker and DNA. When MB binds to the surface of the DNA film, the anionic marker can now gain access to MB bound on the DNA film and electron transfer between MB and the electrode can be enhanced by the redox reaction of anionic marker. Barton's group developed a highly sensitive detecting system of mismatched DNA duplex using this electrocatalytic reduction system [85, 88]. The complex of FND with DNA is regarded as a pseudo-polyferrocene coating the DNA duplex because of the intercalator's binding mode (nearest exclusion model), and hence an electron transfer through this array was envisaged. In fact, such an electron transfer was observed between the electrode and the reduced glucose oxidase generated from the reaction of glucose oxidase and glucose [89]. This result suggested that the electrochemical signal is enhanced further upon DNA duplex formation in this system. The electron transfer occurs through the ferrocene array covering the groove of DNA duplex in Takenaka's system, whereas it occurs through the π-stacked base pairs of DNA duplex in Barton's system.

Tris(2,2'-bipyridine)ruthenium chelate, $Ru(bpy)_3{}^{3+}$, described above, can mediate the oxidation reaction of guanine bases:

$$Ru(bpy)_3{}^{2+} \rightarrow Ru(bpy)_3{}^{3+} + e^-$$

$$Ru(bpy)_3{}^{3+} + guanine \rightarrow Ru(bpy)_3{}^{2+} + (guanine)_{ox}$$

Thorp and co-workers studied the mediation reaction of unpaired guanine bases on DNA duplex through $Ru(bpy)_3{}^{2+}$ using an indium tin oxide (ITO) electrode [38, 90]. Figure 9.12 shows an example of the mediation current of $Ru(bpy)_3{}^{2+}$ for calf thymus DNA. They proved that the mediation of $Ru(bpy)_3{}^{2+}$ derives from the guanine exposed to the solvent on the basis of their observation that this mediation current depended upon the amount of unpaired guanine bases. They developed a DNA-sensing system using a DNA probe-immobilized ITO electrode, on which the mediated current of $Ru(bpy)_3{}^{2+}$ increased with an increase in the amount of target DNA. The magnitude of the mediation current was relatively constant with several DNA samples because of the uniformity of the GC content of natural DNA used.

9.9
Application to DNA Microarray

Recently, DNA microarrays are receiving attention as a means of high-throughput analysis of DNA. This technique, developed by Brown of Stanford University, consists of a range of procedures [91]. Many different probe DNAs are spotted for immobilization on a glass plate with a microarrayer (preparation of a DNA microarray). A labeled DNA sample is allowed to hybridize on the microarray plate. After

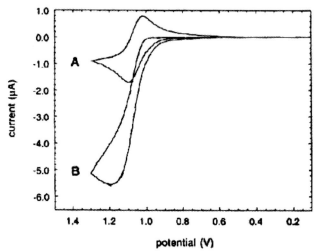

Fig. 9.12. Cyclic voltammograms of $Ru(bpy)_3^{2+}$ in the absence (A) and presence (B) of calf thymus DNA at 25 mV s^{-1} in 700 mM NaCl/50 mM sodium phosphate buffer, pH 6.8.

washing the plate, the sample DNA bound on the plate is detected by the label's signal. Affymetrix Inc. (USA) developed a highly condensed DNA array called the DNA chip by oligonucleotide synthesis on the silicon surface by photolithography using amidate reagents carrying a photoreactive protecting group [91]. An electrochemical integrated multi-electrode system has been developed using the application of the DNA sensor technique. Xanthon Inc. (USA) constructed a multi-ITO electrode on the well of titer plates and applied it to the mediated DNA-detecting system using $Ru(bpy)_3^{2+}$ [92]. TUM Inc. (Japan) developed a multi-gold electrode called electrochemical array (ECA), in which many DNA probes were immobilized on a gold electrode and the hybridization events were detected by FND-1 as a hybridization indicator [81]. Takenaka and co-workers tried to visualize a DNA microarray by scanning electrochemical microscopy (SECM) [93]. When a DNA microarray was dipped in an electrolyte containing FND-1, the ligand was concentrated in the DNA duplex region on the glass plate. The reason for the enhanced SECM signal is that the DNA duplex regions of the spots serve as a conductive polymer. Since this system can be used as an ordinary DNA microarrayer and a fluorescent dye is unnecessary to immobilize and detect the DNA duplex region, it is expected to emerge as the DNA microarray technology of the future.

9.10
Conclusion

As described above, DNA, which is electrochemically inactive under ordinary conditions, can be analyzed electrochemically by using electrochemically active DNA-binding ligands. This method is a sort of a DNA sensor or genosensor based on the use of DNA-binding ligands as a hybridization indicator. This novel genosensor enables not only general DNA analysis but also mismatch detection including analysis of single nucleotide polymorphisms (SNPs). It is expected that this novel methodology will be seen as a breakthrough in the development of highly sensitive, quick, inexpensive, and compact analytical devices for target DNA. In light of an increased interest in DNA chip technology, research on electrochemical chips based on integrated multi-electrodes is being carried out intensively. The fruits of this research, once achieved, will exert a strong impact or even revolutionize diagnosis in medicine in the future.

9.11
Summary

Electrochemical methods have proved to be an excellent technique for the study of interactions of small molecules with DNA and RNA, especially where the molecules are spectrophotometrically silent. Electrochemical molecules called hybridization indicators can be used in DNA sensors or genosensors coupled with a DNA immobilized electrode. In this regard, hybridization indicators with high discrimination ability for double-stranded DNA (hybrid with target DNA) over single-stranded DNA (unreacted DNA probe) are required. Bis-intercalators and threading intercalators seem to be promising candidates for this. Mismatched DNA detection using the electron transfer reaction through a π-stack of DNA duplex or relying on an decrease in the amount of the bound intercalator per mismatched DNA duplex has provided a new technology for single nucleotide polymorphism (SNP) analysis. Electrochemical array (ECA) and its homolog DNA chip have also been developed for electrochemical DNA sensing. Visualization of DNA microarrays on an ordinary glass plate was realized by ferrocenyl naphthalene diimide as threading intercalator coupled with scanning electrochemical microscopy (SECM).

Acknowledgments

The author is grateful to all of his co-workers whose names appear in the papers cited in references for their enthusiasm for research. Special thanks are due to Professors Makoto Takagi of Kyushu University and Hiroki Kondo of Kyushu Institute of Technology for advice and helpful discussions.

References

1 WILSON, W. D., JONES, R. J. Intercation in biological systems, in *Intercalation Chemistry*, eds M. S. Whittingham and A. J. Jacobson, Academic Press, New York, 1982, 445–501.

2 LOBER, G. The fluorescence of dye-nucleic acid complexes. *J. Luminesc.* 1981, 22, 221–265.

3 BERG, H. Polarographic studies on nucleoc acids and nuceleases. I. Polarographic determination of

proteins in the presence of nucleic acids. *Biochem. Z.* **1957**, *329*, 274–276.

4 BERG, H. Electrochemistry of biopolymers, in *Comprehensive Treatise of Electrochemistry*, Vol. 10, eds S. Srinivasan, Y. A. Chizmadzhev, J. O'M. Bockris, B. E. Comway and E. Yeager, Plenum Press, New York, 1985, 189–229.

5 PALECEK, E., FOJTA, M. Detecting DNA hybridization and damage. *Anal. Chem.* **2001**, *73*, 75A–83A.

6 PALECEK, E. Past, present and future of nucleic acids electrochemistry. *Talanta* **2002**, *56*, 809–819.

7 PALANTI, S., MARRAZZA, G., MASCINI, M. Electrochemical DNA probes. *Anal. Lett.* **1996**, *29*, 2309–2332.

8 ELVING, P. J., SMITH, D. L. The graphite electrode. An improved technique for voltammetry and chronopotentiometry. *Anal. Chem.* **1960**, *32*, 1849–1854.

9 DRYHURST, G., ELVING, P. J. Electrochemical oxidation of adenine: reaction products and mechanisms. *J. Electrochem. Soc.* **1968**, *115*, 1014–1020.

10 DRYHURST, G., PACE, G. F. Electrochemical oxidation of guanine at the pyrolytic graphte electrode. *J. Electrochem. Soc.* **1970**, *117*, 1259–1264.

11 BRABEC, V., SCHINDLEROVA, I. Electrochemical behavior of proteins at praphite electrodes. Part III. The effect of protein adsorption. *Boielectrochem. Bioenerg.* **1981**, *8*, 437–449.

12 PANG, D.-W., QI, Y. P., WANG, Z-L., CHENG, J.-K., WANG, J.-W. Electrochemical oxidation of DNA at a gold microelectrode. *Electroanalysis* **1995**, *7*, 774–777.

13 PALECEK, E. Electrochemical behavior of biological macromolecules. *Bioelectrochem. Bioenerg.* **1986** *15*, 275–295.

14 PALECEK, E. From polarography of DNA to microanalysis with nucleic acid-modified electrodes. *Electroanalysis* **1996**, *8*, 7–14.

15 PALECEK, E., JELEN, F., TEIJEIRO, C. Biopolymer-modified electrodes in the voltammetric determination of nucleic acids and proteins at the submicrogram level. *Anal. Chim. Acta* **1993**, *273*, 175–186.

16 KERMAN, K., MERIC, B., OZHAN, D., KARA, P., ERDEM, A., OZSOZ, M. Electrochemical DNA biosensor for the determination of benzo[a]pyrene-DNA adducts. *Anal. Chim. Acta* **2001**, *450*, 45–52.

17 BREGADZE, V. G. Metal ion interactions with DNA: considerations on structure, stability, and effects from metal ion binding. *Metal Ions Biol. Syst.* **1996**, *32*, 419–451.

18 LIEBE, D. C., STUEHR, J. E. Copper(II)-DNA denaturation. I. Concentration dependence of melting temperature and thermal relaxation time. *Biopolymers* **1972**, *11*, 145–166.

19 LIEBE, D. C., STUEHR, J. E. Copper(II)-DNA denaturation. II. The model of DNA denaturation. *Biopolymers* **1972**, *11*, 167–184.

20 BOSENBERG, B., TROSKO, J. E., MANSOUR, V. H. Platinum compounds: a new class of potent antitumour agents. *Nature* **1969**, 385–386.

21 TAKAHARA, P. M., PATRICIA, M., ROSENZWEIG, A. C., FREDERICK, C. A., LIPPARD, S. J. Crystal structure of double-stranded DNA containing the major adduct of the anticancer drug cisplatin. *Nature* **1995**, *377*, 649–652.

22 BARTON, J. K., DANISHEFSKY, A. T., GOLDBERG, J. M. Tris(phenanthroline) ruthenium(II) : Steroselectivity in binding to DNA. *J. Am. Chem. Soc.* **1984**, *206*, 2172–2176.

23 MAHADEVAN, S., PALANIANDAVAR, M. Chiral discrimination in the binding of tris(phenanthroline)ruthenium(II) to calf thymus DNA: An electro-chemical study. *Bioconjug. Chem.* **1996**, *7*, 138–143.

24 CARTER, M. T., BARD, A. J. Voltam-metric studies of the interaction of Tris(1,10-phenanthroline)cobalt (III) with DNA. *J. Am. Chem. Soc.* **1987**, *109*, 7528–7530.

25 CARTER, M. T., RODRIGUEZ, M., BARD, A. J. Voltammetric studies of the interaction of metal chelates with

DNA. 2. Tris-chelated complexes of cobalt (III) and iron(II) with 1,10-phenanthroline and 2,2′-bipyridine. *J. Am. Chem. Soc.* **1989**, *111*, 8901–8911.

26 ZHAO, Y.-D., PANG, D.-W., WANG, Z.-L., CHENG, J.-K., QI, Y.-P. J. DNA-modified electrodes. Part 2. Electro-chemical characterization of gold electrodes modified with DNA. *J. Electroanal. Chem.* **1997**, *431*, 203–209.

27 MISHIMA, Y., MOTONAKA, J., IKEDA, S. Utilization of an osmium complex as a sequence recognizing material for DNA-immobilized electrochemical sensor. *Anal. Chim. Acta* **1997**, *345*, 45–50.

28 FOJTA, M., HAVAN, L., FULNECKOVA, J., KUBICAROVA, T. Adsorptive transfer stripping AC voltammetry of DNA complexes with intercalators. *Electroanalysis* **2000**, *12*, 926–934.

29 ASLANOGLU, M., ISAAC, C. J., HOULTON, A., HORROCKS, B. R. Voltammetric measurements of the interaction of metal complexes with nucleic acids. *Analyst* **2000**, *125*, 1791–1798.

30 MILLAN, K. M., SPURMANIS, A. J., MIKKELSEN, S. R. Covalent immobil-ization of DNA onto glassy carbon electrodes. *Electroanalysis* **1992**, *4*, 929–932.

31 MILLAN, K. M., MIKKELSEN, S. R. Sequence-selective biosensor for DNA based on electroactive hybridization indicators *Anal. Chem.* **1993**, *65*, 2317–2323.

32 MILLAN, K. M., SARAULLO, A., MIKKELSEN, S. R. Voltammetric DNA biosensor for cystic fibrosis based on a modified carbon paste electrode. *Anal. Chem.* **1994**, *66*, 2943–2948.

33 STEEL, A. B., HERNE, T. M., TARLOV, M. J. Electrostatic interactions of redox cations with surface-immobilized and solution DNA. *Bioconjug. Chem.* **1999**, *10*, 419–423.

34 STEEL, A. B., HERME, T. M., TARLOV, M. J. Electrochemical quantitation of DNA immobilized on gold. *Anal. Chem.* **1998**, *70*, 4670–4677.

35 CARTER, M. T., BARD, A. J. Electrochemical investigations of the interaction of metal chelates

with DNA. 3. Electrogenerated chemi-luminescent investigation of the interaction of tris(1,10-phenanthroline)ruthenium(II) with DNA. *Bioconjug. Chem.* **1990**, *1*, 257–263.

36 YAN, F., ERDEM, A., MERIC, B., KERMAN, K., OZSOZ, M., SADIK, O. A. Electrochemical DNA biosensor for the detection of specific gene related to Microcystis species. *Electrochem. Commun.* **2001**, *3*, 224–228.

37 WELCH, T. W., CORBETT, A. H., THORP, H. H. Electrochemical determination of nucleic acid diffusion coefficients through noncovalent association of a redox-active probe. *J. Phys. Chem.* **1995**, *99*, 11757–11763.

38 JOHNSTON, D. H., GLASGOW, K. C., THORP, H. H. Electrochemical measurement of the solvent accessibility of nucleobases using electron transfer between DNA and metal complexes. *J. Am. Chem. Soc.* **1995**, *117*, 8933–8938.

39 KELLY, S. O., JACKSON, N. M., HILL, M. G., BARTON, J. K. Long-range electron transfer through DNA films. *Angew. Chem. Inter. Eng. Ed.* **1999**, *38*, 941–945.

40 RODRIGUEZ, M., BARD, A. J. Electrochemical studies of the interaction of metal chelates with DNA. 4. Voltammetric and electro-generated chemiluminescent studies of the interaction of tris(2,2′-bipyridine)osmium(II) with DNA. *Anal. Chem.* **1990**, *62*, 2658–2662.

41 WELCH, T. W., THORP, H. H. Distribution of metal complexes bound to DNA determined by normal pulse voltammetry. *J. Phys. Chem.* **1996**, *100*, 13829–13836.

42 NAPIER, M. E., LOOMIS, C. R., SISTARE, M. F., KIM, J., ECKHARDT, A. E., THORP, H. H. Probing biomolecule recognition with electron transfer: Electrochemical sensors for DNA hybridization. *Bioconjug. Chem.* **1997**, *8*, 906–913.

43 NAPIER, M. E., THORP, H. H. Modification of electrodes with dicarboxylate self-assembled monolayers for attachment and detection of nucleic acids. *Langumir* **1997**, *13*, 6342–6344.

44 AMISTEAD, P. M., THORP, H. H. Oxidation kinetics of guanine in DNA molecules adsorbed onto indium tin oxide electrodes. *Anal. Chem.* **2001**, *73*, 558–564.

45 RODRIGUEZ, M., KODAEK, T., TORRES, M., BARD, A. J. Cleavage of DNA by electrochemically activated Mn^{II} and Fe^{III} complexes of meso-tetrakis(N-methyl-4-pyridiniumyl)porphine. *Bioconjug. Chem.* **1990**, *1*, 123–131.

46 LIU, S., YE, J., HE, P., FANG, Y. Voltammetric determination of sequence-specific DNA by electroactive intercalator on graphite electrode. *Anal. Chim. Acta* **1996**, *335*, 239–243.

47 QU, F., ZHU, Z., LI, N.-Q. Electrochemical studies of the hemin-DNA interaction. *Electroanalysis* **2000**, *12*, 831–835.

48 MOLINIER-JUMEL, C., MALFOY, B, REYNAUD, J. A., AUBEL-SADRON, G. Electrochemical study of DNA-anthracyclines interaction. *Biochem. Biophys. Res. Commun.* **1978**, *84*, 441–449.

49 YANG, M., YAU, H. C. M., CHAN, H. L. Adsorption kinetics and ligand-binding properties of thiol-modified double-stranded DNA on a gold surface. *Langmuir* **1998**, *14*, 6121–6129.

50 BERG, H., HORN, G., LUTHARDT, U., IHN, W. Interaction of anthracycline antibiotics with biopolymers. Part V. Plarographic behavior and complexes with DNA. *Bioelectrochem. Bioenerg.* **1981**, *8*, 537–553.

51 HASHIMOTO, K., ITO, K., ISHIMORI, Y. Novel DNA sensor for electrochemical gene detection. *Anal. Chim. Acta* **1994**, *286*, 219–224.

52 SUN, X., HE, P., LIU, S., YE, J., FANG, Y. Immobilization of single-stranded deoxyribonucleic acid on gold electrode with self-assembled aminoethanethiol monolayer for DNA electrochemical sensor applications. *Talanta* **1998**, *47*, 487–495.

53 CHU, X., SHEN, G.-L., JIANG, J.-H., KANG, T.-F., XIONG, B., YU, R.-Q.

54 WANG, J., OZSOZ, M., CAI, X. *et al.* Interactions of antitumor drug daunomycin with DNA in solution and at the surface. *Bioelectrochem. Bioenerg.* **1998**, *45*, 33–40.

55 MARRAZZA, G., MASCINI, C. M. Disposable DNA electrochemical sensor for hybridization detection. *Biosensors Bioelectronics* **1999**, *14*, 43–51.

56 MARRAZA, G., CHITI, G., MASCHINI, M., ANICHINI, M. Detection of human apolipoprotein E genotypes by DNA electrochemical biosensor coupled with PCR. *Clin. Chem.* **2000**, *46*, 31–37.

57 IBRAHIM, M. S. Voltammetric studies of the interaction of nogalamycin antitumor drug with DNA. *Anal. Chim. Acta* **2001**, *443*, 63–72.

58 ZHANG, H.-M., LI, N.-Q. Electrochemical studies of the interaction of adriamycin to DNA. *J. Pharmac. Biomed. Anal.* **2000**, *22*, 67–73.

59 KELLY, J. M., LYONS, M. E., VAN DER PUTTEN, W. J. M. An electrochemical method to determine binding constants of small ligands to nucleic acids. *Anal. Chem. Symp. Ser. (Electrochem., Sens. Anal.)* **1986**, *25*, 205–213.

60 KELLEY, S. O., BARTON, J. K. Electrochemistry of methylene blue bound to a DNA-modified electrode. *Bioconjug. Chem.* **1997**, *8*, 31–37.

61 EEDEM, A., KERMAN, K., MERIC, B., AKARCA, U.S., OZSOZ, M. Novel hybridization indicator methylene blue for the electrochemical detection of short DNA sequences related to the hepatitis B virus. *Anal. Chim. Acta* **2000**, *422*, 139–149.

62 TANG, T.-C., HUANG, H.-J. Electrochemical studies of intercalation of ethidium bromide to DNA. *Electroanalysis* **1999**, *11*, 1185–1190.

63 HASHIMOTO, K., ITO, K., ISHIMORI, Y. Sequence-specific gene detection with a gold electrode modified with DNA probes and an electrochemically active dye. *Anal. Chem.* **1994**, *66*, 3830–3833.

64 WANG, J., RIVAS, G., CAI, X. *et al.* Accumulation and trace measurements of phenothiazine drugs at DNA-modified electrodes. *Anal. Chim. Acta* **1996**, *332*, 139–144.

65 BRETT, A. M. O., MACEDO, T. R. A., RAIMUNDO, D., MARQUES, M. H., SERRANO, S. H. P. Voltammetric behavior of mitoxantrone at a DNA-biosensor. *Biosensors Bioelectronics* **1998**, *13*, 861–867.

66 RADI, A. Voltammetric study of khellin at a DNA-coated carbon paste electrode. *Anal. Chim. Acta* **1999**, *386*, 61–68.

67 PANG, D.-W., ABRUNA, H. D. Interactions of benzyl viologen with surface-bound single- and double-stranded DNA. *Anal. Chem.* **2000**, *72*, 4700–4706.

68 JELEN, F., ERDEM, A., PALECEK, E. Cyclic voltammetry of echinomycin and its complexes with DNA. *J. Biomol. Struct. Dyn.* **2000**, *17*, 1176–1177.

69 ERDEM, A., OZSOZ, M. Interaction of the anticancer drug epirubicin with DNA. *Anal. Chim. Acta* **2001**, *437*, 107–114.

70 TAKENAKA, S., IHARA, T., TAKAGI, M. Bis-9-acridinyl derivative containing a viologen linker chain: Electrochemicallyactive intercalator for reversible labeling of DNA. *Chem. Commun.* **1990**, 1485–1487.

71 TAKENAKA, S., SHIGEMOTO, N., KONDO, H. Involvement of nucleic bases in the quenching of the fluorescence of acridine by methylviologen. *Supramol. Chem.* **1988**, *9*, 47–56.

72 TAKENAKA, S. SATO, H., IHARA, T., TAKAGI, M. Synthesis and DNA binding properties of bis-9-acridinyl derivatives containing mono-, di-, and tetra-viologen units as a connector of bis-intercalators. *J. Heterocyclic Chem.* **1997**, *34*, 123–127.

73 TAKENAKA, S., SATO, H., IHARA, T., TAKAGI, M. DNA-binding behavior of viologen-containing, electrochemically

active intercalators, *Anal. Sci.* **1991**, *7* supplement, 1385–1386.

74 TAKENAKA, S., UTO, Y., SAITA, H., YOKOYAMA, M., KONDO, H., W. D. Wilson Electrochemically active threading intercalator with high double stranded DNA selectivity. *Chem. Commun.* **1998**, 1111–1112.

75 TAKENAKA, S., YAMASHITA, K., TAKAGI, M., UTO, Y., KONDO, H. DNA sensing on a DNA probe-modified electrode using ferrocenylnaphthalene diimide as the electrochemically active ligand. *Anal. Chem.* **2000**, *72*, 1334–1341.

76 SATO, S., FUJII, S., YAMASHITA, K., TAKAGI, M., KONDO, H., TAKENAKA, S. Ferrocenyl naphthalene diimide can bind to DNA RNA hetero duplex: potential use in an electrochemical detection of mRNA expression. *J. Organomet. Chem.* **2001**, *637–639*, 476–483.

77 WANG, J. Towards genoelectronics: Electrochemical biosensing of DNA hybridization. *Chem. Eur. J.* **1999**, *5*, 1681–1685.

78 TAKENAKA, S. Highly sensitive probe for gene analysis by electrochemical approach. *Bull. Chem. Soc, Jpn* **2001**, *74*, 217–224.

79 HASHIMOTO, K., ISHIMORI, Y. Preliminary evalution of electrochemical PNA array for detection of single base mismatch mutations. *Lab on a Chip* **2001**, *1*, 61–63.

80 WANG, J., RIVAS, G., CAI, X., PALECEK, E., NIELSEN, P., SHIRAISHI, H., DONTHA, N., LUO, D., PARRAADO, C., CHICHARRO, M. *et al.* DNA electrochemical biosensors for environmental monitoring. *Anal. Chim. Acta* **1997**, *347*, 1–8.

81 MIYAHARA, H., YAMASHITA, K., TAKAGI, M., KONDO, H., TAKENAKA, S. Electrochemical array (ECA) as an integrated multi-electrode DNA sensor. *Trans. IEE Jpn* **2001**, *121-E*, 187–191.

82 TAKENAKA, S., MIYAHARA, H., YAMASHITA, K., TAKAGI, M., KONDO, H. Base mutation analysis by a ferrocenyl naphthalene diimide drivative. *Nucleosides Nucleotides Nucleic Acids* **2001**, *20*, 1429–1432.

83 MIYAHARA, H., YAMASHITA, K., KANAI, M. *et al.* Electrochemical analysis of single nucleotide polymorphisms of p53 gene. *Talanta* **2002**, *56*, 829–835.

84 YAMASHITA, K., TAKAGI, M. KONDO, H., TAKENAKA, S. Electrochemical detection of base pair mutation. *Chem. Lett.* **2000**, 1038–1039.

85 KELLY, S. O., BOON, E. M., BARTON, J. K., JACKSON, N. M., HILL, M. G. Single-base mismatch detection based on charge transduction through DNA. *Nucleic Acids Res.* **1999**, *27*, 4830–4837.

86 KELLY, S. O., BARTON, J. K., JACKSON, N. M. *et al.* Orientating DNA helices on gold using applied electric fields. *Langmuir* **1998**, *24*, 6781–6784.

87 WANG, J., PALECEK, E., NIELSEN, P. E. *et al.* Peptide nucleic acid probes for sequence-specific DNA biosensors. *J. Am. Chem. Soc.* **1996**, *118*, 7667–7670.

88 BOON, E. M., CERES, D. M., DRUMMOND, T. G., HILL, M. G., BARTON, J. K. Mutation detection by electrocatalysis at DNA-modified electrodes. *Nature Biotechnol.* **2000**, *18*, 1096–1100.

89 TAKENAKA, S., UTO, Y., TAKAGI, M., KONDO, H. Enhanced electron transfer from glucose oxidase to DNA-immobilized electode aided by ferrocenyl naphthalene diimide, a threading intercalator. *Chem. Lett.* **1988**, 989–990.

90 AMISTEAD, P. M., THORP, H. H. Modification of indium tin oxide electrodes with nucleic acids: detection of attomole quantities of immobilized DNA by electrocatalysis. *Anal. Chem.* **2000**, *72*, 3764–3770.

91 SCHENA, M. *DNA Microarrays*, Oxford University Press, New York, 1999.

92 POPOVICH, N. D. Mediated electrochemical detection of nucleic acids for drug discovery and clinical diagnostics. *IVD Technol* **2001**, *7*, 36–42.

93 YAMASHITA, K., TAKAGI, M., UCHIDA, K., KONDO, H., TAKENAKA, S. Visualization of DNA microarrays by scanning electrochemical microscopy (SECM). *Analyst* **2001**, *126*, 1210–1211.

10
Design and Studies of Abasic Site Targeting Drugs: New Strategies for Cancer Chemotherapy

Jean-François Constant and Martine Demeunynck

10.1
Introduction

10.1.1
Importance of Abasic Sites in Cells

The emergence of resistance is one of the factors that limits the efficacy of anti-tumor drugs. Some DNA-repair processes have been shown to be implicated in resistance to DNA-damaging agents [1] and an increasing number of studies are being carried out to design DNA repair inhibitors, with the ultimate goal of improving the antitumour effects of DNA-damaging drugs [2, 3]. Among the numerous lesions produced by these drugs on DNA, modified bases are processed by a specific process: the base excision repair (BER) pathway, which can be summarized as consisting of three major steps (Fig. 10.1):

1. Recognition and excision of the damaged base by a specific glycosylase, leaving a 2′-deoxyribose residue, the abasic site (apurinic/apyrimidinic or AP-site).
2. DNA breakage at the abasic site by a specific AP endonuclease (usually HAP1 enzyme in humans).
3. DNA resynthesis by two possible mechanisms named the short patch and long patch repair pathways.

The short patch repair pathway replaces only one or two nucleotides and involves DNA polymerase β (pol β), XRCC1 (X-ray cross-complementation protein 1), and a ligase. In the long patch repair pathway up to 10 nucleotides are replaced, involving other enzymes such as FEN1 (flap endonuclease 1), PCNA (proliferating cell nuclear antigen), a polymerase (β, δ, or ε) and a ligase [1]. Base modifications that result from the action of alkylating agents are handled by the BER pathway. Thus a deficiency of this repair pathway could result in increased cellular sensitivity to alkylating agents [4]. BER has not been fully investigated as a potential actor in resistance to antitumor drugs and very few reports have addressed the potential for targeting abasic site processing.

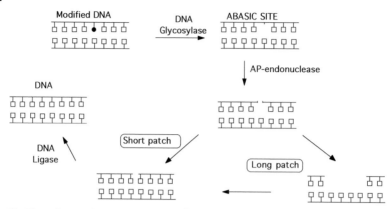

Fig. 10.1. Base excision repair (BER) pathway.

As mentioned above, AP-sites are key intermediates in the BER pathway. They are formed by enzymatic cleavage of the C1'-N glycosidic bond of a damaged or inappropriate nucleic base [1]. Many BER glycosylases have been isolated and characterized [5, 6].

Recent work has shown that abasic sites may interfere with other vital enzymes in cells such as ligases [7] and topoisomerases [8]. Apurinic sites appear to stimulate DNA breakage up to 20-fold when located within the four-base overhang produced by topoisomerase II-mediated cleavage [9–11]. This lesion appears to be 1000 times more potent than specific topoisomerase II inhibitors [12]. Topoisomerase I activity is also affected by the presence of abasic sites. Depending on its position, a single abasic site is able to induce new topoisomerase I cleavage sites or to trap the topoisomerase I-cleavable complex [13]. Abasic sites and other DNA damage may therefore be considered to be site-specific topoisomerase poisons [14].

All these studies demonstrate the central role played by abasic sites as key intermediates in BER process and as potential poisons for other DNA-modifying enzymes. In our group we came up with the idea that targeting this lesion by specific drugs might give rise to new pharmacological activities (Fig. 10.2):

- Drugs that bind strongly at abasic site might mask the lesion to the hydrolytic activity of AP endonucleases, the enzymes that initiate the AP-site repair by incising DNA at the adjacent phosphodiester bond.
- A similar effect might be expected from drugs that react with AP-sites in their open aldehydic form, leading to covalent adducts, the repair of which would be less efficient.
- Drugs able to catalyze single-strand cleavage at abasic sites might trigger other enzymatic activities such as polyADP ribose polymerase (PARP). PARP production is induced by high levels of single-strand breaks in DNA [15]. This enzyme interacts at the cleavage site and catalyzes the synthesis of NAD oligomers. The cell depletion of this essential cofactor might participate in the numerous processes leading to γ-irradiated cell death.

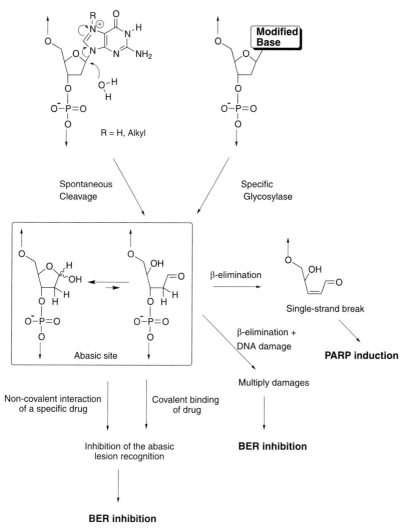

Fig. 10.2. Importance of abasic sites as a target for specific drugs.

- Another class of drugs is derived from the single-strand breakers mentioned above. These compounds also possess a DNA-damaging motif. When binding at abasic sites, such compounds might induce multiply-damaged sites (single-strand breaks + other damage) known to be more difficult to repair.

10.1.2
Structure of Abasic DNA

The chemical structure of abasic sites was studied by NMR spectroscopy analysis of oligodeoxynucleotides containing ^{17}O- and ^{13}C-labeled abasic sites [16–18].

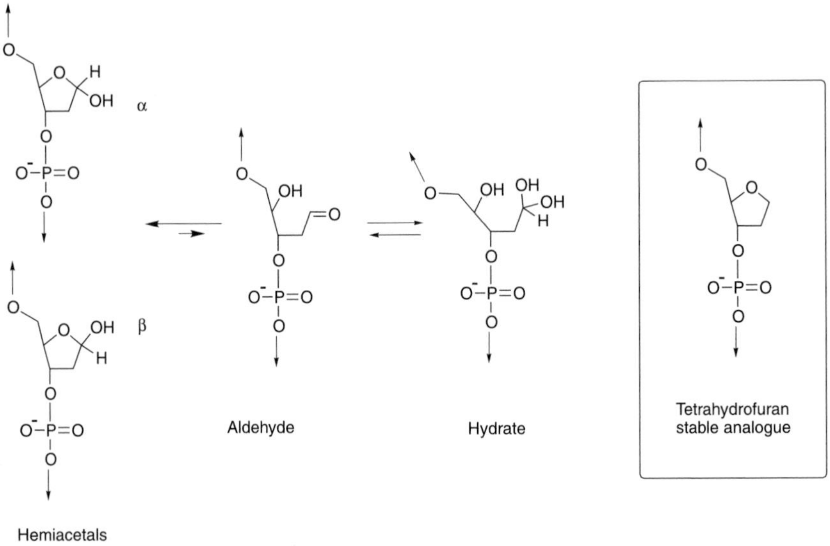

Hemiacetals

Fig. 10.3. Chemical structure of the abasic lesion.

Abasic sites can exist in three different forms in equilibrium: open-chain aldehyde, α and β hemiacetals, or hydrate. NMR studies of this equilibrium in solution indicated that the mixture of α/β cyclic hemiacetals is the predominant form in the duplex and that the aldehyde represents only 1% of the total (Fig. 10.3).

The use of synthetic oligonucleotides containing stable analogs of the abasic lesion, such as the tetrahydrofuran (Fig. 10.3) [19–23] allowed studies of the conformational influence of an abasic site [24–27]. Different types of abasic duplexes were examined by high-field NMR spectroscopy, calorimetry, and restrained molecular dynamics. The influence of the nature of the base opposite the abasic site, a purine or a pyrimidine (i.e. respectively an apyrimidinic or an apurinic site), and the nature of the flanking bases was studied. The main results that emerged were as follows: (1) the backbones of the abasic duplex DNAs are regular with right-handed B-form helices, (2) the conformational changes due to the presence of the abasic site depend on the nature of the flanking and opposite bases. In the duplexes containing unpaired purines (apyrimidinic sites) the purine remains stacked within the helix in an intrahelical conformation [22, 23, 25, 27–29]. The situation is more complex for apurinic sites, the pyrimidine opposite the abasic site may be either stacked inside the helix or expelled outside, or in equilibrium between the two situations [20, 26, 29]. However, some differences are observed between the "true" abasic site (also called the "aldehydic" abasic site) and the stable tetrahydrofuran analog. In the case of the "true" abasic site, which exists predominantly as a mixture of α- and β-anomers, the deoxyribose residue is within the helix for the β-anomer, while the sugar in the α-configuration is out of the helix [25].

Another feature that distinguishes the structure of an abasic duplex from that of the parent unmodified oligonucleotide, is formation of a kink at the site of the lesion. The bending value into the major groove was estimated to be approximately 10° in an apyrimidinic undecamer [28] or approximately 30° for an apuric dodecamer [30]. Furthermore the abasic lesion considerably increases the flexibility of duplex DNA [28].

10.1.3
Abasic Site Reactivity

In its opened aldehydic form, deoxyribose may undergo two types of reaction (Fig. 10.4): (1) strand cleavage in alkaline conditions and (2) condensation with amino nucleophiles. In mild alkaline conditions, abasic sites undergo a β-elimination reaction with formation of an α,β-unsaturated aldehyde [31, 32]. In more drastic alkaline medium, a second elimination reaction (δ-elimination) may occur. Similar reactivity was observed in the presence of polyamines [32, 33]. These β- and δ-elimination reactions are implicated in the cleavage of AP-sites by class I AP endonucleases or AP lyases (see Section 10.1.4).

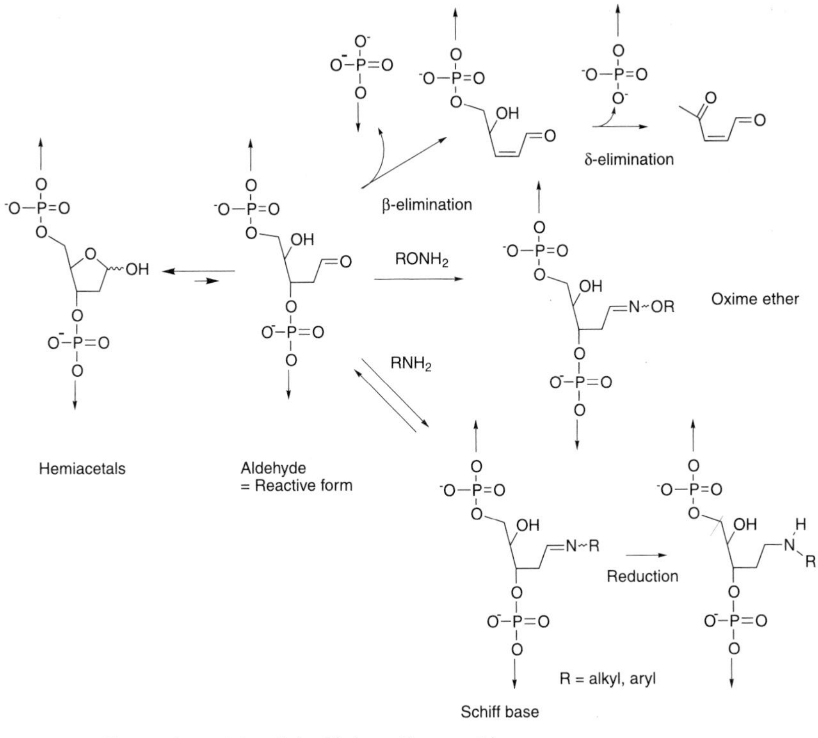

Fig. 10.4. Chemical reactivity of the 2′-deoxyribose residue.

The β-elimination process occurs on the open aldehydic form of the deoxyribose residue. NMR spectroscopy experiments confirmed the stereospecific abstraction of the 2′-pro-S hydrogen atom [34–36].

The reactivity of the abasic site with amino nucleophiles has been used by several groups to derivatize oligonucleotides either by formation of a Schiff base between the electrophilic aldehydic abasic site and amino group containing ligands, followed by reduction (reductive amination) or by reaction with alkoxyamines leading to stable oxime ethers. This reaction has been largely used for detection and titration of abasic sites.

10.1.4
Enzymology of the Abasic Site

As shown in Fig. 10.1 the very first step in the abasic site-repair process is an incision of DNA at one of the phosphodiester bonds adjacent to the lesion. AP-sites can be processed following two enzymatic pathways (Fig. 10.5). Class II enzymes named AP endonucleases, such as exonuclease III from *Escherichia coli* [37] and APE-I, its human homolog [38], cleave the DNA backbone at the 5′ side of the AP-site via hydrolytic process [39], thus generating a 3′-OH group and a 5′-deoxyribose 5′-phosphate residue. Class I AP endonucleases, also termed AP lyases [40], are associated with glycosylase activities and proceed via a β-elimination mechanism leaving a 3′-α,β unsaturated aldehyde residue and a 5′-phosphate group. The formation of a Schiff base between an amino residue of the enzyme and the aldehyde group of the abasic site has also been postulated [41]. δ-Elimination has also been proposed to follow β-elimination cleavage of AP-sites during enzymatic repair of damaged DNA [42].

If not repaired, the abasic site was shown to be mutagenic [43–46] or lethal [47]. If replication proceeds through the abasic lesion, *in vitro* experiments found that that DNA polymerases from various organisms preferentially (but not exclusively) incorporate dAMP opposite abasic lesions [48–51]. This is known as the "A rule." Precise kinetics studies were made with synthetic gapped matrix containing a site-specific abasic analog (tetrahydrofuran). The bypass synthesis by DNA polymerases was studied by varying the position of the 3′ and 5′ terminus of the primer relative to the abasic site [52, 53]. In all cases the "A rule" was obeyed. It was also shown that the persistence of the non-coding AP-site in the cell may lead to the blockage of DNA replication [54].

10.2
Drug Design

10.2.1
Introduction

Various compounds have been shown to interact or react with DNA at abasic sites. As already mentioned in Section 10.1.3, alkoxyamine derivatives bind covalently to

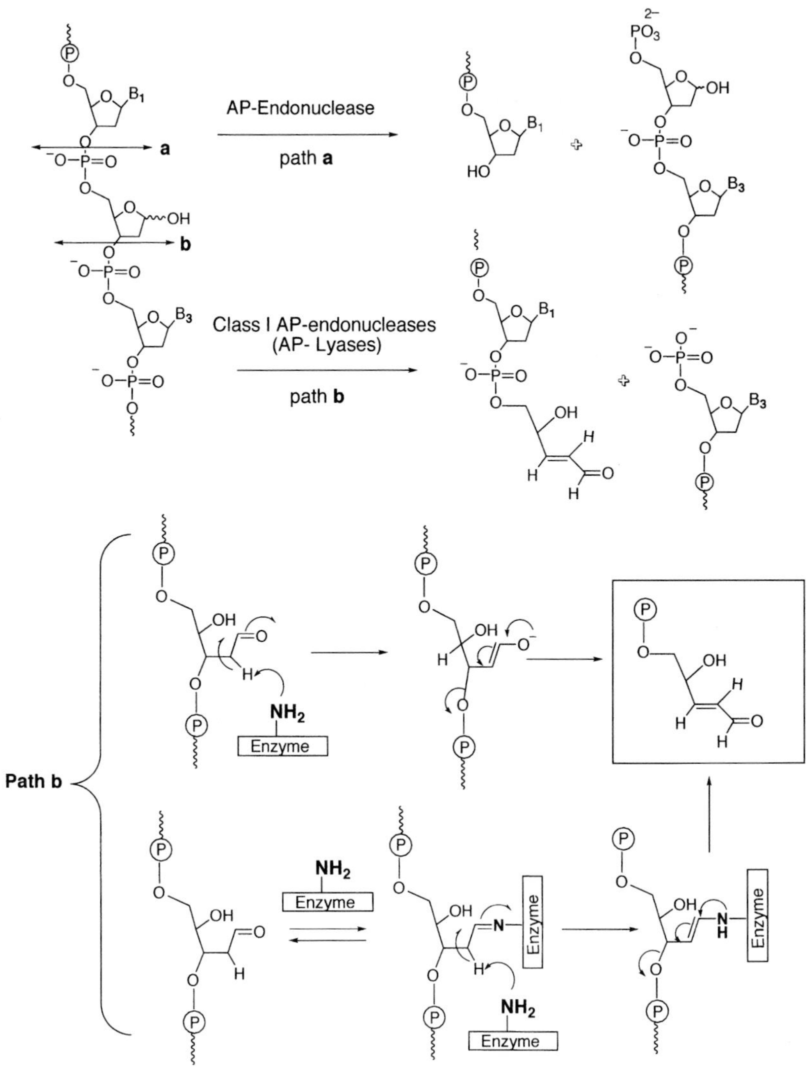

Fig. 10.5. Enzymatic cleavage of the abasic lesion.

the abasic site. Several derivatives were designed for the detection and quantification of the lesion. The simplest one, ^{14}C-labeled methoxyamine, allows quantification of AP-sites by measuring radioactivity of labeled DNA resulting from the formation of ^{14}C-labeled oxime methylether [55, 56]. Other probes have also been prepared (Fig. 10.6), such as ARP, a biotin derivative used in an ELISA-like assay [57–60], and the lissamine probe for fluorescence detection [61, 62].

Two intercalating drugs, 9 amino-elipticine (9-AE) and 3-aminocarbazole, have been shown to interact with AP-sites and induce DNA cleavage at these sites at low

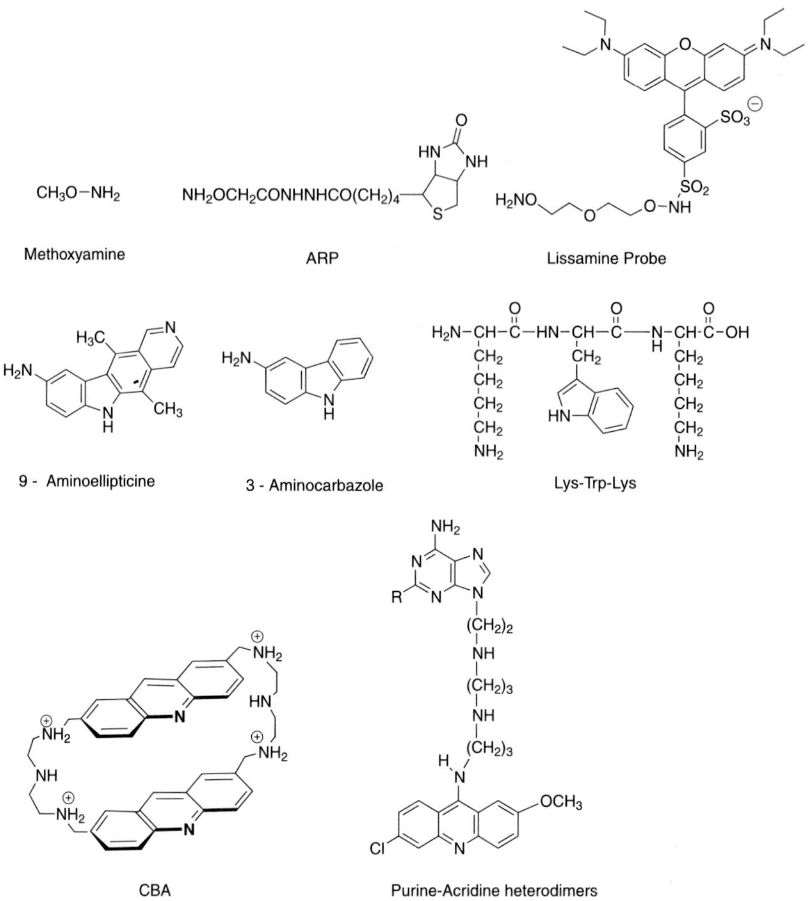

Fig. 10.6. Molecules showing specificity for abasic lesion.

doses [63–65]. The mechanism of cleavage by 9-AE has been thoroughly examined; it is proposed to involve formation of a Schiff base by reaction of the amino group of the drug with the aldehyde function of the AP-site, followed by β-elimination reaction. This results in the formation of a 2′,3′-unsaturated deoxyribose and cleavage of the DNA strand at the 3′-side of the site. In presence of a reducing agent, the Schiff base is converted into a stable covalent adduct, and the fluorescence of 9-amino-ellipticine can be exploited to detect and quantify the abasic lesions [66].

Similar reactivities have been observed with oligopeptides containing aromatic and basic amino acids. The interaction of the tripeptide Lys-Trp-Lys with abasic sites containing DNA has been studied in detail. The tripeptide catalyzes the cleavage of the phosphodiester bond adjacent to the AP-site via a β-elimination reaction [67–70]. Specific binding of the tripeptide at the AP-site places the amino group of a lysine residue in close proximity to the aldehydic function of the ring-opened deoxyribose residue. Proton abstraction at the C'-2 carbon, occurring either on the free aldehyde and/or on the imine formed with an amino group of the Lys-Trp-Lys results in β-elimination of the 3'-phosphate [35].

More recently the macrocyclic bisacridine CBA, which belongs to the cyclo-bisintercaland family [71], was reported to interact at abasic sites in a very specific mode known as threading intercalation. Combined high-field NMR spectroscopy and restrained molecular dynamics calculations suggest the formation of a major complex in which CBA penetrates the double helix at the abasic site, sandwiching the base pair 3' to the lesion between the two acridine rings, and leaving one protonated linking chain in each groove [72]. Furthermore, CBA induced selective photocleavage in the vicinity of the abasic lesion on both strands of a 23-mer duplex containing tetrahydrofuran analog [73]. For more on the cyclobisintercaland family, see Chapter 11.

With the ultimate goal of developing molecules capable of interfering with the repair processes in the cell, we designed a series of heterodimers for specific recognition and binding to abasic sites in DNA [74, 75]. These tailor-made molecules include: (1) a recognition unit, i.e. a nucleic base that inserts into the abasic cavity and gives hydrogen bonds with the complementary base in the opposite strand; (2) an intercalator for strong but non-sequence-specific binding to DNA; and (3) a linking chain endowed with binding and/or cleavage activities. These molecules may act as "artificial nucleases" and/or DNA-repair inhibitors (Fig. 10.7).

Following this hypothesis, a large series of heterodimers was prepared [74–79]. The most representative compounds are depicted in Figs 10.8–10.10.

Compounds **1–9** that contain aliphatic amines in the linker behaved as "artificial nucleases" (see Section 10.2.4). They are able to cleave the abasic strand very efficiently via a β-elimination catalysis. A second DNA-modifying motif was also introduced on the intercalating moiety as exemplified by molecules **10–16** (Fig. 10.9). Their nuclease properties are described in Section 10.2.3.

In a third group of compounds, the linker containing amide and/or guanidi-

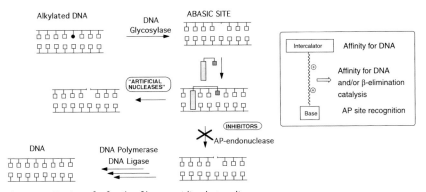

Fig. 10.7. Design of a family of base-acridine heterodimers.

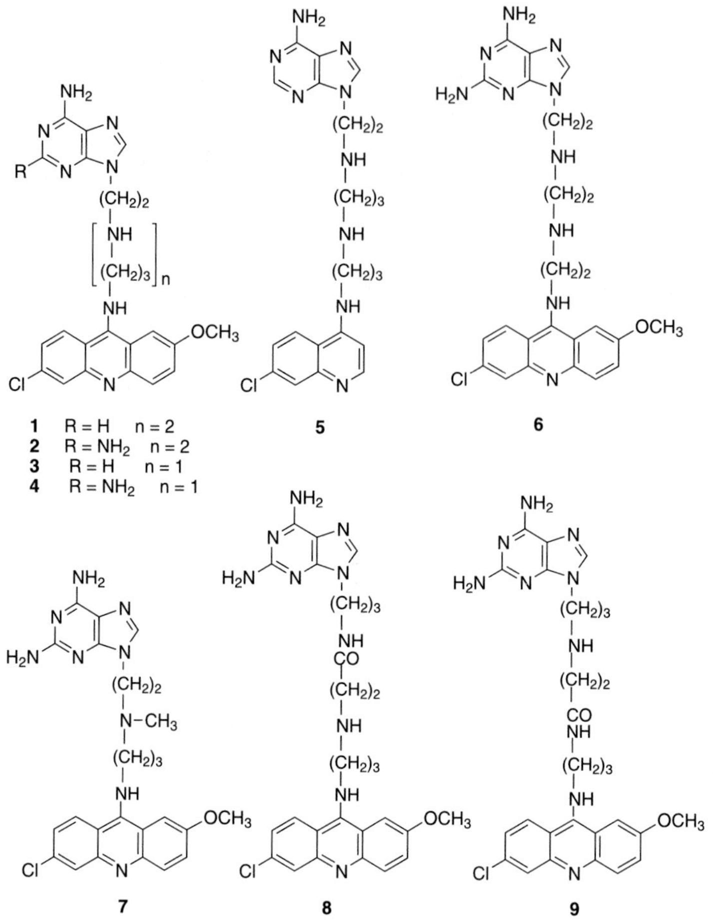

Fig. 10.8. "Artificial nucleases."

nium functions is totally devoid of cleavage properties (compounds **17–28**, Fig. 10.10). The biological properties of these compounds are discussed in Sections 10.2.6 and 10.3.

10.2.2
Synthesis of the Heterodimers

As shown in Schemes 10.1 and 10.2, two main strategies were designed to prepare the different series of heterodimers. Molecules containing polyamines as linkers were synthesized in three steps (Scheme 10.1): (1) *N*-9 alkylation of the nucleic base, adenine or 2,6-diaminopurine, by 1,2-dibromoethane; (2) chain elongation by reaction with the polyamine; and (3) introduction of the intercalating nucleus, ac-

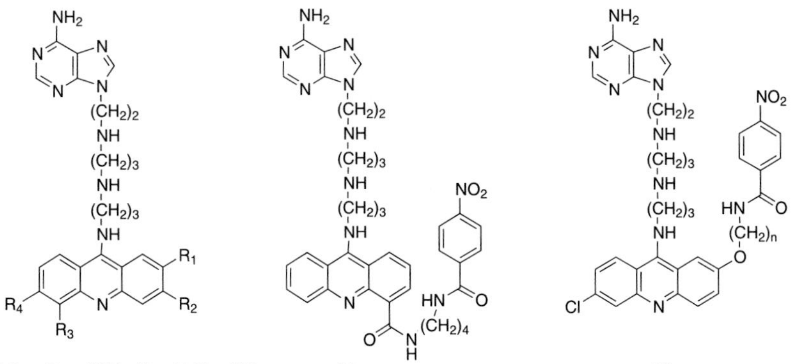

10 $R_1 = R_2 = OCH_3$, $R_3 = H$, $R_4 = NO_2$
11 $R_1 = OEt$, $R_2 = R_3 = H$, $R_4 = NH_2$
12 $R_1 = OEt$, $R_2 = H$, $R_3 = I$, $R_4 = NH_2$

13

14

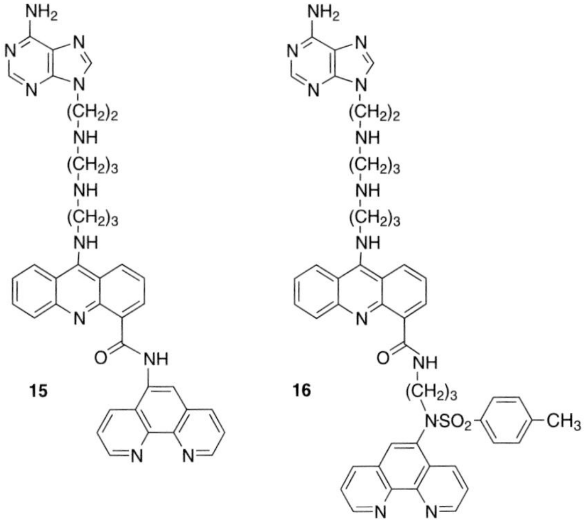

15

16

Fig. 10.9. Bifunctional "nucleases."

ridine or quinoline, by reaction of the terminal primary amine of the linker with the phenoxy or chloro heterocycle [75]. Post-functionalization of the secondary amines of compounds **4** and **1**, afforded the tertiary amino-substituted compound **7** [80] and the bis-guanidino derivative **18** [79], respectively.

For the preparation of the bifunctional nucleases **10–16**, two strategies were used: (1) the acridine nucleus itself was converted into a photocleavage agent by introduction of various substituents (R_2, $R_3 = NO_2$, NH_2, or iodine) on the heterocycle, (2) a modifying agent, i.e. nitrobenzamide or phenanthroline, was tethered to the acridine nucleus. In both strategies, the purine linker synthon is introduced on the functionalized 9-chloro or 9-phenoxy acridines in the last step.

Compounds **20–28**, in which the two chromophores are tethered via *N,N′*-dialkylguanidine, were prepared by the step-by-step construction of the guanidine

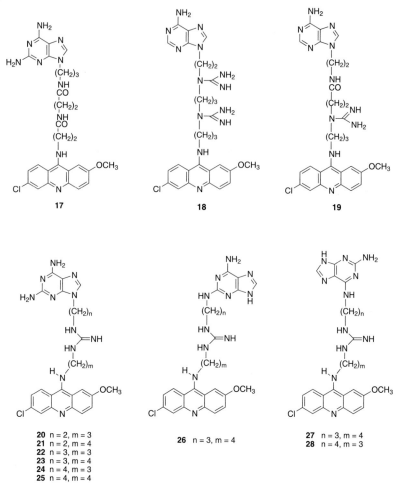

Fig. 10.10. DNA repair inhibitors.

linker (Scheme 10.2) [76, 77]. The key step involves coupling of an ω-alkylamino-substituted purine with a protected amino thiourea.

The linker was introduced on various positions of the 2,6-diaminopurine and different pathways were adopted to prepare the ω-alkylamino-substituted purine key intermediates (Scheme 10.3) [77].

A slightly modified route was used to prepare molecules possessing amido or mixed amino-amido linkers [74, 78]. The key step is the synthesis of the poly-functionalized chains $Cl\text{-}(CH_2)_n\text{-}X\text{-}(CH_2)_m\text{-}Y\text{-}(CH_2)_p\text{-}NHP$, bearing a halogen atom at one end for purine alkylation, a protected primary amine at the other end for reaction with the activated acridine, and including amido and protected secondary amino groups (X, Y). A typical synthesis is shown in Scheme 10.4.

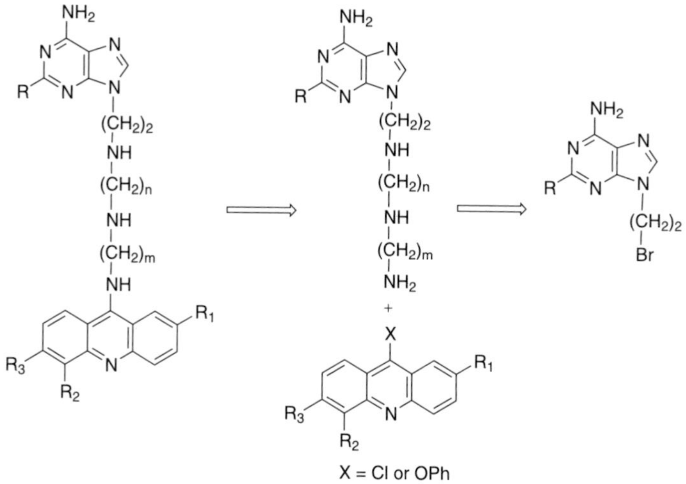

Scheme 10.1. General strategy for the synthesis of the "artificial nucleases."

10.2.3
Nuclease Properties

In a first approach to optimize the selectivity of the heterodimers for abasic sites, we used their ability to cleave the DNA strand at AP-sites. The cleavage efficiency was evaluated by agarose gel electrophoresis using supercoiled pBR322 plasmids containing an average of two abasic sites per DNA molecule [75]. Single-strand cleavage at the abasic site converts the supercoiled form into a relaxed circular form, which migrates more slowly. To illustrate the methodology, the results obtained with Lys-Trp-Lys and compound **2** are shown in Fig. 10.11.

We also measured the association constant of each compound with calf thymus DNA. The data are collected in Tab. 10.1. All molecules bind strongly to DNA. The three parts of the molecules are necessary for cleavage, the "half molecules" **29**

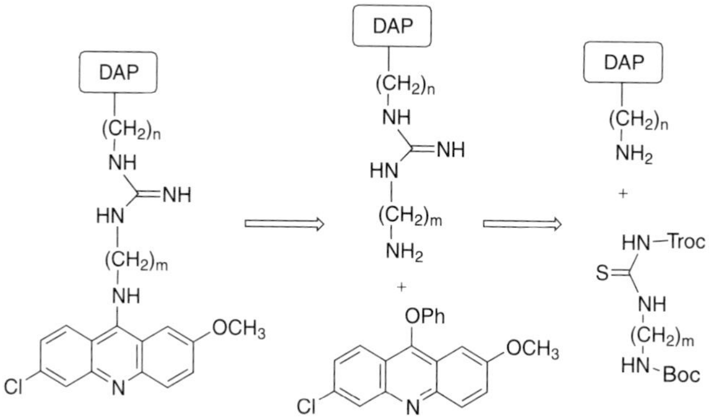

Scheme 10.2. Synthesis of guanidinium containing heterodimers.

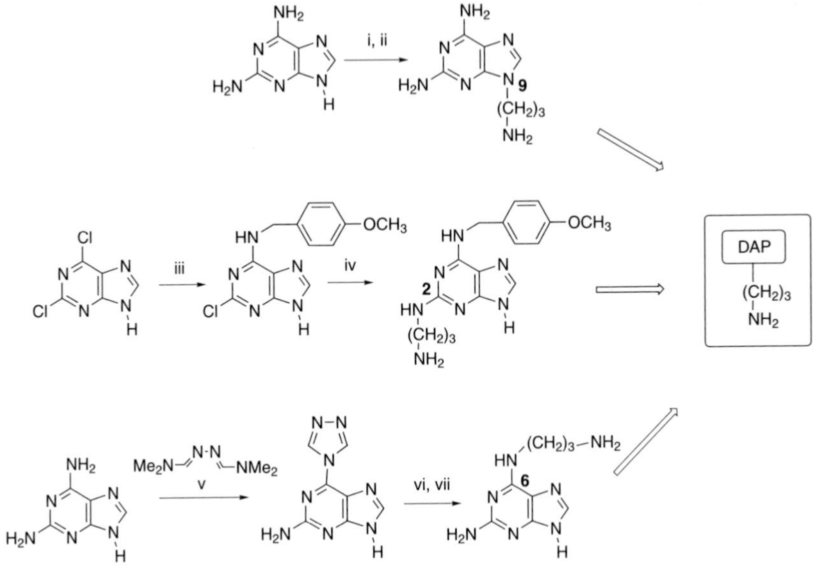

i) 3-Bromopropylphthalimide, DMF, NaH, rt, 6 days; ii)12N HCl/H $_2$O/AcOH 1/1/1 mixture, 100°C, 24h; iii) *para*-methoxybenzylamine, NEt$_3$, DMF, 80°C, 2h; iv) 1,3-diaminopropane, 140°C, 2h; v) DMF, 180°C, 18h ; vi) BocNH-(CH$_2$)$_3$-NH$_2$, 140°C, 1.5 days; vii) 1N HCl in AcOH, rt, 1h.

Scheme 10.3. Regioselective functionalization at position 2, 6, or 9 of the 2,6-diaminopurine [76, 77]. (i) 3-Bromopropylphthalimide, DMF, NaH, rt, 6 days. (ii) 12N HCl/H$_2$O/AcOH 1/1/1 mixture, 100°C, 24 h. (iii) *para*-

Methoxybenzylamine, Net$_3$. DMF, 80°C, 2 h. (iv) 1,3-diaminopropane, 140°C, 2 h. (v) DMF, 180°C, 18 h. (vi) BocNH-(CH$_2$)$_3$-NH$_2$, 140°C, 1.5 days. (vii) 1N HCl in AcOH, rt, 1 h.

and **30** (Fig. 10.12) being almost totally inactive in the conditions of the experiments, and for molecules possessing identical linker a relationship is observed between binding and cleavage. The presence of at least one secondary aliphatic amine in the linking chain is necessary for cleavage activity.

The most efficient molecule, **2**, possesses 2,6-diaminopurine as nucleic base, 9-aminoacridine nucleus as intercalator, and two secondary amines in the linker. This molecule is able to cleave plasmid DNA at nanomolar concentrations, and can therefore be considered as the most potent abasic selective artificial nuclease. All data are in agreement with formation of a specific complex to which all three modules participate, prior to cleavage, as was confirmed by high-field NMR experiments (see Section 10.2.5) [30].

It is interesting to note the critical effect of the protonation state of the amines present in the linker. For the most active molecules, **1** and **2**, which possess the same linking chain, one of the amino groups of the linker that is essentially protonated at neutrality (pK_a 8.5), participates in the formation of the specific complex by ionic interaction with the DNA phosphates, while the second amino group of

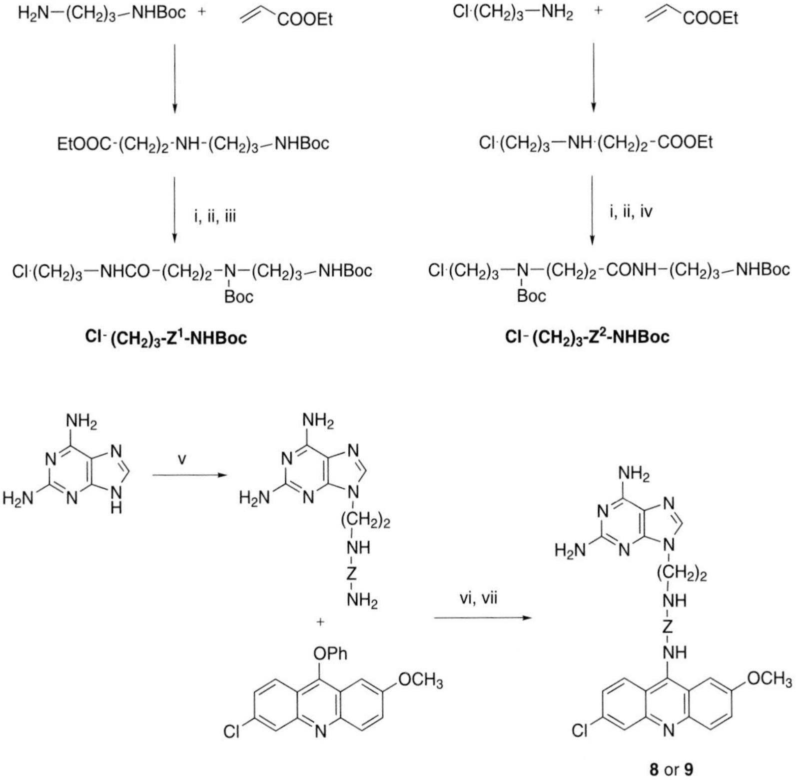

i) NaOH, EtOH, rt, 24h; ii) Boc_2O, dioxane/H_2O; iii) Chloropropylamine hydrochloride, isobutylchloroformate, NEt_3, THF, 0°C; iv) t-butoxycarbonylamino-propylamine, isobutylchloroformate, NEt_3, THF, 0°C; v) Cl-(CH $_2$)$_3$-Z-NHBoc, K_2CO_3, DMF, 80°C, 48h; vi) 1N HCl in AcOH, rt; vii) PhOH, 80°C

Scheme 10.4. Synthesis of compounds 8 and 9 containing amine/amide linkers [74]. (i) NaOH, EtOH, rt, 24 h. (ii) Boc_2O, dioxane/H_2O. (iii) Chloropropylamine hydrochloride, isobutylchloroformate, NEt_3, THF, 0°C. (iv) *t*-Butoxycarbonylamino-propylamine, isobutylchloroformate, NEt_3, THF, 0°C. (v) Cl-$(CH_2)_3$-Z-NHBoc, K_2CO_3, DMF, 80°C, 48 h. (vi) 1N HCl in AcOH, rt. (vii) PhOH, 80°C.

the linker, being partially unprotonated (pK_a 6.7), catalyzes the cleavage. Changing the number of methylenes in the bridge as in molecule **6** dramatically lowers the pK_a values of the aliphatic amines (pK_a 4.7 and 7.8), and the resulting molecule exhibits a very different pH/activity profile, with the highest cleaving activity observed at pH 4 [80].

Molecules such as **18** that contain highly basic guanidine groups (p$K_a > 11$) display the highest affinity for DNA but do not induce any cleavage as they are fully protonated at physiological pH. The most important features that emerge from this study are summarized in Fig. 10.13.

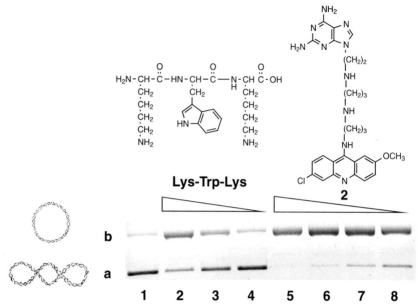

Fig. 10.11. Agarose gel analysis of the nuclease activity of Lys-Trp-Lys and compound **2**. (a) pBR322 supercoiled plasmid containing two AP sites/plasmid $2 \cdot 10^{-9}$ M. (b) Relaxed circular form. Lane 1: Reference; lanes 2–4: Lys-Trp-Lys 10^{-3}, 10^{-4}, and 10^{-5} M; lanes 5–8: Compound **2** 10^{-5}, 10^{-6}, 10^{-7}, and 10^{-8} M.

10.2.4
Molecules Inducing Multiple DNA Damage

Clustered lesions, or locally multiply damaged sites (LMDS), in tracks of a few base pairs appear to be very toxic as they present a challenging repair problem for the cell [81, 82]. Hence the production of new lesions in the vicinity of the abasic site seems an attractive strategy to increase the biological activity. This can be done by introducing a second DNA-damaging or cleavage group on the artificial AP nuclease **1** (Fig. 10.14).

The DNA-modifying activity of these bifunctional compounds (Fig. 10.9) was tested on a 23-oligomer containing the tetrahydrofuran analog (Fig. 10.15).

In compounds **10–12** the acridine nucleus bears amino-, nitro-, or iodo- substituents. The modification of the acridine does not alter the specificity of the drugs for the abasic site and their cleavage efficiency. The three molecules show photo-damaging activity that occurs mainly in close proximity to the lesion [83].

In compounds **13–16**, a reporter group capable of inducing DNA damage or cleavage is tethered to the acridine 2- or 4-positions [84]. No DNA-damaging activity was observed with **13**, but nitrobenzamide-acridine conjugates **14** ($n = 4$ or 6) appeared to be AP-site-specific photodamaging agents [84]. As illustrated by molecular modeling calculations, the molecules interact with DNA in a very simi-

Tab. 10.1. Binding constants (K_a) for native calf thymus DNA, measured by ethidium bromide displacement and cleavage activity on pBR322 DNA plasmid containing approximately two apurinic sites.

Molecules	Affinity $K_a \times 10^{-5}$ (M^{-1})	Cleavage activity	
		a (%)	b
1 Ade-C_2-NH-C_3-NH-C_3-NHAcr	2	90	5×10^{-9}
2 DAP-C_2-NH-C_3-NH-C_3-NHAcr	11	100	1×10^{-9}
5 Ade-C_2-NH-C_3-NH-C_3-NHClQ	0.1	90	8×10^{-8}
6 DAP-C_2-NH-C_2-NH-C_2-NHAcr	4	80	7×10^{-7}
3 Ade-C_2-NH-C_3-NHAcr	0.7	23	$>10^{-5}$
4 DAP-C_2-NH-C_3-NHAcr	2	26	$>10^{-5}$
5 DAP-C_2-N(CH_3)-C_3-NHAcr	0.4	10	$>10^{-5}$
17 DAP-C_2-NHCO-C_2-NHCO-C_2-NHAcr	1.2	0	
8 DAP-C_2-NHCO-C_2-NH-C_3-NHAcr	5	23	$>10^{-5}$
9 DAP-C_2-NH-C_2-CONH-C_3-NHAcr	3.5	23	$>10^{-5}$
18 Ade-C_2-Gua-C_3-Gua-C_3-NHAcr	20	0	
29 Ade-C_2-NH-C_3-NH-C_3-NH_2		0	
30 AcrNH-C_3-NH-C_3-NH_2		0	

Plasmid concentration 2×10^{-9} M, incubation time 20 min, pH $= 7.4$, $T = 37°C$.
(a) Observed cleavage rate percentage for 1×10^{-5} M drug concentration.
(b) Drug concentration (M) inducing 50% cleavage.
Abbreviations: C_2, $(CH_2)_2$; C_3, $(CH_2)_3$; Acr, 6-chloro-2-methoxyacridine; ClQ, 7-chloroquinoline; Gua, guanidine.

lar way to that of the original artificial nucleases **1** or **2** [85]. Both side chains of the intercalated acridine are positioned in the minor groove.

The acridine–phenanthroline conjugate **16** is very promising as it cleaves efficiently and selectively the DNA strand at the base opposite the abasic site (Fig. 10.16). This molecule probably interacts with DNA by threading intercalation, i.e.

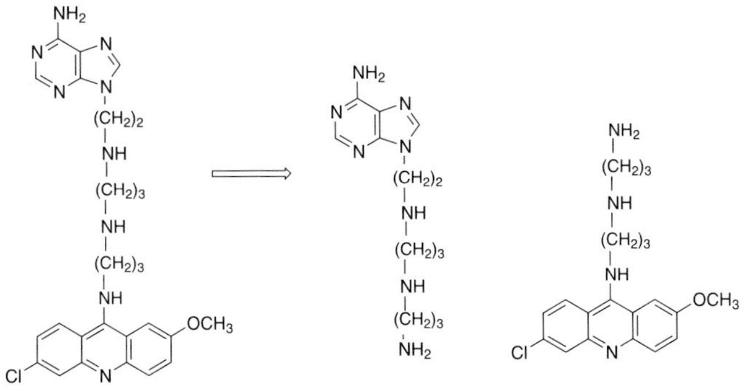

Fig. 10.12. "Half-molecules" **29** and **30**.

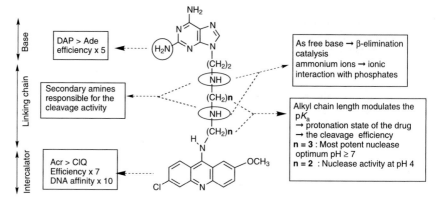

Fig. 10.13. Optimization of the AP site cleavage activity.

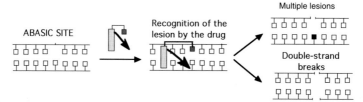

Fig. 10.14. Induction of multiply-damaged sites.

by positioning a side chain in each groove of the DNA [86, 87]. The presence of the chain joining the phenanthroline to the acridine 4-position confers to the conjugate the degree of freedom required for adequate positioning of the DNA-cleavage agent inside one groove. This compound thus seems to be a good candidate for the generation of double-stranded DNA cleavage (via its intrinsic β-elimination cleaving activity on the abasic strand and radical cleavage on the opposite strand).

10.2.5
Drug–DNA Interaction

The mode of recognition of the abasic sites by drugs and the structure of the drug/DNA complexes were investigated by studying the interaction of the drugs with a

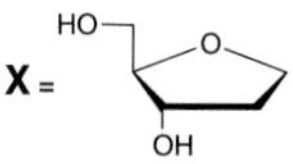

$$X = \text{[structure: 2-deoxyribose tetrahydrofuran with HO— and OH groups]}$$

Strand 1: $^{5'}C_1\ G_2\ C_3\ G_4\ T_5\ A_6\ C_7\ G_8\ C_9\ A_{10}C_{11}\mathbf{X_{12}}C_{13}A_{14}C_{15}G_{16}C_{17}A_{18}T_{19}G_{20}C_{21}G_{22}C_{23}\ ^{3'}$

Strand 2: $^{3'}G_{46}C_{45}G_{44}C_{43}A_{42}T_{41}G_{40}C_{39}G_{38}T_{37}G_{36}\mathbf{T}_{35}G_{34}T_{33}G_{32}C_{31}G_{30}T_{29}A_{28}C_{27}G_{26}C_{25}G_{24}\ ^{5'}$

Fig. 10.15. A 23-oligomer containing the tetrahydrofuran analog.

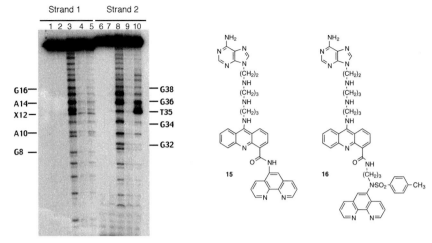

Fig. 10.16. DNA cleavage induced by phenanthroline-acridines **15** and **16**. Lanes 1 and 6: DNA alone; lanes 2 and 7: Cu; lanes 3 and 8: phenanthroline-Cu; lanes 4 and 9: **15** and Cu; lanes 5 and 10: **16** and Cu. (Mercaptopropionic acid, MPA, was used as reducing agent.)

short (11-mer) synthetic DNA duplex containing the stable tetrahydrofuran analog of the abasic site. The sequences are shown in Fig. 10.17.

Different methods such as thermal denaturation studies, ^{1}H NMR, and electron paramagnetic resonance (EPR) were used. UV Spectrophotometry is a simple method used to study the thermal denaturation of DNA duplexes. We determined the T_m values (melting temperatures) of the abasic site-containing duplex (TX) and of the conventional TA duplex in the absence and presence of interacting drugs [88]. The loss of a base leads to a significant decrease of T_m value (37°C for the TX duplex versus 55°C for the TA duplex in 10 mM phosphate buffer at pH 7, containing 20 mM NaCl). We investigated the effects of drugs such as **1**, **2**, **30**, and ethidium bromide (BET) on TA and TX duplexes. All four molecules stabilized the TA duplex quite efficiently due to non-specific binding of the intercalating moiety and ionic interaction contribution of the amino chain. The stabilizing effect of these drugs on the abasic duplex was considerably increased. At low drug-to-AP-site ratios (less than 1), the specific contribution of the recognition of the abasic pocket by the nucleobase of the drug could be evaluated (the T_m increase was more significant for **1** and **2** than for **30** and BET). At higher drug concentrations, the

CGCACXCACGC   CGCACACACGC
GCGTGTGTGCG   GCGTGTGTGCG

Abasic site containing   Natural duplex (TA)
duplex (TX)

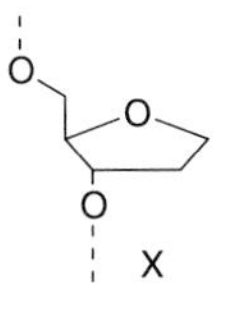

Fig. 10.17. Sequence of the synthetic oligonucleotide used for interaction studies.

Fig. 10.18. CPK representations of the abasic oligomer (left) and of its complex with **1**. The strand containing the abasic lesion is shown in red, the complementary strand in blue, and the interacting drug **1** in green [30].

specific contribution of the recognition moiety was overcome by the ionic interaction contribution of the protonated amino groups of the aliphatic chain (the T_m increases for **2** and **30** became similar). Other compounds, e.g. guanidinium-containing molecules **20–28**, were also tested, and similar T_m increases were measured [76].

High-field NMR spectroscopy coupled to molecular modeling revealed a similar mode of binding for molecules **1** and **2** [30]. Two complexes are formed in which the purine base of the drug inserts into the abasic pocket and pairs with thymine in the opposite strand, most probably in the Hoogsteen mode. The acridine intercalates at a two-base-pair distance 5′ to the abasic site, and the polyamino chain lies in the minor groove (Fig. 10.18). The two complexes only differ by a 180° rotation of the acridine ring around the C9–N bond.

Compound **18**, the analog of **1** that possesses two guanidinium functions in the linker [79], also forms specific 1:1 complexes with the stable model abasic duplex (TX), as indicated by thermal denaturation experiments and high-field NMR spectroscopy. Two types of specific complexes were identified by ^{1}H NMR. In both complexes the adenine moiety of the drug is inserted into the abasic cavity. Watson–Crick base-pairing of this adenine with thymine in the opposite strand was definitely established by measurement of characteristic NOEs between protons of the complementary bases (T17–H3 and Ade–H2 of the drug).

EPR studies of the interaction were done with compound **32**, a spin-labeled analog of **1** in which a nitroxide probe was grafted on the acridine 2-position (Scheme 10.5). Its binding to abasic oligonucleotide provided additional data on the kinetics of complex formation [89, 90].

The observed correlation time is significantly increased by the interaction with the abasic duplex ($\tau_C = 1.3 \times 10^{-10}$ s for the free probe in solution and $\tau_C = 6 \times 10^{-9}$ s in the presence of the abasic duplex). More interestingly, it was shown

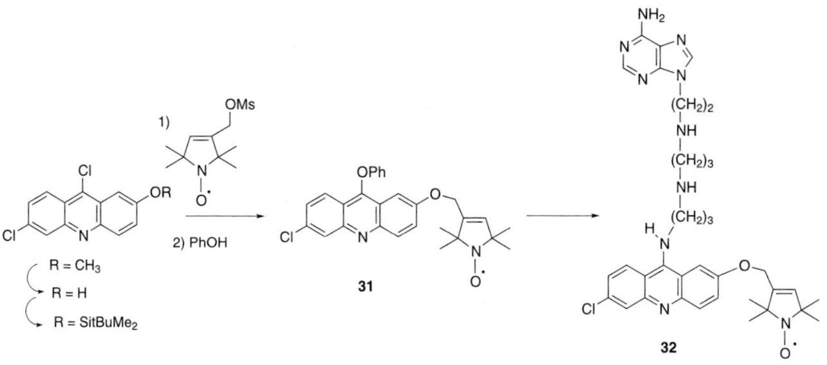

Scheme 10.5. Synthesis of the nitroxide labeled compound **32** [89].

by this technique that the affinity of **32** is much higher for the abasic duplex than for the natural duplex ($K_{aff} = 1.5 \times 10^6$ M^{-1} and 4×10^4 M^{-1} respectively).

10.2.6
Enzyme Inhibition

As discussed in the introduction, abasic sites are formed as intermediates during the repair of modified bases produced by anticancer alkylating agents through the BER mechanism. Molecules that bind specifically and strongly to abasic sites may inhibit this repair process and thus sensitize tumor cells by potentiating the activity of alkylating drugs.

The BER pathway has only been recently suggested as a target for such a potentiation and very few BER inhibitors have been reported. The first inhibitor mentioned in the literature was methoxyamine, a molecule that reacts with abasic sites and inhibits AP endonucleases *in vitro*, but at very high doses (50 mM) [91]. Malvy *et al.* reported that 9-aminoellipticine (9-AE) and the structurally related iso-propyl-oxazolopyridocarbazole potentiate the cytotoxic effect of alkylating drugs such as dimethyl sulfate in *E. coli* through a mechanism involving apurinic sites [92]. They later showed that, *in vitro*, 9-AE inhibits 65% of the endonuclease activity of *E. coli* exonuclease III [93]. It was suggested that inhibition might result from irreversible adduct formation at the 3′-side of the apurinic site following cleavage [26, 63, 94].

In our laboratory we have been developing for the last decade a new concept of inhibition in relation to the AP-site. This strategy is based on specific recognition of the abasic lesion by the synthetic heterodimeric compounds described above. By masking the AP-site to AP endonucleases these compounds may inhibit their repair activity. Even the heterodimers acting as artificial endonucleases might increase the toxicity of DNA-damaging agents whose adducts are repaired by the BER pathway, since this chemical cleavage activity occurs irrespective of the co-regulation proteins involved in the BER pathway. The BER inhibition was investigated [95] using the 3-D assay, a chemiluminescence microplate assay summarized by the following four steps:

1. A plasmid DNA is adsorbed in the wells of sensitized microplates.
2. DNA is treated with methyl methane sulfonate (MMS).
3. The compounds to be tested are incubated with a mixture containing all the constituents necessary for the *in vitro* DNA repair reaction [96]. During this reaction, damaged DNA fragments are excised and replaced by neo-synthesized DNA fragments incorporating digoxygenylated dUTPs.
4. The DNA repair activity is quantified with an anti-digoxygenin antibody conjugated with alkaline phosphatase. The conjugated enzyme dephosphorylates a substrate (lumigen) which becomes fluorescent.

The results indicated that **2** and **1** were potent inhibitors of BER with IC_{50} values of 70 and 62 μM respectively. These compounds were clearly more efficient than methoxyamine (480 μM), a BER inhibitor used as a reference [97].

Since the BER pathway involves key steps such as abasic site incision by specific AP endonucleases, some of the heterodimeric compounds described above were tested for their ability to inhibit the AP-endonucleolytic activity of exonuclease III *in vitro*. Exonuclease III is a class II AP endonuclease, which accounts for more than 80% of the cellular AP endonuclease activity in *E. coli* [98]. The substrates of exonuclease III were synthetic oligonucleotides (23-mers) containing the abasic analog X (Fig. 10.19).

The sequences were chosen in order to form duplexes containing the four different abasic sites (A, C, G, or T facing the lesion).

We tested the effect of **1** and **2** on the activity of exonuclease III. For comparison, we also investigated molecules devoid of recognition motif: **30** and BET, a classical intercalator. In our conditions, the four drugs decreased very slightly the cleavage activity of exonuclease III at apyrimidic sites (A or G opposite X). In contrast, a significant inhibitory effect was observed with **1**, **2** and even **30** at the apurinic site where a thymine is facing the lesion. At a drug-to-AP-site ratio of 2, the percentage of cleavage by the enzyme decreased from 30% to 15% after an incubation of 30 min (unpublished results). All drugs were ineffective on the other apurinic site where a cytosine faces the abasic lesion.

The specific effect of **1** and **2** was interpreted in terms of a specific interaction in which the nucleobase of the drug (adenine or diaminopurine) pairs with the complementary single thymine of the baseless duplex. The surprising inhibitory effect

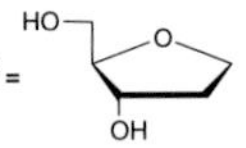

Strand 1: $^{5'}C_1\ G_2\ C_3\ G_4\ T_5\ A_6\ C_7\ G_8\ C_9\ A_{10}C_{11}\mathbf{X_{12}}C_{13}A_{14}C_{15}G_{16}C_{17}A_{18}T_{19}G_{20}C_{21}G_{22}C_{23}\ ^{3'}$

Strand 2: $^{3'}G_{46}C_{45}G_{44}C_{43}A_{42}T_{41}G_{40}C_{39}G_{38}T_{37}G_{36}N_{35}G_{34}T_{33}G_{32}C_{31}G_{30}T_{29}A_{28}C_{27}G_{26}C_{25}G_{24}\ ^{5'}$

N = A, G, C, T

Fig. 10.19. 23-mer duplexes used for enzyme inhibition experiments.

(a decrease of the cleavage activity from 30% to 20% in the presence of two drugs per lesion) of **30** (devoid of abasic site recognition unit) at both apurinic sites was attributed to a sequence context effect. In our duplex, the local environment at apurinic sites (GpyrG facing CXC) might be more flexible than at apyrimidic sites (GpurG facing CXC). This flexibility might allow the acridine insertion in the apurinic pocket with a significant contribution of the protonated amino chain to the binding. Similar observations were done when investigating the specific recognition of apurinic sites by these same compounds using melting temperature (T_m) measurements (see Section 10.2.3).

In a similar approach, **19** was tested for exonuclease III inhibition on the natural abasic site (2′-deoxyribose). Compound **19** (Fig. 10.10) shows no cleavage activity since the amino groups of the linker in **1** were replaced by one amido and one guanidine functions (fully protonated at pH 7). At a concentration of one molecule of **19** for six base pairs, a 50% inhibition of the AP endonucleolytic activity of exonuclease III was observed on a depurinated plasmid DNA (pBR322) containing two apurinic sites per DNA molecule [78]. The overall effects of these drugs on mammalian cells have been also investigated (see Section 10.3).

In addition, more recently, it has been reported that genomic damage, such as abasic sites, may influence the activities of topoisomerases [12, 13]. This represents an additional impulse for designing abasic site-specific drugs.

10.3
Pharmacological Data

Compounds **1** and **2** tested against L1210 cultured cells are moderately cytotoxic (IC_{50} 1.5 and 2 µM respectively) but more interestingly potentiate the DNA-methylating agent methyl methane sulfonate (MMS) as the IC_{50} drops from 350 µM with MMS alone to 10 µM with MMS combined with either **1** (0.7 µM) or **2** (1 µM) [95]. In the same study, sensitization to bis-chloroethylnitrosourea (BCNU), a clinically useful anticancer drug was also observed. In the presence of BCNU the IC_{50} value shifts from 5.6 µM with BCNU alone to 0.43 or 0.85 µM with a combination of BCNU and either **1** (0.7 µM) or **2** (1 µM). A synergistic effect was also observed with thiotepa used in combination with **2**, but none with alkylating agents such as mitomycin C and streptozocin [95].

Following this study, the heterodimers **18–28** specifically designed as DNA-repair inhibitors (devoid of DNA-cleavage activity) were also investigated and compared with the artificial nucleases **1** and **2**. Intrinsic cytostatic activity was estimated by measuring the dose-dependent growth inhibition of L1210 murine leukemia cells. Cytostatic/cytotoxic properties as well as synergistic potency with BCNU were determined by measuring the clonogenicity of A549 human pulmonary carcinoma cells. Results are given in Tab. 10.2 [76, 77, 79].

Compared with molecules **1** and **2**, the cytostatic activity of **18–28** alone on L1210 cells appeared to be in the same range and rather weak, and cytotoxicity on A549 was moderate.

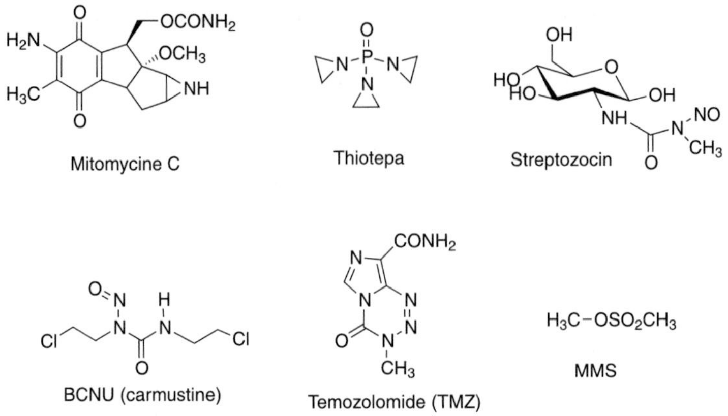

Mitomycine C

Thiotepa

Streptozocin

BCNU (carmustine)

Temozolomide (TMZ)

$H_3C-OSO_2CH_3$

MMS

Fig. 10.20. Alkylating agents used for potentiation studies.

Tab. 10.2. Binding constants (K_a) and cytostatic/cytotoxic properties (IC$_{50}$ alone and in association with BCNU) of the drugs tested as DNA-repair inhibitors.

Compound	IC_{50} L1210 (μM)[a] (Growth inhibition)	IC_{50} A549 (μM)[a] (Clonogenicity inhibition)	Synergy[b]	K_a ($\times 10^{-5}$ M^{-1})
1	>10[d]	0.1	+[c]	2
2	>10[d]	0.05	−	11
18	33	10	+94%	20
19	>100	4	−	2
20	74	3	−	6
21	26	6	−	8
22	>100	28	+74%	0.6
23	60	7	+16%	3.7
24	>100	31	−	2.1
25	>100	43	+49%	4
26	36	5	−	2
27	>100	5	−	3
28	70	1	−	2.2

[a] Tested compounds were added at various concentrations and incubated for 24 h.
[b] Apparent synergy was measured by simultaneous exposition of A549 cells to 10 μM of the tested compound and 10 μM BCNU and expressed as the % of toxicity increase with regard to the addition of the toxic effects of both drugs alone (theoretical combined toxicity) taken as 100%.
[c] +25% synergy observed after exposition to 0.05 μM of 1 and 10 μM BCNU. Cytotoxicity was too strong at higher doses of 1.
[d] IC$_{50}$ of respectively 1.6 and 2.2 μM have been previously observed after 48 h incubation with L1210 cells [95].

Tab. 10.3. Activity of compound **18** alone or in association with BCNU on the P388 murine leukemia.

Treatment	Average survival (days ± s.d.)	T/C (Survival of treated animals/Survival of controls) × 100	Number of surviving animals at 60 days
Controls	14.9 ± 8	–	0/7
18 (2 mg kg^{-1})	16.2 ± 3	109	0/6
BCNU (6 mg kg^{-1})	46 ± 17	309	3/6
18 (2 mg kg^{-1}) + BCNU (6 mg kg^{-1})	–	–	6/6

Compounds **18**, **22**, **23**, and **25** displayed a significant synergy with BCNU. Interestingly, **22** and **25** are among the less toxic compounds on both cell lines. The most active compound, **18**, was also tested *in vivo* on the murine leukemia P388 (Tab. 10.3). Compound **18** showed no antitumor property by itself but definitely potentiated the action of BCNU (treatment with BCNU alone led to a 60-day survival rate of 50%, while association with **18** resulted in 100% survival). However this compound elicited curare-like acute toxicity, possibly due to the presence of two guanidinium cations in the linker, that severely limited further *in vivo* experiments [79].

Recently, Taverna *et al.* reported sensitization of resistant cell lines to chemotherapeutic alkylating agents by methoxyamine (MX) [99]. Methoxyamine increases significantly the cytotoxicity of temozolomide (TMZ) against HCT116 colon cancer cells and SW480 cell line (Tab. 10.4). HCT116 cells are deficient in DNA mismatch repair (MMR) and are therefore resistant to methylating agents such as TMZ and nitrosourea.

Due to the extreme complexity of the system, these biological data cannot be directly interpreted in terms of mechanism of action. Nevertheless, most of the compounds that were designed to bind specifically and strongly to the abasic site present the most interesting activities *in vitro* and even *in vivo*, most particularly in association with alkylating and/or anticancer drugs that create multiple abasic lesions in the cell. It thus seems that targeting the abasic site by specific drugs may be a pertinent and useful strategy to improve the clinical efficiency of currently available DNA-damaging agents.

Tab. 10.4. Synergistic activity of methoxyamine.

Cell line	TMZ IC$_{50}$ (μM)	MX IC$_{50}$ (mM)	MX + TMZ IC$_{50}$ (μM)
SW480	395	50	172
HCT116	950	28	306

Cells are exposed to 0–1500 μM TMZ for 2 h or 6 mM MX and TMZ for 2 h.

Dedication

We dedicate this chapter to Professor Jean Lhomme on the occasion of his retirement. Together with the many colleagues who participated in the long and stimulating story of abasic site recognition, we wish Professor Jean Lhomme an enjoyable retirement. He helped us to progress in the fascinating world of bioorganic chemistry, and many young scientists benefited from his teaching and advice.

Acknowledgments

Many parts of this work were financed by the Ligue Nationale contre le Cancer and Association pour la Recherche sur le Cancer (ARC). We also thank the Région Rhône-Alpes for financial support. The results contained in this article summarize the work of many other colleagues; we wish to thank them and all students who participated to this project.

References

1 LINDAHL, T., KARRAN, P., WOOD, R. D. DNA excision repair pathways. *Curr. Opin. Genet. Devel.* **1997**, *7*, 158–169.

2 BARRET, J. M., HILL, B. T. DNA repair mechanisms associated with cellular resistance to antitumor drugs: potential novel targets. *Anti-Cancer Drugs* **1998**, *9*, 105–123.

3 BERTHET, N., BOTURYN, D., CONSTANT, J.-F. DNA repair inhibitors. *Exp. Opin. Ther. Patents* **1999**, *9*, 401–415.

4 ENGELWARD, B. P., DRESLIN, A., CHRISTENSEN, J. Repair-deficient 3-methyladenine DNA glycosylase homozygous mutant mouse cells have increased sensitivity to alkylation-induced chromosome damage and cell killing. *EMBO J.* **1996**, *15*, 945–952.

5 DAVID, S. S., WILLIAMS, S. D. Chemistry of glycosylases and endonucleases involved in base-excision repair. *Chem. Rev.* **1998**, *98*, 1221–1261.

6 KROKAN, H. E., STANDAL, R., SLUPPHAUG, G. DNA glycosylases in the base excision repair of DNA. *Biochem. J.* **1997**, *325*, 1–16.

7 BOGENHAGEN, D. F., PINZ, K. G. The action of DNA ligase at abasic sites in DNA. *J. Biol. Chem.* **1998**, *273*, 7888–7893.

8 KINGMA, P. S., OSHEROFF, N. The response of eukariotic topoisomerases to DNA damage. *Biochim. Biophys. Acta* **1998**, *1400*, 223–232.

9 KINGMA, P. S., OSHEROFF, N. Spontaneous DNA damage stimulates topoisomerase II-mediated DNA cleavage. *J. Biol. Chem.* **1997**, *272*, 7488–7493.

10 KINGMA, P. S., OSHEROFF, N. Apurinic sites are position-specific topoisomerase II poisons. *J. Biol. Chem.* **1997**, *272*, 1148–1155.

11 KINGMA, P. S., OSHEROFF, N. Topoisomerase II-mediated DNA cleavage and religation in the absence of base-pairing. Abasic lesions as a tool to dissect enzyme mechanism. *J. Biol. Chem.* **1998**, *273*, 17999–18002.

12 KINGMA, P. S., CORBETT, A. H., BURCHAM, P. C., MARNETT, L. J., OSHEROFF, N. Abasic sites stimulate double-stranded DNA cleavage mediated by topoisomerase II.DNA lesions as endogenous topoisomerase II poisons. *J. Biol. Chem.* **1995**, *270*, 21441–21444.

13 POURQUIER, P., UENG, L.-M., KOHLHAGEN, G., MAZUMDER, A., GUPTA, M., KOHN, K. W., POMMIER, Y. Effects of uracil incorporation, DNA

mismatches, and abasic sites on cleavage and religation activities of mammalian topoisomerase I. *J. Biol. Chem.* **1997**, *272*, 7792–7796.

14 WILSTERMANN, A. M., OSHEROFF, N. Base excision repair intermediates as topoisomerase II poisons. *J. Biol. Chem.* **2001**, *276*, 46290–46296.

15 ZIEGLER, M., OEI, S. L. A cellular survival switch: poly(ADP-ribosyl)ation stimulates DNA repair and silences transcription. *Bioessays* **2001**, *23*, 543–548.

16 MANOHARAN, M., GERLT, J. A. Coexistence of conformations revealed by site specific labeling with ^{13}C-labeled nucleotides. *J. Am. Chem. Soc.* **1987**, *109*, 7217–7219.

17 MANOHARAN, M., RANSOM, S. C., MAZUMDER, A. *et al.* The characterization of abasic sites in DNA heteroduplexes by site specific labeling with carbon-13. *J. Am. Chem. Soc.* **1988**, *110*, 1620–1622.

18 WILDE, J. A., BOLTON, P. H., MAZUMDER, A., MANOHARAN, M., GUERLT, J. A. Characterization of the equilibrating forms of the aldehydic abasic site in duplex DNA by ^{17}O NMR. *J. Am. Chem. Soc.* **1989**, *111*, 1894–1896.

19 RAAP, J., DREEF, C. E., VAN DER MAREL, G. A., VAN BOOM, J. H., HILBERS, C. W. Synthesis and proton-NMR studies of oligonucleotides containing an apurinic (AP) site. *J. Biomol. Struct. Dyn.* **1987**, *5*, 219–247.

20 COPPEL, Y., BERTHET, N., COULOMBEAU, C., GARCIA, J., LHOMME, J. Solution conformation of an abasic DNA undecamer duplex d(CGCACXCACGC) x d(GCGTGTGTGCG): the unpaired thymine stacks inside the helix. *Biochemistry* **1997**, *36*, 4817–4830.

21 CUNIASSE, P., SOWERS, L. C., ERITJA, R. *et al.* Abasic frameshift in DNA. Solution conformation determined by proton NMR and molecular mechanics calculations. *Biochemistry* **1989**, *28*, 2018–2026.

22 CUNIASSE, P., SOWERS, L. C., ERITJA, R. *et al.* An abasic site in DNA. Solution conformation determined by proton NMR and molecular mechanics calculations. *Nucleic Acids Res.* **1987**, *15*, 8003–8022.

23 KALNIK, M. W., CHANG, C. N., GROLLMAN, A. P., PATEL, D. J. NMR studies of abasic sites in DNA duplexes: deoxyadenosine stacks into the helix opposite the cyclic analogue of 2-deoxyribose. *Biochemistry* **1988**, *27*, 924–931.

24 GOLJER, I., WITHKA, J. M., KAO, J. Y., BOLTON, P. H. Effects of the presence of an aldehydic abasic site on the thermal stability and rates of helix opening and closing of duplex DNA. *Biochemistry* **1992**, *31*, 11614–11619.

25 GOLJER, I., KUMAR, S., BOLTON, P. H. Refined solution structure of a DNA heteroduplex containing an aldehydic abasic site. *J. Biol. Chem.* **1995**, *270*, 22980–22987.

26 SINGH, M. P., HILL, G. C., PEOC'H, D., RAYNER, B., IMBACH, J. L., LOWN, J. W. High-field NMR and restrained molecular modeling studies on a DNA heteroduplex containing a modified apurinic abasic site in the form of covalently linked 9-aminoellipticine. *Biochemistry* **1994**, *33*, 10271–10285.

27 WITHKA, J. M., WILDE, J. A., BOLTON, P. H., MAZUMDER, A., GERLT, J. A. Characterization of conformational features of DNA heteroduplexes containing aldehydic abasic sites. *Biochemistry* **1991**, *30*, 9931–9940.

28 CLINE, S. D., JONES, R. W., STONE, M. P., OSHEROFF, N. DNA abasic lesions in a different light: solution structure of an endogenous topoisomerase II poison. *Biochemistry* **1999**, *38*, 15500–15507.

29 CUNIASSE, P., FAZAKERLEY, G. V., GUSCHLBAUER, W., KAPLAN, B. E., SOWERS, L. C. The abasic site as a challenge to DNA polymerase. A nuclear magnetic resonance study of G, C and T opposite a model abasic site. *J. Mol. Biol.* **1990**, *213*, 303–314.

30 COPPEL, Y., CONSTANT, J.-F., COULOMBEAU, C., DEMEUNYNCK, M., GARCIA, J., LHOMME, J. NMR and molecular modeling studies of the interaction of artificial AP lyases with a DNA duplex containing an apurinic abasic site model. *Biochemistry* **1997**, *36*, 4831–4843.

31 BAILLY, V., VERLY, W. G. Possible roles of β-elimination and δ-elimination reactions in the repair of DNA containing AP (apurinic/apyrimidinic) sites in mammalian cells. *Biochem. J.* **1988**, *253*, 553–559.

32 BAILLY, V., DERYDT, M., VERLY, W. G. Delta-elimination in the repair of AP (apurinic/apyrimidinic) sites in DNA. *Biochem. J.* **1989**, 261, 707–713.

33 McHUGH, P. J., KNOWLAND, J. Novel reagents for chemical cleavage at abasic sites and UV photoproducts in DNA. *Nucleic Acids Res.* **1995**, 23, 1664–1670.

34 MANOHARAN, M., MAZUMDER, A., RANSOM, S. C., GERLT, J. A., BOLTON, P. H. Mechanism of UV endonuclease V cleavage of abasic sites in DNA determined by carbon-13 labeling. *J. Am. Chem. Soc.* **1988**, *110*, 2690–2691.

35 MAZUMDER, A., GERLT, J. A., ABSALON, M. J., STUBBE, J., CUNNINGHAM, R. P., WITHKA, J., BOLTON, P. H. Stereochemical studies of the beta-elimination reactions at aldehydic abasic sites in DNA: endonuclease III from *Escherichia coli*, sodium hydroxide, and Lys-Trp-Lys. *Biochemistry* **1991**, *30*, 1119–1126.

36 MAZUMDER, A., GERLT, J. A., RABOW, L., ABSALON, M. J., STUBBE, J.-A., BOLTON, P. H. UV endonuclease V from bacteriophage T4 catalyzes DNA strand cleavage at aldehydic abasic sites by a syn β-elimination reaction. *J. Am. Chem. Soc.* **1989**, *111*, 8029–8030.

37 MOL, C. D., KUO, C. F., THAYER, M. M., CUNNINGHAM, R. P., TAINER, J. A. Structure and function of the multifunctional DNA-repair enzyme exonuclease III. *Nature* **1995**, *374*, 381–386.

38 GORMAN, M. A., MORERA, S., ROTHWELL, D. G. et al. The crystal structure of the human DNA repair endonuclease HAP1 suggests the recognition of extra-helical deoxyribose at DNA abasic sites. *EMBO J.* **1997**, *16*, 6548–6558.

39 DOETSCH, P. W., CUNNINGHAM, R. P. The enzymology of apurinic/apyrimidinic endonucleases. *Mutat. Res.* **1990**, *236*, 173–201.

40 BAILLY, V., VERLY, W. G. AP endonucleases and AP lyases. *Nucleic Acids Res.* **1989**, *17*, 3617–3618.

41 McCULLOUGH, A. K., SANCHEZ, A., DODSON, M. L., MARAPAKA, P., TAYLOR, J. S., LLOYD, R. S. The reaction mechanism of DNA glycosylase/AP lyases at abasic sites. *Biochemistry* **2001**, *40*, 561–568.

42 LATHAM, K. A., LLOYD, R. S. Delta-elimination by T4 endonuclease V at a thymine dimer site requires a secondary binding event and amino acid Glu-23. *Biochemistry* **1995**, *34*, 8796–8803.

43 DOGLIOTTI, E., FORTINI, P., PASUCCI, B. Mutagenesis of abasic sites, in *Base Excision Repair of DNA Damage*, Landes Bioscience, Austin, Texas, USA, 1997, 81–101.

44 KUNKEL, T. A. Mutational specificity of depurination. *Proc. Natl Acad. Sci. USA* **1984**, *81*, 1494–1498.

45 LOEB, L., PRESTON, B. Mutagenesis by apurinic/apyrimidinic sites. *Annu. Rev. Genet.* **1986**, *20*, 201–230.

46 SCHAAPER, R. M., KUNKEL, T. A., LOEB, L. A. Infidelity of DNA synthesis associated with Bypass of apurinic sites. *Proc. Natl Acad. Sci. USA* **1983**, *80*, 487–491.

47 GENTIL, A., CABRAL-NETO, J. B., MARIAGE-SAMSON, R. et al. Mutagenicity of a unique apurinic/apyrimidinic site in mammalian cells. *J. Biol. Chem.* **1992**, *227*, 981–984.

48 BOITEUX, S., LAVAL, J. Coding properties of poly (deoxycytidilic acid) templates containing uracil or apyrimidinic sites: in vitro modulation of mutagenesis by deoxyribonucleic acid repair enzymes. *Biochemistry* **1982**, *21*, 6746–6751.

49 CAI, H., BLOOM, L. B., ERITJA, R., GOODMAN, M. F. Kinetics of deoxyribonucleotide insertion and extension at abasic template lesions in different sequence contexts using HIV-1 reverse transcriptase. *J. Biol. Chem.* **1993**, *268*, 23567–23572.

50 RANDALL, S. K., ERITJA, R., KAPLAN, B. E., PETRUSKA, J., GOODMAN, M. F. Nucleotide insertion kinetics opposite abasic lesions in DNA. *J. Biol. Chem.* **1987**, *262*, 6864–6870.

51 TAKESHITA, M., CHANG, C. N., JOHNSON, F., WILL, S., GROLLMAN, A. P. Oligodeoxynucleotides containing synthetic abasic sites. *J. Biol. Chem.* **1987**, *262*, 10171–10179.

52 PAZ-ELIZUR, T., TAKESHITA, M., LIVNEH, Z. Mechanism of bypass synthesis through an abasic site analog by DNA polymerase I. *Biochemistry* **1997**, *36*, 1766–1773.

53 SHIBUTANI, S., TAKESHITA, M., GROLLMAN, A. P. Translesional synthesis on DNA templates containing a single abasic site. A mechanistic study of the "A rule". *J. Biol. Chem.* **1997**, *272*, 13916–13922.

54 HEVRONI, D., LIVNEH, Z. Bypass and termination at apurinic sites during replication of single-stranded DNA in vitro: a model for apurinic sites mutagenesis. *Proc. Natl Acad. Sci. USA* **1988**, *85*, 5046–5050.

55 TALPAERT-BORLÉ, M., LIUZZI, M. Reaction of apurinic/apyrimidinic sites with [^{14}C] methoxyamine, a method for the quantitative assay of AP sites in DNA. *Biochem. Biophys. Acta* **1983**, *740*, 410–416.

56 TALPAERT-BORLÉ, M., LIUZZI, M. A process for directly determining apurinic and apyrimidinic sites in DNA, in *Eur. Patent Application*, O122 507 A2, 1984.

57 NAKAMURA, J., SWENBERG, J. A. Endogenous apurinic/apyrimidinic sites in genomic DNA of mammalian tissues. *Cancer Res.* **1999**, *59*, 2522–2526.

58 NAKAMURA, J., WALKER, V. E., UPTON, P. B., CHIANG, S. Y., KOW, Y. W., SWENBERG, J. A. Highly sensitive apurinic/apyrimidinic site assay can detect spontaneous and chemically induced depurination under physiological conditions. *Cancer Res.* **1998**, *58*, 222–225.

59 ASAEDA, A., IDE, H., TANO, K., TAKAMORI, Y., KUBO, K. Repair kinetics of abasic sites in mammalian cells selectively monitored by the aldehyde reactive probe (ARP). *Nucleosides Nucleotides* **1998**, *17*, 503–513.

60 ASAEDA, A., IDE, H., TERATO, H., TAKAMORI, Y., KUBO, K. Highly sensitive assay of DNA abasic sites in mammalian cells-optimization of the aldehyde reactive probe method. *Anal. Chim. Acta* **1998**, *365*, 35–41.

61 BOTURYN, D., BOUDALI, A., CONSTANT, J.-F., DEFRANCQ, E., LHOMME, J. Synthesis of fluorescent probes for the detection of abasic sites in DNA. *Tetrahedron* **1997**, *53*, 5485–5492.

62 BOTURYN, D., CONSTANT, J. F., DEFRANCQ, E., LHOMME, J., BARBIN, A., WILD, C. P. A simple and sensitive method for in vitro quantitation of abasic sites in DNA. *Chem. Res. Toxicol.* **1999**, *12*, 476–482.

63 BERTRAND, J. R., VASSEUR, J.-J., RAYNER, B. *et al.* Synthesis, thermal stability and reactivity towards 9-aminoellipticine of double-stranded oligonucleotides containing a true abasic site. *Nucleic Acids Res.* **1989**, *17*, 10307–10319.

64 BERTRAND, J. R., VASSEUR, J. J., GOUYETTE, A. *et al.* Mechanism of cleavage of apurinic sites by 9-aminoellipticine. *J Biol. Chem.* **1989**, *264*, 14172–14178.

65 MALVY, C., PREVOST, P., GANSSER, C., VIEL, C., PAOLETTI, C. Efficient breakage of DNA apurinic sites by the indoleamine related 9-aminoellipticine. *Chem. -Biol. Interactions* **1986**, *57*, 41–53.

66 BERTRAND, J.-R., MALVY, C., PAOLETTI, C. Quantification by fluorescence of apurinic sites in DNA. *Biochem. Biophys. Res. Commun.* **1987**, *143*, 768–774.

67 BEHMOARAS, T., HÉLÈNE, C. A tryptophan-containing peptide recognizes and cleaves DNA at apurinic sites. *Nature* **1981**, *292*, 858–859.

68 BEHMOARAS, T., TOULME, J. J., HELENE, C. Specific recognition of apurinic sites in DNA by a tryptophan-containing peptide. *Proc. Natl Acad. Sci. USA* **1981**, *78*, 926–930.

69 DUKER, N. J., HART, D. M. Cleavage of DNA at apyrimidinic sites by lysyl-tryptophyl-alpha-lysyl. *Biochem. Biophys. Res. Commun.* **1982**, *105*, 1433–1439.

70 PIERRE, J., LAVAL, J. Specific nicking of DNA at apurinic sites by peptides containing aromatic residues. *J. Biol. Chem.* **1981**, *256*, 10217–10220.

71 TEULADE-FICHOU, M. P., VIGNERON, J. P., LEHN, J. M. Molecular recogition of nucleosides and nucleotides by a water-soluble cyclo-bis-intercaland receptor based on acridine subunits. *J. Supramol. Chem.* **1995**, *5*, 139–147.

72 JOURDAN, M., GARCIA, J., LHOMME, J., TEULADE-FICHOU, M.-P., VIGNERON, J.-P., LEHN, J.-M. Threading bis-intercalation of a macrocycle Bisacridine at abasic sites in DNA: ¹H NMR and molecular modeling study. *Biochemistry* **1999**, *38*, 14205–14213.

73 BERTHET, N., MICHON, J., LHOMME, J., TEULADE-FICHOU, M.-P., VIGNERON, J.-P., LEHN, J.-M. Recognition of abasic sites in DNA by a cyclobisacri-dine molecule. *Chem. Eur. J.* **1999**, *5*, 3625–3630.

74 FKYERAT, A., DEMEUYNCK, M., CONSTANT, J.-F., LHOMME, J. Synthesis of purine-acridine hybrid molecules related to artificial endonucleases. *Tetrahedron* **1993**, *49*, 11237–11252.

75 FKYERAT, A., DEMEUYNCK, M., CONSTANT, J.-F., MICHON, P., LHOMME, J. A new class of artificial nucleases that recognize and cleave apurinic sites in DNA with great selectivity and efficiency. *J. Am. Chem. Soc.* **1993**, *115*, 9952–9959.

76 ALARCON, K., DEMEUYNCK, M., LHOMME, J., CARREZ, D., CROISY, A. Potentiation of BCNU cytotoxicity by molecules targeting abasic lesions in DNA. *Bioorg. Med. Chem.* **2001**, *9*, 1901–1910.

77 ALARCON, K., DEMEUYNCK, M., LHOMME, J., CARREZ, D., CROISY, A. Diaminopurine-acridine heterodimers for specific recognition of abasic site containing DNA. Influence on the biological activity of the position of the linker on the purine ring. *Bioorg. Med. Chem. Lett.* **2001**, *11*, 1855–1858.

78 BELMONT, P., DEMEUYNCK, M., CONSTANT, J.-F., LHOMME, J. Synthesis and study of a new adenine-acridine tandem, inhibitor of exonuclease III. *Bioorg. Med. Chem. Lett.* **2000**, *10*, 293–295.

79 BELMONT, P., JOURDAN, M., DEMEUYNCK, M. et al. Abasic site recognition in DNA as a new strategy to potentiate the action of anticancer alkylating drugs? *J. Med. Chem.* **1999**, *42*, 5153–5159.

80 BELMONT, P., BOUDALI, A., CONSTANT, J.-F. et al. Efficient and versatiles chemical tools for cleavage of abasic sites in DNA. *N. J. Chem.* **1997**, *21*, 47–54.

81 CHAUDHRY, M. A., WEINFELD, M. Reactivity of human apurinic/apyrimidinic endonuclease and *Escherichia coli* exonuclease III with bistranded abasic sites in DNA. *J. Biol. Chem.* **1997**, *272*, 15650–15655.

82 HARRISON, L., HATAHET, Z., PURMAL, A. A., WALLACE, S. S. Multiply damaged sites in DNA: interactions with Escherichia coli endonucleases III and VIII. *Nucleic Acids Res.* **1998**, *26*, 932–941.

83 MARTELLI, A., BERTHET, N., CONSTANT, J. F., DEMEUYNCK, M., LHOMME, J. The abasic site as a new target for generation of locally multiply damaged sites. *Bioorg. Med. Chem. Lett.* **2000**, *10*, 763–766.

84 MARTELLI, A., CONSTANT, J. F., DEMEUYNCK, M., LHOMME, J., DUMY, P. Design of site specific DNA damaging agents for generation of multiply damaged sites. *Tetrahedron* **2002**, in press.

85 AYADI, L., FORGET, D., MARTELLI, A., CONSTANT, J. F., DEMEUYNCK, M., LHOMME, J. Molecular modeling study of DNA abasic site. *Theor. Chem. Acc.* **2000**, *104*, 284–289.

86 HÉNICHART, J.-P., WARING, M. J., RIOU, J. F., DENNY, W. A., BAILLY, C. Copper-dependent oxidative and Topoisomerase II-mediated DNA cleavage by a netropsin/4'-(9-acridinylamino)methanesulfon-m-anisidine combilexin. *Mol. Pharmacol.* **1997**, *51*, 448–461.

87 WAKELIN, L. P. G., DENNY, W. A. Kinetic and equilibrium binding studies of a series of intercalating agents that bind by threading a sidechain through the DNA helix, in *Molecular Basis of Specificity in Nucleic Acid–Drug Interactions*, Kluwer Academic, Dordrecht, 1990, 191–206.

88 BERTHET, N., CONSTANT, J.-F., DEMEUNYNCK, M., MICHON, P., LHOMME, J. Search for DNA repair inhibitors: selective binding of nucleic bases-acridine conjugates to a DNA duplex containing an abasic site. *J. Med. Chem.* **1997**, *40*, 3346–3352.

89 BELMONT, P., CHAPELLE, C., DEMEUNYNCK, M., MICHON, J., MICHON, P., LHOMME, J. Introduction of a nitroxide group on position 2 of 9-phenoxyacridine: easy access to spin labelled DNA-binding conjugates. *Bioorg. Med. Chem. Lett.* **1998**, *8*, 669–674.

90 THOMAS, F., MICHON, J., LHOMME, J. Interaction of a spin-labeled adenine-acridine conjugate with a DNA duplex containing an abasic site model. *Biochemistry* **1999**, *38*, 1930–1937.

91 LIUZZI, M., WEINFELD, M., PATERSON, M. C. Selective inhibition by methoxyamine of the apurinic/apyrimidinic endonuclease activity associated with pyrimidine dimer-DNA glycosylase from *Micrococcus luteus* and bacteriophage T4. *Biochemistry* **1987**, *26*, 3315–3321.

92 MALVY, C., SAFRAOUI, H., BLOCH, E., BERTRAND, J. R. Involvement of apurinic sites in the synergistic action of alkylating and intercalating drugs in *Escherichia coli*. *Anti-Cancer Drug Design* **1988**, *2*, 361–370.

93 LEFRANÇOIS, M., BERTRAND, J. R., MALVY, C. 9-Amino-ellipticine inhibits the apurinic site-dependent base excision-repair pathway. *Mutat. Res.* **1990**, *236*, 9–17.

94 VASSEUR, J.-J., RAYNER, B., IMBACH, J.-L. *et al.* Structure of the adduct formed between 3-aminocarbazole and the apurinic site oligonucleotide moded d[Tp(Ap)pT]. *J. Org. Chem.* **1987**, *52*, 4994–4998.

95 BARRET, J. M., ETIEVANT, C., FAHY, J., LHOMME, J., HILL, B. T. Novel artificial endonucleases inhibit base excision repair and potentiate the cytotoxicity of DNA-damaging agents on L1210 cells. *Anticancer Drugs* **1999**, *10*, 55–65.

96 BARRET, J. M., SALLES, B., PROVOT, C., HILL, B. T. Evaluation of DNA repair inhibition by antitumor or antibiotic drugs using chemiluminescence microplate assay. *Carcinogenesis* **1997**, *18*, 2441–2445.

97 LIUZZI, M., TALPAERT-BORLÉ, M. A new approach to study of the base-excision repair pathway using methoxyamine. *J. Biol. Chem.* **1985**, *260*, 2552–2558.

98 KOW, Y. W. Mechanism of action of *Escherichia coli* exonuclease III. *Biochemistry* **1989**, *28*, 3280–3287.

99 TAVERNA, P., LIU, L., HWANG, H.-S., HANSON, A. J., KINSELLA, T. J., GERSON, S. L. Methoxyamine potentiates DNA single strand breaks and double strand breaks induced by temozolomide in colon cancer cells. *Mutat. Res.* **2001**, *485*, 269–281.

11
Interactions of Macrocyclic Compounds with Nucleic Acids

Marie-Paule Teulade-Fichou and Jean-Pierre Vigneron

11.1
Introduction

The selective recognition of single- and double-stranded nucleic acids, as well as their many secondary and tertiary structures, constitutes a challenge to chemists and biologists. Among the many DNA ligands discussed in the literature, the macrocyclic compounds are the most intriguing ones. Their very structure implies that it would be very difficult for them to bind to double-stranded (ds) DNA, at least in the intercalative mode. However, natural antibiotics of the triostin family, for instance, which are composed of a cyclic peptide with two intercalator quinoline rings, are able to associate with double-stranded DNA by inducing a strong local distortion of the double helix. This means that the double-stranded DNA matrix can accommodate large complex synthetic molecules. Moreover, increasing knowledge on nucleic acids structure, particularly on the conformational flexibility of double helices, makes it possible to envisage new binding modes for synthetic molecules.

In the past decade, a number of macrocyclic synthetic ligands have been designed and synthesized, inspired by both the cyclic structure of many natural compounds and the numerous studies centered on bisintercalators. Indeed, in addition to their high affinity for dsDNA, the latter, such as bisanthracyclines and ditercalinium, elicited significant antitumor activity which strongly stimulated research in this field. More or less simultaneously, several groups synthesized topologically constrained macrocyclic bisintercalators in order to test the ability of DNA to accommodate such structures and possibly to point up new binding modes or unprecedented structural binding selectivities.

In many cases, synthetic strategies and the design of the cyclic structures are rooted in studies on the complexation of nucleotides by synthetic receptors that were carried out in the past two decades. Indeed, work on the molecular recognition of nucleosides and nucleotides by well-defined synthetic models can shed light on the non-covalent interactions between nucleic acids and various important biological compounds. Moreover, the nucleotides not only participate in the fundamental mechanisms of storage, replication and transcription of genetic information as building blocks of nucleic acids, but also play a role in a variety of biological processes as chemical energy sources. This research topic led to the rapid development of various molecular architectures essentially based on acyclic dimeric aromatics (molecular tweezers) and macrocyclic polyamines. Noticeable synthetic difficulties came from the need for compounds with large aromatic moieties that were water soluble at a physiological pH. It is important to briefly summarize these studies before reviewing the macrocycle–nucleic acids interactions themselves.

11.2
Nucleotide Complexation

The selective complexation of nucleotides is the result of an interplay between many parameters; of these, electrostatic forces, hydrogen bonding, stacking interactions, and hydrophobic effects play the most important roles. Many multifunctional receptors, such as macrocyclic polyamines, azoniacyclophanes, and cyclo-intercaland compounds, have been designed so as to optimize the participation of each of these different contributions. So far, most of them make use of electrostatic and hydrophobic forces only, hydrogen bonding being highly disfavored in water.

11.2.1
Macrocyclic Polyamines

Like biological polyamines (i.e. putrescine, spermidine, and spermine), macrocyclic polyamines, when protonated, bind to nucleotides strongly and selectively by electrostatic interactions between the cationic ammonium groups of the receptor and the negatively charged phosphate groups [1]. The measured affinities are comparable to those found in enzyme–substrate complexes; for instance, when it is fully protonated, the [24]-N_6O_2 macrocycle **1** binds AMP^{2-}, ADP^{3-}, and ATP^{4-} with log K_s values of 6.95, 8.30, and 11.00, respectively [2, 3]. But, in order to improve the selectivity of the nucleotide recognition, macrocycles also need to contain other binding sites capable of interactions with the other parts of the molecule, the sugar moiety, and/or the nucleic base. Nucleic base recognition may be achieved either by stacking interactions or by hydrogen bonding; in the former case the distinction between different nucleic bases rests on differences in stacking energy, while in the latter it results from the presence in the receptor of sites capable of forming complementary hydrogen-bonding patterns. A combination of both is necessary to obtain strong recognition and good selectivity.

1   X = Y= R = H

2a   X = Y= H, R = (CH$_2$)$_3$NHZ

2b   X= H, Y = R = (CH$_2$)$_3$NHZ

Compound **2a** was the first synthetic multifunctional structure able to bind nucleotides by an interplay of electrostatic and stacking interactions [1, 4]. It combines a macrocyclic polyammonium moiety as anion-binding site and an acridine side arm for stacking interactions. Due to this double recognition, protonated **2a** forms stronger complexes with the nucleotides than the parent macrocycle **1**, which does not possess the acridine side arm. However, protonated **2a** is unable to discriminate between the different nucleic bases of the nucleotides. To perform molecular recognition of nucleic bases more specific information features must be introduced in the structure. Of note, **2a** binds strongly to the supercoiled circular dsDNA plasmid pBR322, probably via double interaction involving both intercalation of the acridine subunit within the base pairs and electrostatic attraction between the ammonium groups of the receptor and the phosphate groups of DNA.

The bis-functional analog **2b** possessing two acridine subunits was synthesized subsequently [5]. Unexpectedly, this structural modification did not provide the macrocycle with higher affinities either for nucleotides or for dsDNA as compared with **2a**.

11.2.2
Azoniacyclophanes

Adding stacking interactions to coulombic forces was also the purpose of the design of azoniacyclophanes **3–8**, but the binding of AMP^{2-}, ADP^{3-}, and ATP^{4-} by these ligands results essentially from electrostatic interactions, although NMR studies provided unambiguous evidence for participation of π-stacking to the stabilization of the various complexes [6–9].

Compared with the synthetic receptors studied so far, the **CPnn** azoniacyclophanes **9–12** delineate more structured cavities. These lipophilic compounds are polyfunctional macrocycles composed of two apolar diphenylmethane subunits connected by two polar bridges having variable length and bearing positively charged nitrogen atoms. The geometrical features of the diphenylmethane group make it a suitable structural element for the design of molecular cavities capable of including substrates, particularly aromatic flat compounds [10]. The number of methylene groups in the linkers, *n*, varies from three to six, which allows the cavity to be tailored to fit specific guests, the ammonium groups providing the receptor with water solubility. They were prepared according to Koga by cyclization of bis-tosylated bis(4-aminophenyl)methane with α,ω-dibromoalkanes followed by removal of the tosyl groups and permethylation [11].

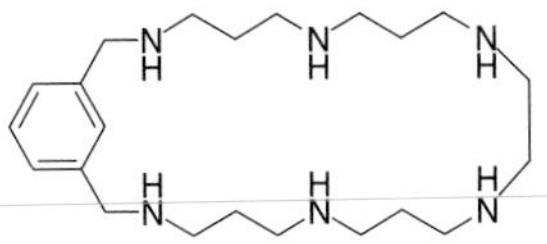

3 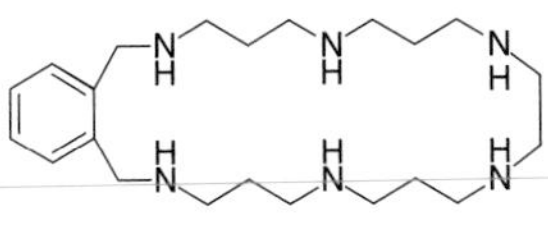   4 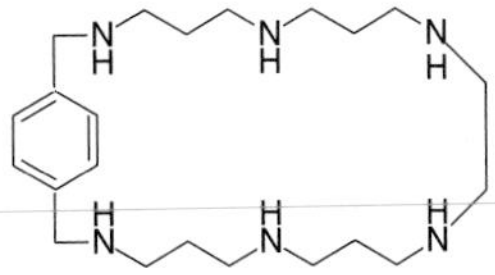   5

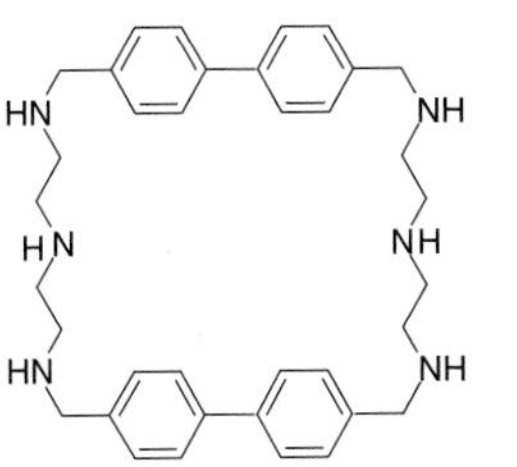

6 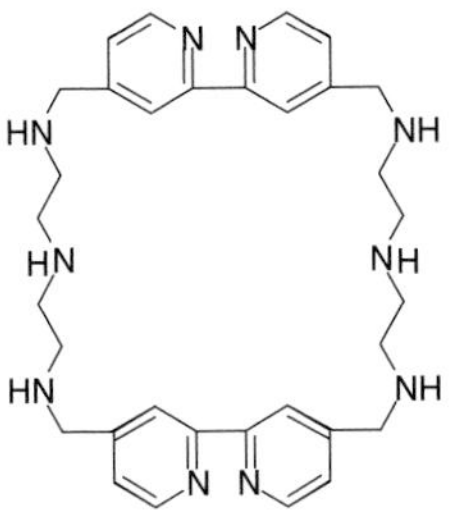   7 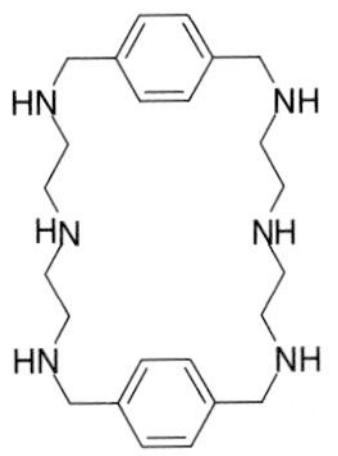   8

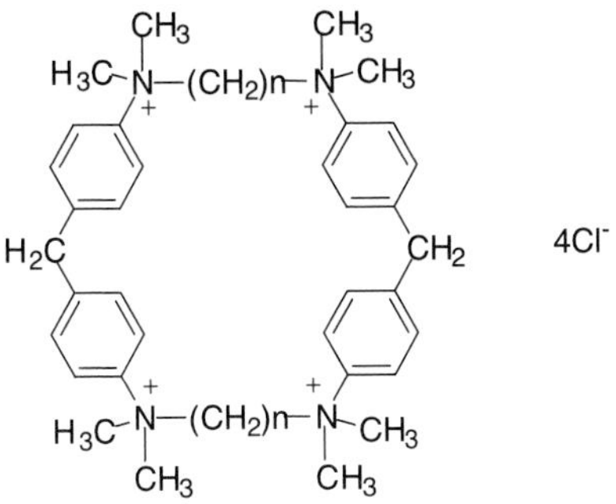

9   CP33   n=3
10   CP44   n=4
11   CP55   n=5
12   CP66   n=6

Complexation of nucleosides and nucleotides by **CP66**, in aqueous solution, has been extensively studied [12]. It binds purine derivatives strongly and the complexation-induced NMR shift values, up to -1.7 ppm, are in favor of the inclusion of the adenine moiety into the cavity of **CP66** with the sugar ring outside. NMR titration curves showed that the stoichiometry of the complexes is 1:1 for every substrate tested. Among purines, the selectivity for adenine derivatives is remarkable: for example, the association constant is 5 times higher for adenosine than for guanosine; the complexation-induced NMR shifts are also smaller for all the other nucleobases. **CP66** gives only loose associations with pyrimidine derivatives. Clearly, on the route towards the search of selective receptors for nucleotides, a great step was accomplished with the design of the **CPnn** azoniacyclophanes.

11.2.3
Cyclobisintercalands

The concept of cyclointercalation, developed by Lehn's group, was a breakthrough towards the design of selective synthetic receptors for nucleotides and nucleic acids. In a manner reminiscent of the binding of intercalators between the plateaux of base pairs in double-stranded nucleic acids, the incorporation of flat subunits into macropolycyclic structures gives receptor molecules which may be expected to display molecular recognition of flat substrates. These planar subunits must be of sufficient surface and positioned at a distance suitable for the intercalation of planar substrates such as nucleic bases; moreover the bridges which link them must be sufficiently rigid in order to prevent the collapse of the cavity because of hydrophobic effects in water. Of special interest, as structural groups, are the planar molecules known to interact with nucleic acids by intercalation. In that case, receptors of cyclointercaland type are obtained; they may be expected to form supramolecular complexes with every flat substrates, especially with the nucleic bases of nucleosides and nucleotides, and to present selective properties towards nucleic acids due to their bulky structure. Of note, this generic name refers to structural factors but not to intercalative properties [13, 14].

11.2.3.1 Acridinium derivatives
The triply bridged bisintercalands **13** and **14**, based on acridine subunits, were among the first to be prepared and studied in the context of nucleosides and nucleotides binding in water [14–16].

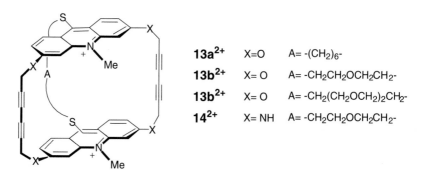

13a^{2+}	X=O	A= -(CH$_2$)$_6$-
13b^{2+}	X= O	A= -CH$_2$CH$_2$OCH$_2$CH$_2$-
13b^{2+}	X= O	A= -CH$_2$(CH$_2$OCH$_2$)$_2$CH$_2$-
14^{2+}	X= NH	A= -CH$_2$CH$_2$OCH$_2$CH$_2$-

The addition of increasing amounts of a series of nucleosides and nucleotides to their aqueous solutions resulted in a decrease of the intensity of the band located around 400 or 470 nm in their electronic absorption spectrum. Stability constants of 10^3 to 10^4 were determined and the stoichiometry of the complexes formed by **13** and **14** was found to be 1:1 for all substrates, indicating that a well-defined species is generated. The marked hypochromism observed in all cases reveals the formation of stacked structures between π-systems. Taken together, the 1:1 stoichiometry and the hypochromism suggest a sandwich-type structure for the complexes, the substrate being located between the two flat acridine units. Several features indicate that substrate binding by **13** and **14** is dominated by stacking effects that involve both van der Waals and hydrophobic effects: there is no significant increase of the stability constants with increase of the charge of the substrate, the binding constants for AMP^{2-}, ADP^{3-}, and ATP^{4-} being the same; likewise the stability constants measured for the neutral nucleosides are comparable to those of the doubly charged nucleotides.

17   X = O,   A= -CH$_2$CH$_2$CH$_3$

18   X = NH,   A= -CH$_2$CH$_2$CH$_3$

15a^{2+}   X= O,   A= -(CH$_2$)$_6$-

15b^{2+}   X= O,   A= -CH$_2$CH$_2$OCH$_2$CH$_2$-

15c^{2+}   X= O,   A= -CH$_2$(CH$_2$OCH$_2$)$_2$CH$_2$-

16^{2+}   X= NH,   A= -CH$_2$CH$_2$OCH$_2$CH$_2$-

Although the stability of the complexes formed in aqueous solutions are about two orders of magnitude higher than for the monomers reference compounds **17** and **18**, they are however smaller than those measured with the acyclic analogs **15** and **16** by factors of 5–10. This could indicate that the macrobicycle is not well tailored for the intercalation of planar substrates. Indeed, the crystal structure of **13** indicates that the two acridine walls are too far apart to perfectly accommodate flat compounds, particularly since the rigidity of the diacetylenic bridges prevent any adaptation of the size of the cavity to allow a close contact with the substrate [17].

11.2.3.2 Phenanthridinium derivatives

Diacetylenic bridges have also been used to connect two planar phenanthridinium units in a series of cyclobisintercaland receptors [18–20]. The addition of increasing amounts of nucleotide to aqueous solutions of each of compounds **19–21** resulted in the quenching of their fluorescence emission, which allowed the determination of the stability constants by fluorometric titrations; very high values, ranging from 10^5 to 10^6, were measured for various nucleotides and the neutral adenosine. These constants were found to be practically charge-independent, which indicates that the stacking interactions are the driving force for the binding. The high stability of the complexes is, thus, remarkable for receptors which bind only the nucleic base of the nucleotide.

	R
19	$(CH_2)_4$
20	$(CH_2)_6$
21	$p-C_6H_4$

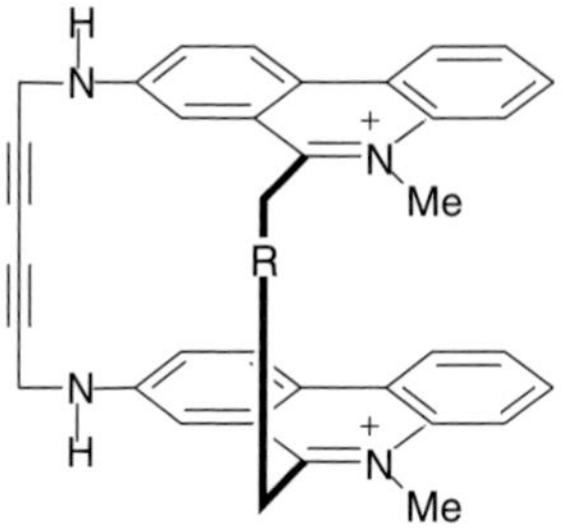

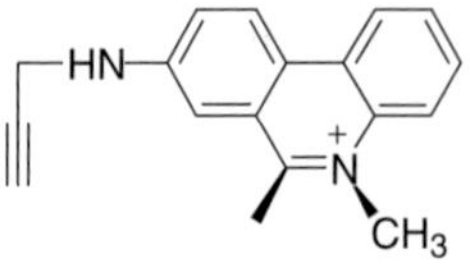

22

The observed 1:1 stoichiometry for the complexes and the fact that the K_s values for the corresponding acyclic monomer **22** are more than one order of magnitude lower than those of macrocyclic compounds **19–21**, are in accordance with an intercalative type of binding.

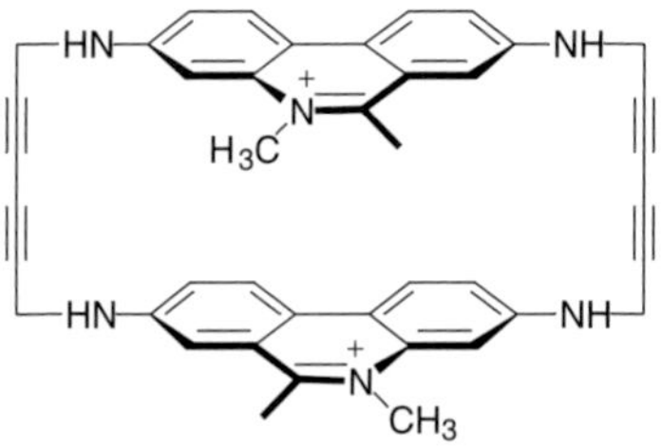

23

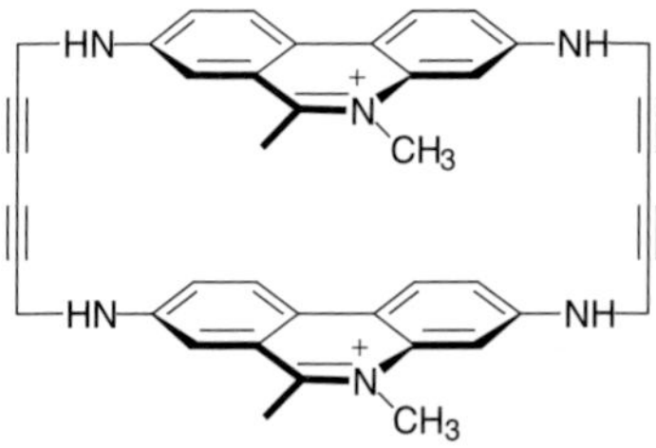

24

Macrocyclic ligands **23** and **24** are remarkable since they differentiate AMP^{2-} from GMP^{2-} or UMP^{2-}: an increase of the fluorescence is observed upon complexation of the former whereas only a very slight emission change is seen upon the addition of GMP^{2-} and UMP^{2-} [20].

In summary, except for the K_s values, which are higher for the phenanthridinium compounds, acridinium and phenanthridinium cyclobisintercalands behave in the same way. The binding of nucleotides by this first series of cyclointercalands is mainly due to stacking and hydrophobic effects and is virtually not selective. This is why a second generation of compounds, provided with positive sites able to interact with phosphate groups, was designed.

11.2.3.3 Polyamino naphthalenophanes and acridinophanes

According to the results obtained with the rigid acridinium compounds (Section 11.2.3.1), it appeared necessary to render the structure more flexible to help the fit of interactions with nucleotidic substrates and also to introduce binding sites for electrostatic interactions. The macrocycles **BisNP** and **BisA**, in which the two aromatic units are bridged by two flexible diethylene triamine (DIEN) chains, have thus been designed [21, 22]. In these compounds the length of the triamine allows a positioning of the two aromatic units at a distance compatible with the trapping of an aromatic nucleus (∼3.5 Å) and the four positive charges, globally developed at physiological pH (5–7), greatly help the interaction with phosphate groups;

these charges also prevent the flattening of the cavity in water. **BisNP** and **BisA** were obtained via the very efficient $[2 + 2]$ condensation between DIEN and the corresponding naphthalene or acridine dialdehydic derivative (Scheme 11.1). The naphthalene dialdehyde was prepared in two steps from the dimethylester derivative and the 2,7-acridinedicarboxaldehyde was synthesized through a four-step pathway using usual acridine reactions (Scheme 11.2). Crystals of complexes of **BisA** and **BisNP** with flat aromatic dicarboxylates have been obtained [23, 24]. Structural X-ray analysis showed that the macrocycles adopt a semi-closed conformation with an interchromophoric distance of ∼7 Å suitable for π-sandwiching of an aromatic nucleus (Fig. 11.1).

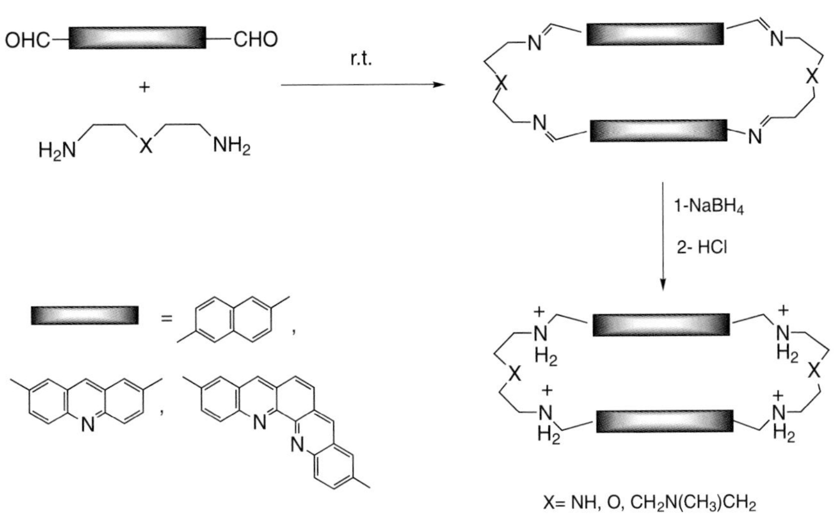

Scheme 11.1. General synthetic pathway of cyclobisintercaland macrocycles.

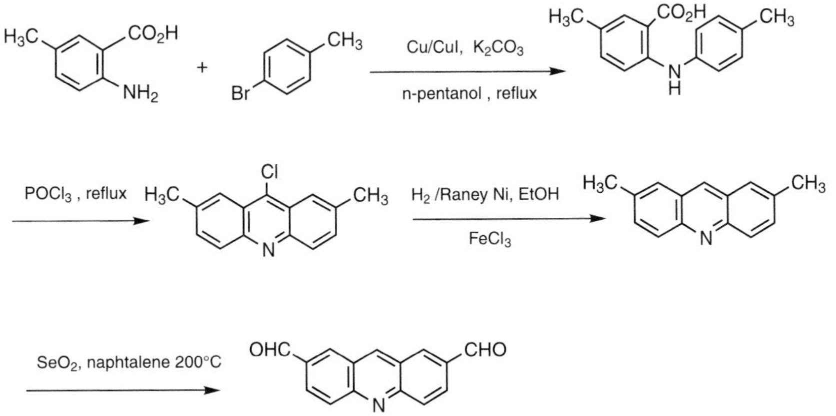

Scheme 11.2. Synthetic scheme of 2,7-acridine dicarboxaldehyde.

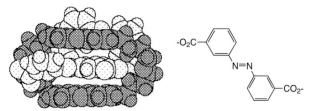

Fig. 11.1. Space-filling representation of the X-ray structure of the complex between **BisA** (gray) and *trans*-3,3′-azobenzene dicarboxylate (white).

BisNP

BisA

The affinity of **BisNP** and **BisA** for various nucleotides has been measured by NMR and fluorimetric titrations. In all cases 1/1 complexes were formed and remarkably high stability constants were determined (Tab. 11.1). In particular **BisA** displays one of the highest affinity constants for ATP^{4-} ever measured in water (K_s for $ATP^{4-} > 10^8$ M^{-1}). For both macrocycles the same trends in selectivity were observed: purinic derivatives are bound more strongly than pyrimidinic ones and the binding constant increases with the anionic charge (one order of magnitude per phosphate group). These features indicate clearly the respective contributions of electrostatic forces and stacking/hydrophobic interactions between the macrocyclic polycationic structure and the nucleotide. A more detailed comparison of the binding properties of the two macrocycles shows that **BisA** is much more efficient than **BisNP**. Indeed, the replacement of the bicyclic naphthalene by the larger tricyclic acridine produces a large increase of the binding constants specially for the purine derivatives (an increase of 2–3 orders of magnitude is observed

Tab. 11.1. Stability constants (log K_s) calculated for the complexes of macrocycles **BisA** and **BisNP** with nucleotides

Nucleotide	BisA	BisNP
AMP^{2-}	6.1	4.1
ATP^{4-}	8.4	5.2
GMP^{2-}	4.6	3.6
UMP^{2-}	3.8	3.8

for AMP^{2-}, GMP^{2-}, and ATP^{4-}, see Tab. 11.1). This might simply be due to the larger size of the cavity delineated by the bisacridine derivative but it might also reflect the considerable influence of the van der Waals and solvophobic effects on the complexation since the strength of the stacking interaction is strongly dependent on the area of the aromatic surfaces in contact.

The variations of the fluorescence properties of **BisA** when binding to nucleotides are worth noting. The macrocycle displays a remarkable behavior with two types of response: strong quenching of the fluorescence intensity (70–80%) with the purine derivatives and considerable enhancement of the emission (70–130%) with pyrimidine ones. On the other hand, when **BisNP** associates with nucleotides a strong quenching of the emission of the naphthalene units is observed in any case. The ability of **BisA** to discriminate between the two types of nucleobases is remarkable both qualitatively and quantitatively and will be used further to monitor interactions with DNA hairpins (see Section 11.3.4.2).

In conclusion, thanks to an interplay of electrostatic and stacking/hydrophobic interactions, the naphthalenophane and acridinophane macrocycles are among the most efficient synthetic compounds designed for complexation of nucleotides in physiological conditions.

In an effort to improve the strength and the selectivity of the nucleotides binding again and, also to mimic the base pairing that occurs in biology, larger macrocycles **BisQ**$_{1-3}$ were designed. They consist of crescent-shaped quinacridine subunits linked by the same linkers as in **BisNP** and **BisA** and they were synthesized according the general Scheme 11.1 [25].

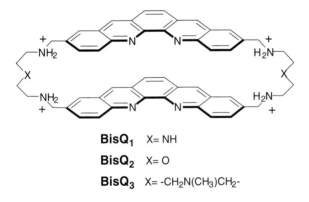

BisQ$_1$ X= NH

BisQ$_2$ X= O

BisQ$_3$ X= -CH$_2$N(CH$_3$)CH$_2$-

Absorption and fluorimetric experiments showed that **BisQ**$_{1-3}$ display noncovalent associations with nucleotides and revealed a new recognition pattern for such host–guest systems [26]. As indicated by fluorimetric titrations of **BisQ$_1$** by GMP^{2-} and also by mass spectrometry studies, the stoichiometry of the complexes is 1:2 (host/guest) with nucleoside monophosphates. This unusual binding of a nucleoside monophosphate pair involves π-stacking/hydrophobic effects between the nucleobases and the quinacridine subunits of the macrocycles as schematically represented in Fig. 11.2. The formation of hydrogen bonding between the two

Fig. 11.2. Schematic representation of the 1/2 complex formed between **BisQ** and GMP²⁻.

nucleobases within the complex could also contribute to the high stability of the ternary association and to the selectivity for GG base pairs.

11.3
Nucleic Acids Complexation

On the way to nucleotide recognition, efforts have mainly been devoted to the design of ligands able to selectively bind isolated planar substrates, including the nucleic bases. In the case of nucleic acids the situation is totally different since the monomer nucleotides are engaged in much more complex architectures with a variety of secondary and tertiary structures. However, in such structures, the efficient ligands for nucleotide recognition should be able to bind isolated nucleic bases when present. Thus, some azoniacyclophanes and cyclointercaland compounds, exhibiting interesting binding properties for nucleotides, have been tested on nucleic acids.

11.3.1
Azoniacyclophanes

The tetracationic azoniacyclophanes **CPnn** (Section 11.2.2) also interact strongly with nucleic acids [27, 28]. Competitive displacements of ethidium bromide intercalated into calf thymus DNA were used to estimate the interactions of the various macrocycles with DNA (calf thymus). The effect of the four **CP** compounds, **9–12**, was up to 15 times higher than that of the corresponding open-chain analogs. Moreover, the stabilization against thermal denaturation brought by **CP44** is higher than with the other cyclophanes: $\Delta T_m = +5.1°C$ instead of $+2.2$, $+1.4$, $+3.0$ for **CP33**, **CP55**, and **CP66** respectively. NMR and viscosity measurements ruled out the participation of intercalation and, thus, electrostatic interactions remain the main factor to explain the strong binding of azoniacyclophanes. Molecular modeling showed that **CP44** fits particularly well in the large groove of the DNA double helix, thereby allowing very efficient coulombic interactions between the positively charged ammonium groups of the macrocycle and the negatively charged phosphate groups of the DNA. Because of their size, other azoniacyclophanes do not fit so easily in the major groove and cannot develop the same electrostatic interactions as **CP44**.

More interestingly, the interactions of all these azoniacyclophanes with DNA and

RNA duplexes of the same sequence have also been investigated, by thermal denaturation and viscosity measurements as well as by CD and NMR studies [27]. A surprising behavior was observed: all the compounds stabilize DNA but, depending on the size of the macrocycles and on the experimental conditions, they can either stabilize or destabilize RNA duplexes. With DNA duplexes, as was already seen with DNA from calf thymus, **CP44** forms the most stable complex and **CP66** the least stable one. It is very different with RNA duplexes: whereas a strong stabilization is observed with **CP33**, the RNA ΔT_m values measured with **CP44**, **CP55**, and **CP66** are smaller than the corresponding variations of melting temperature on DNA. With **CP66**, ΔT_m even becomes negative, indicating that RNA is actually destabilized in the conditions of the experiment. Thermal melting curves studies, viscosity, and CD results showed that there are two types of complexes for **CP66** depending on the ratio of macrocycle to RNA. At lower ratios, the major complex involves mainly electrostatic interactions with little deformation of the double helix, whereas at higher ratios, CD studies suggested base-pair opening with inclusion of one or more bases into the cavity of the macrocycle; in addition to the ionic interactions this process also involves van der Waals and hydrophobic forces. Further information provided by NMR experiments at different temperatures confirmed this mechanism and showed that only adenine bases are inserted in the cavity of **CP66**. Two effects can explain the different behaviors of **CP66** with DNA and RNA: stabilization with the former, destabilization with the second. First, the grooves in DNA and RNA are quite different and so **CP66** may fit well in the major groove of DNA but not in that of RNA. Second, NMR experiments clearly demonstrated a stronger interaction of **CP66** with an adenine base in an RNA strand than the same in a DNA strand and only the cavity of **CP66** is large enough to accommodate a purine base.

The interactions of compounds **6–8** and of some analogs with nucleic acids have also been examined [29]. It is worth noting that **6**, which was shown to be large enough to allow insertion of a purine base into the macrocycle cavity, displays the same behavior as **CP66**. The RNA-melting temperature increases at low macrocycle/RNA ratio and then decreases as more **6** is added. These results reveal an initial electrostatic stabilization of the double-stranded RNA followed by base-pair opening and ultimately by denaturation of the double helix.

These surprising observations are reminiscent of biological mechanisms such as base flipping by DNA enzymes, which is a common event in biology, but the fact that a small organic molecule can do the same selectively with RNA is amazing.

11.3.2
Porphyrin Derivatives

The first cyclointercaland compounds were described as early as 1984 [30]. They are macrotetracyclic and macropentacyclic compounds which combine one or two porphyrin rings and [18]-N_2O_4 aza-oxamacrocycles. In the first derivative, **25**, one porphyrin subunit is linked by two macrocycles to a biphenyl group; the amine groups being protonated around pH 7 in water, it may be expected to interact with

DNA via coulombic interactions. Moreover its bulky structure may allow discrimination between single-stranded and double-stranded nucleic acids, the former being much more flexible. Actually, studies using a set of photophysical methods indicate that it is the case [31].

The binding of **25** to polynucleotides occurs in two steps: the first one is a highly non-specific cooperative binding of the macrocycle along the polymer, whereas the second one is the binding to isolated sites. Using absorption and fluorescence spectroscopy, binding and competition studies on the affinity of **25** for isolated sites on single-stranded and double-stranded polymers indicated that it binds more strongly to single-stranded polynucleotides than to double-stranded ones.

25   **26**

Thermal denaturation measurements showed that it binds efficiently to denatured DNA whereas there is no significant stabilization of the double helix. This selectivity is pH-dependent, the highest affinity being observed at pH 6.6 for poly(dA) and the ss/ds selectivity is larger at pH 4.6 than at pH 6.6. Moreover, the binding constants of **25** for poly[d(A-T)]$_2$ and poly(dA)-poly(dT) obtained from fluorescence measurements are comparable, while known intercalators bind more tightly to the former [32]. The similarity between the binding constants of **25** for these two polynucleotides and the absence of stabilization of the double helix upon binding of **25** show that it does not intercalate into double helices like planar aromatic cations. In addition, fluorescence anisotropy results and the fluorescence quenching by the bromine atom of poly[d(A-^{5}BrU)] suggested that it binds double helices into the major groove. The binding selectivity and the binding location may be attributed to the macrocyclic cryptand cage structure into which the porphyrin subunit is incorporated. The rigidity of the cage walls causes a steric hindrance which prevents the intercalative insertion of the porphyrin ring between the base pairs and the overall bulky structure of the molecule cannot accommodate the narrow minor groove of the double helix.

The porphyrin derivatives, like most of cyclobisintercalands based on dye molecules, present another interesting feature. Since the porphyrin is a photoactive moiety, it may induce DNA photocleavage when irradiated in visible light. This has been shown using the macropentacyclic compound **26** which is fitted with two porphyrin rings (Section 11.3.4.2).

11.3.3
Phenanthridinium Derivatives

The affinities of phenanthridinium derivatives **23** and **24** (Section 11.2.3.2) for single- and double-stranded polynucleotides of DNA and RNA have been studied by fluorescence, viscometric, and thermal denaturation measurements and the results have been compared to those obtained with the corresponding monomer **22** [20]. With regard to the structural features of these molecules, particularly the position of the connecting bridges, which are unfavorable for insertion in double-stranded nucleic acids, preferred binding to single-stranded nucleic acids may be expected.

Actually, the two macrocyclic ligands bind single-stranded polynucleotides more strongly than double-stranded ones; for instance, their affinity for poly(dG) is 25 times higher than that for poly(dG)-poly(dC). The K_s values for **23** and **24** with ss-poly(dA) and poly(dG) are two orders of magnitude higher than those of monomer **22**. A similar result was already found with bis- and monointercalators; this is in favor of a participation of both phenanthridinium subunits, with a single base inserted between them and validates the concept that has led to the design of cyclo-bisintercaland compounds. However, other studies would be necessary to assert that true bisintercalation is the mode of binding. In contrast, complementary studies indicated nonintercalative binding of **23** and **24** with double-stranded polynucleotides and were in favor of groove binding driven by hydrophobic and electrostatic interactions.

11.3.4
Acridinium Derivatives

Two bisacridine compounds, **SDM** and **BisA** (Section 11.2.3.3), have been extensively studied by the groups of Zimmerman and Lehn respectively [22, 33, 34]. They differ by the nature and the position of the bridges that connect the two acridine subunits. Depending on the position of attachment of the linkers on the acridine rings, different modes of intercalation into the double helix of DNA can be envisaged (Fig. 11.3):

1. When the two linkers are closely attached on the same side of the ring, bis-intercalation can occur in a normal way with both chains in a single groove; this situation is schematized in Fig. 11.3a.
2. But when these linking chains are positioned on opposing sides of the major axis of the acridine ring three complexes are theoretically possible. In Fig. 11.3b, the two acridine rings are partially inserted between the base pairs with the linkers on the same groove as in Fig. 11.3a and with the long axes of the acridines almost perpendicular to those of the base pairs; this complex can be supposed to be weak because of the small overlap with the base pairs.

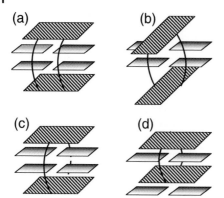

Fig. 11.3. Various binding models of cyclobisintercalation.

3. In Fig. 11.3c, a catenated complex is formed, with the long axis of the intercalator and that of the base pairs approximately parallel and with the linkers in the opposite grooves, which requires a transient disruption of the Watson–Crick base pairs. This case is reminiscent of the threading intercalators and polyintercalators which bear two substituents at the extremities of the diagonal line of an aromatic ring [35].

4. In these three cases, the bisintercalation is supposed to obey the neighbor exclusion principle [36], although violation of this principle, as in Fig. 11.3d, with less favorable bisintercalation at adjacent sites, cannot be excluded. It is also possible that the cyclobisintercaland compound does not bind as an intercalator at all or that it behaves as a monointercalator with one single ring intercalating between the base pairs.

SDM X:

SPDA X: H,H

11.3.4.1 **SDM Macrocycle**

The compound studied by Zimmerman's group is composed of two acridine rings connected by two different linking chains: a spermine unit linking the 9-positions and a 2-aminoethanethiol succinamide motif connecting the 4-positions of the ac-

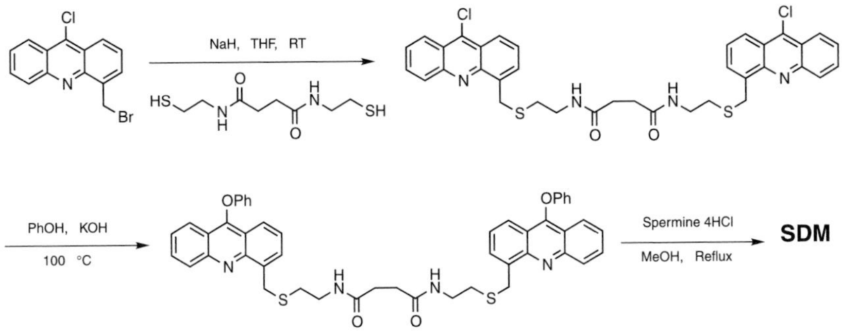

Scheme 11.3. Synthesis of **SDM** macrocycle.

ridine parts [33]. It was prepared, in three steps, by reaction of 4-(bromomethyl)-9-chloroacridine with *N,N′*-bis(2-mercapto-ethyl)succinamide, in presence of base, followed by conversion of the resulting dichloride into bis(9-phenoxyacridine), which reacted with spermine tetrahydrochloride under high dilution conditions (Scheme 11.3). In **SDM** structure, the 14 atoms linking chains provide water solubility and are long enough (about 16 Å in their fully extended conformation) to allow bisintercalation according to the neighbor exclusion principle. Binding studies performed with various nucleic acids indicated a high affinity. Thermal denaturation measurements, spectroscopic data, viscometric analysis, ability to unwind a closed circular supercoiled DNA and comparaison with the known bisintercalator spermine bisacridine **SPDA**, all were in favor of a bifunctional intercalator behavior [33]. However further experiments were necessary to try to discriminate, if possible, between the binding models discussed above. In addition to visible spectroscopy and stopped-flow kinetics, NMR studies confirmed that **SDM** binds DNA according to an intercalation mechanism with the long axes of the acridine rings parallel to the long axes of the base pairs. All the data were analyzed for consistency with the different binding models. The attachment of the linkers in the 4- and 9-positions of the acridine ring rules out the participation of a binding mechanism of type in Fig. 11.3a and all the accumulated data, included modeling studies, suggest that the model in Fig. 11.3b as highly unlikely. In contrast, all the results fit with the models in Fig. 11.3c and d, in which side chains are in opposite grooves. At this stage it is difficult to distinguish between these two models although NOESY NMR spectra of the **SDM** complex with d(CGCG)$_2$ seem more consistent with Fig. 11.3d, that is to say bisintercalation with violation of the neighbor exclusion principle [36].

In fact, the interactions of **SDM** with dsDNA might involve an even more complicated binding pattern as shown by a recent study in which X-ray diffraction analysis of the complex between **SDM** and CGTACG was performed [37]. The refined crystal structure, at 1.1 Å resolution, presents several unexpected features. The terminal base pairs, C1.G12 and G6.C7, are unravelled, thus leaving only four central base pairs in the duplex. Only one acridine ring of the cyclobisintercaland is intercalated between the C5pG6 step while the other acridine ring and the linkers are completely disordered. A number of additional interactions generate a very complex structure in which two acridine rings, belonging to different **SDM** molecules, are intercalated between two sets of GC base pairs. Moreover, NMR studies of the binding of **SDM** to the AACGATCGTT sequence showed that more likely it binds to two different duplexes by intercalating each acridine to a CpG site from different duplexes. A crosslinking would result from such a mechanism which is consistent with the precipitation that occurs during the titration. Of course, this precipitation is also facilitated by the charge neutralization between the bisacridine and the DNA.

In conclusion, these studies show that non-conventional intercalation structures are possible and provide new insights into the design of new intercalating compounds.

11.3.4.2 BisA Macrocycle

The macrocycle **BisA** has been the most extensively studied compound amongst the macrocyclic series of DNA ligands. Its interaction with various DNA conformations has been characterized qualitatively and quantitatively in terms of binding affinities, selectivities and also at the structural level by NMR studies.

Hairpin recognition and duplex destabilization   Hairpins, one of the simplest DNA and RNA secondary structures, result from an intramolecular folding of a partially complementary sequence. They are thus composed by a double helix, the stem, linked to a single-stranded loop. Such structures are found at various regulatory sites such as gene transcription regions and in origins of DNA replication; they are therefore possibly involved in biological processes.

The binding of **BisA** with short oligonucleotides (9–11-mer, referred to as sA_3, sA_5, sT_5, see Fig. 11.4) has been investigated using melting temperature experiments, fluorescence measurements, and gel-filtration experiments [38]. Throughout these studies a monochromophoric compound **MonoA** was used as a reference for comparison with the binding behavior of the dimeric macrocyclic structure.

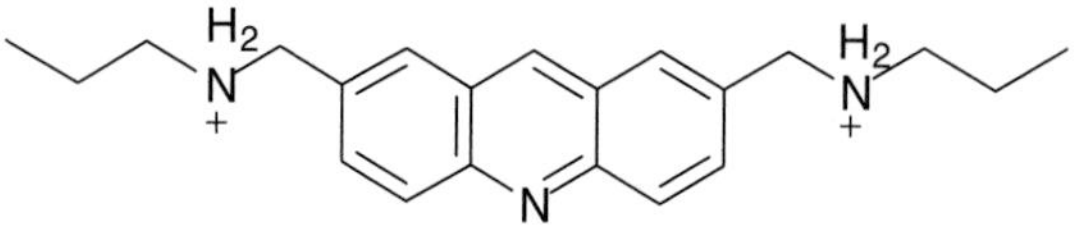

MonoA

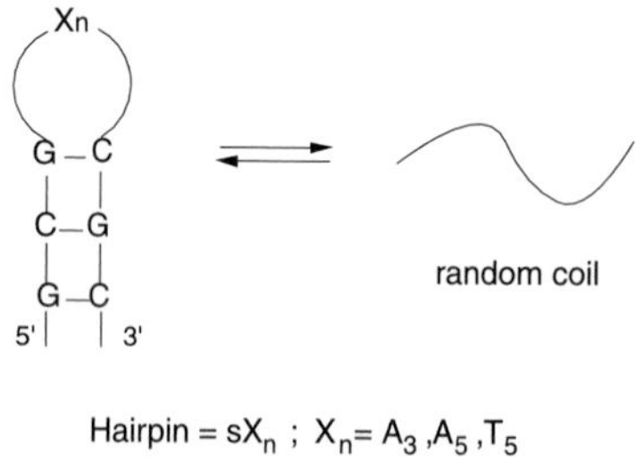

random coil

Hairpin = sX_n ; X_n = A_3 ,A_5 ,T_5

Fig. 11.4.   Intramolecular transition from a hairpin to a random coil.

Tab. 11.2. T_m values of the hairpin-to-coil transition of sA$_3$ in the presence of **BisA**.

BisA/sA$_3$	T_m (°C)	ΔT_m (± 2°C)
free sA$_3$	47 $\pm$ 1	
1/4	47 $\pm$ 1	
1/2	49 $\pm$ 1	2
1/1	57 $\pm$ 2	10
2/1	67 $\pm$ 2	20
4/1	70 $\pm$ 2	23
5/1	75 $\pm$ 2	28

BisA strongly stabilizes the hairpin structure which is indicated by a strong increase of the melting temperature of the hairpin to random coil intramolecular transition (Tab. 11.2). For instance a high affinity constant, $K_s = 4.5 \times 10^7$ M^{-1}, was calculated for the 1/1 complex of **BisA** with sA$_3$. In contrast, the melting temperature of a short duplex, d(GCGCGC)$_2$, mimicking the stem of the hairpin, is not affected by the presence of the macrocycle. This demonstrates that **BisA** exhibits a low affinity for short duplexes and that the binding should arise in the loop part of the hairpin. Compound **MonoA** does not significantly alter the melting of the hairpin, showing that the macrocyclic structure is responsible for the selective binding into the loop. The affinity of **BisA** for the loops appeared dependent on both their sequence and size, small loops sA$_3$ and sA$_5$ being preferred. The fluorescence signature of the macrocycle that distinguishes the heterocyclic base in interaction with the acridine ring (see Section 11.2.3.3) and the comparison of the fluorescence relative yields with various oligo- and polynucleotides allowed the assignment of the binding site to the loop. Gel-filtration experiments performed on the complexes of **BisA** with the sA$_3$ oligonucleotide confirmed the 1/1 stoichiometry but led also to the detection of a 1/2 species that is much less soluble and thereby was not detected by UV-vis spectroscopy. Comparison of the binding constants of **BisA** for a hairpin loop and a single-stranded dA$_3$ ($K_s = 6 \times 10^5$ M^{-1}) demonstrated that the hairpin loop structure was preferred to single strands.

In conclusion, the whole experimental work demonstrated the selectivity of **BisA** for hairpins compared with both double helices and single-stranded oligomers. It seems that 1/1 associations are formed predominantly and that the binding could occur in the vicinity of the loop. However more structural data would be required to gain information on the interaction at the molecular level.

Complementary work based on electrophoresis gel-shift experiments [39] using two DNA hairpins forming a fully paired duplex showed that **BisA** is able to shift the equilibrium towards the hairpin conformation (Fig. 11.5), which confirms the higher affinity of the macrocycle for DNA hairpins compared with double-stranded oligonucleotides. In addition, the premelting of poly[d(A-T)]$_2$ induced in the presence of the macrocycle indicates its preference for single- versus double-stranded oligonucleotides. These observations suggested that the selective binding of **BisA** could be potentially applied for destabilizing DNA double helices. It appeared interesting to further investigate this property since DNA opening processes play a

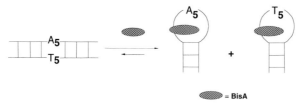

= BisA

Fig. 11.5. Equilibrium between the complementary hairpins sA₅, sT₅ and the fully paired duplex.

crucial role in the initiation of key step events such as replication, transcription, and recombination. The ability of **BisA** to destabilize duplexes has been examined using hybrid systems constituted of a primer (17-mer) hybridized to its complementary sequence on the circular ss DNA of phage M13 (Fig. 11.6).

These DNA constructions were initially designed to evaluate the DNA helix-destabilizing ability of single-strand-binding proteins SSB [40]. By analogy with the behavior of these proteins, this simple assay consisted in measuring the ability of the macrocycle to displace the short ³²P-labeled primer [41]. The analysis is carried out by gel electrophoresis, which allows an easy quantification of the free/bound oligonucleotide ratio due to the large difference in mobility of the two species (Fig. 11.7). When this assay is carried out at increasing temperature with samples taken at regular intervals, denaturation curves are obtained that allow the measurement of a melting temperature with a satisfying accuracy. In these conditions, a strong destabilizing effect of **BisA** ($\Delta T_m = -20°C$) could be detected at low **BisA**/DNA ratio and at high salt concentration (100 mM NaCl). The very low ionic strength dependence of the effect suggests a specific and strong interaction of the macrocycle with DNA.

This led to further investigation of the ability of the macrocycle to compete with SSB proteins for binding to DNA. Competitive binding experiments were monitored by nitrocellulose binding assays using the SSB protein from *E. coli* (*Eco* SSB), which forms 1/1 complexes with a single-stranded 36-mer. The protein was immobilized on a nitrocellulose membrane and incubated with the labeled 36-mer oligonucleotide in the presence of **BisA** and control compounds. The activity of the compounds were evaluated by quantification of the radioactivity retained on the membrane after washing (Fig. 11.8). In the concentration range examined, the formation of the complex SSB/36mer was inhibited by the macrocycle, whereas spermine and the reference monomer **MonoA** elicited no effect. This clearly demonstrated the inhibitory effect of **BisA** likely through a competitive binding to the

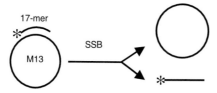

Fig. 11.6. Schematic representation of the hybrid M13:17-mer and of its dissociation by an SSB protein.

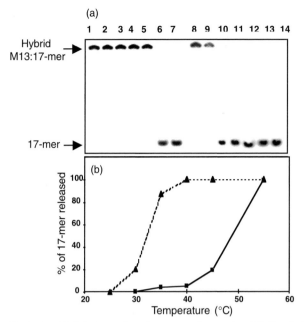

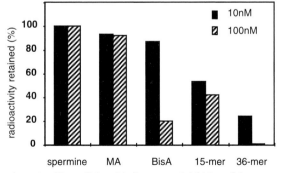

Fig. 11.7. (a) Gel electrophoresis pattern after thermal denaturation of M13:17-mer hybrid alone (lanes 1–7) and in presence of **BisA** (lanes 8–14). Temperature varies from 25°C up to 55°C in 5°C steps. [DNA] in phosphate unit = [**BisA**] = 1 μM. (b) Thermal denaturation curves of M13:17-mer hybrid alone (squares) and in presence of **BisA** (triangles), Tris–HCl buffer (pH 7), 10 mM, NaCl 100 mM.

oligonucleotide. This method also provides a rough evaluation of the affinity of **BisA** for binding to the 36-mer. Comparison with the inhibitory effect of unlabeled oligonucleotides (15-mer and 36-mer) shows that the 15-mer and **BisA** have similar efficiencies. This is clearly indicative of a high affinity of **BisA** for the target oligonucleotide since the apparent binding constant of 15-mer to SSB is $>10^7$ M^{-1}.

Fig. 11.8. Nitrocellulose binding assay. Inhibition of the formation of complex 36-mer/SSB by various competitors at two different concentrations (10 and 100 nM).

The capacity of **BisA** to discriminate between single-stranded and double-stranded oligonucleotides is likely responsible for the helix-destabilizing effect through equilibrium displacement. This also leads to disruption of protein/oligonucleotides complexes. Whether this property could be applied to a cellular context is not known and is under current investigation.

Selective photocleavage As previously mentioned, acridine and porphyrins are photochemically active groups, a property widely applied for inducing redox damages to nucleic acids, suitable for detecting the binding site of compounds and of potential interest for photodynamic therapy. Moreover, photoactive ligands, able to act as charge injectors via direct electron transfer from guanine bases, have raised an extensive curiosity from many groups through the past 5 years (see Chapter ▌). The ability of **BisA** and compound **26** to induce photocleavage of DNA was examined to confirm the selectivity for single-stranded nucleic acids regions shown by the two cyclobisintercalands. To this end, a simple assay has been set up which consists of comparing the ability of the macrocycles to cleave circular dsDNA (plasmids) and circular ssDNA (DNA of phage M13) [42]. The two forms of DNA are commercially available and their cleavage is easily detected on agarose gel. Photocleavage experiments were conducted on circular supercoiled ds plasmids pBR322 and pUC18 and on circular ssDNA M13mp19 and M13mp18. A single cut of either type of DNA results in the formation of new species with very different electrophoretic migration properties. The double-stranded supercoiled DNA yields a relaxed circular form which may be followed by a linear double-stranded one after multiple cleavage events, whereas single-stranded circular DNA gives a linear single-stranded species after a single cut (Fig. 11.9).

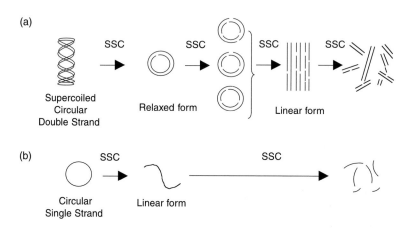

SSC = single strand cleavage

Fig. 11.9. Schematic representation of the species generated by single-strand cleavage reactions of (a) supercoiled circular dsDNA and (b) circular ssDNA.

Tab. 11.3. Photocleavage of equimolar mixtures of supercoiled ds pBr322 and circular ss M13mp19 by cyclobisintercaland **26** (DNA concentration = 0.06 mM in phosphate units).

Concentration of 26 (µM)	0.8	1.6	2	3	6.25	12.5	25	50
ds PBR322 remaining (%)	100	100	100	75	80	60	40	0
ss M13mp19 remaining (%)	90	50	12	0	0	0	0	0

Both compounds induced efficient and selective cleavage of the ssDNA at a low compound/DNA ratio (see Tab. 11.3 and Fig. 11.10). This preferential degradation was observed when irradiation was conducted on each form separately or on mixtures of single- and double-stranded DNA. In both cases significant cleavage of the double-stranded species requires much higher concentration in macrocycles and/or longer irradiation times. Again the reference acridine **MonoA** induces equal degradation of both single- and double-stranded species, thereby confirming that the selectivity of the cyclobisintercalands is provided by the macrocyclic structure. Irradiation experiments were also carried out with **26** on a transfer RNA (tRNAasp) which presents both double- and single-stranded domains. After analysis by electrophoresis, the cleavage patterns have been related to the X-ray crystallographic data of tRNAasp and this revealed that the preferred binding sites are almost exclusively at the ssRNA domains.

According to these experiments **BisA** and **26** display similar level of photocleavage activity which could reflect their similar affinities for single-stranded conformations. However, it is difficult to completely parallel the effects since the efficiency of the cleavage is strongly dependent on both the photophysical properties of each compound (quantum yield) and on the cleavage mechanism. The undisputed advantage offered by **BisA** is that it is active in physiological conditions whereas **26** requires a slightly acidic pH (pH ~4.5) for efficient cleavage.

Altogether these results point to the potential of cyclobisintercaland compounds to function as structural probes for single-stranded domains such as loops, bulges, hairpins, or for local pairing defects in complex nucleic acids.

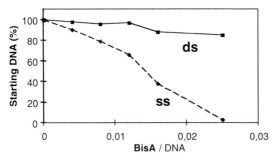

Fig. 11.10. Photocleavage of mixtures of pUC18 and M13mp18 by **BisA**: variation of ds and ssDNA as a function of the concentration of **BisA**. [dsDNA] = [ssDNA] = 0.3 mM in phosphate unit.

Abasic site recognition The loss of a nucleobase is one of the most frequent lesions in DNA and is highly cytotoxic and mutagenic. Abasic sites are produced *in vivo* through the base excision repair pathway (BER) and might also result from the action of ionizing radiations and of alkylating agents, in particular anticancer drugs [43]. It is thus of interest to design molecules capable of interacting specifically with abasic sites either as probes for the detection of damage or as drugs to interfere with the repair process. This fundamental and challenging topic is reviewed in Chapter ▌. The structure of an abasic site depends on its flanking sequences and on the nature of the missing base (apurinic and apyrimidinic); flipped out and flipped in positioning of the unpaired base have therefore been described. Abasic sites can be regarded as particular cases of monobase bulges. Consequently, with regard to these structural features and to the selectivity of **BisA** for unpaired bases in DNA, this compound appeared as a potential new candidate ligand for abasic lesion.

The binding of the drug was investigated with short DNA duplexes (11-mer and 23-mer) oligonucleotides containing a THF cycle facing a thymine in the middle of the sequence [44]. This synthetic modification is commonly used as a stable analog of apurinic (AP)-site (see Fig. 11.11, upper part). A whole set of physico-chemical methods including thermal denaturation, photocleavage experiments, displacement of an abasic site-specific RPE probe, and NMR have been used to demon-

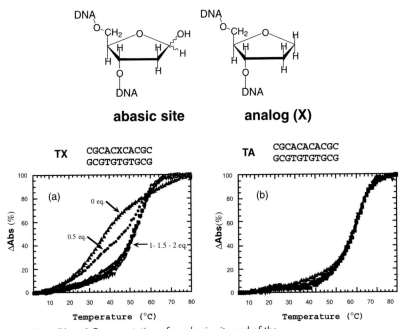

Fig. 11.11. (Upper) Representation of an abasic site and of the THF analog X. (Lower) Melting temperature curves (a) of the undecamer TX and (b) of the regular duplex TA in the presence of **BisA** (0–2 eq.).

strate the specificity and the mode of binding of **BisA** [44]. Thermal denaturation experiments were carried out with duplex TX (11-mer) in the presence of **BisA**. A strong stabilization ($\Delta T_m = +14°C$) was induced whereas no modification of the melting temperature could be detected for the parent unmodified duplex (Fig. 11.11, lower part). This effect levels off at a 1/1 drug/DNA ratio, indicating the existence of only one binding site. In addition, irradiation of a mixture of ^{32}P-labeled oligonucleotide (23-mer) containing the model abasic site X and **BisA** (drug/DNA ratio = 4/1) leads to selective photocleavage in the vicinity of the abasic site on both strands of the duplex (Fig. 11.12). Finally, EPR measurements involving dis-

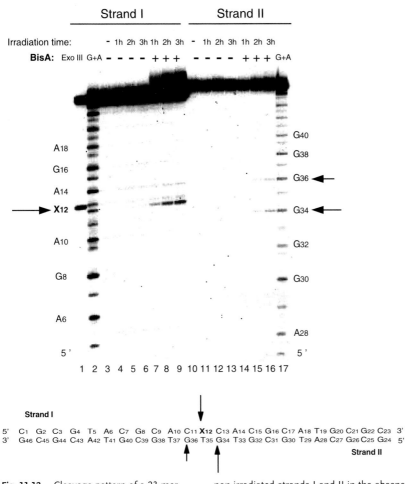

Fig. 11.12. Cleavage pattern of a 23-mer duplex containing an apurinic site X. Lane 1: duplex incubated with exonuclease III to localize the abasic lesion; lanes 2 and 17: G + A for strand I and II; lanes 3 and 10: non-irradiated strands I and II in the absence of **BisA**; lanes 4–6 and 11–13: duplex irradiated in the absence of **BisA** for 1, 2, and 3 h; lanes 7–9 and 14–16: duplex irradiated in the presence of **BisA** (4 μM) for 1, 2, and 3 h.

placement by **BisA** of an abasic site probe labeled by a nitroxide (**ATAC-NO˙**) confirmed the specificity of the binding.

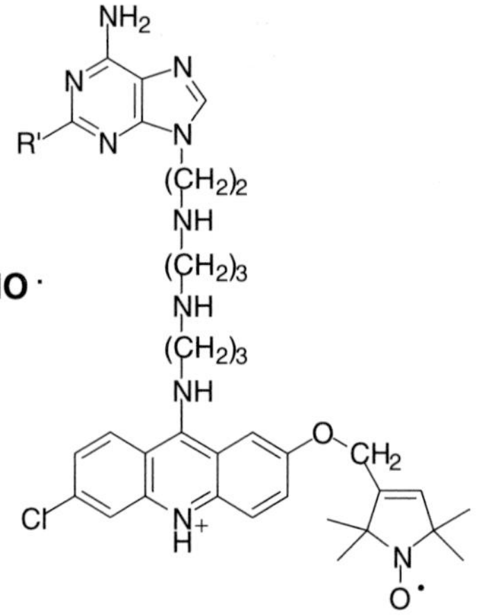

These results obtained by three different methods demonstrate unambiguously that **BisA** binds specifically and cleaves AP-sites.

Study of the interaction between the drug and the undecamer TX by ^{1}H NMR confirmed the binding site [45]. The greatest DNA proton chemical shifts were detected for the two GC base pairs flanking the unpaired thymine and also for the neighboring AT base pair. Data revealed that **BisA** forms two different intercalation complexes with the abasic site containing undecamer. These complexes are present in a 80/20 ratio. The NMR results for the major complex showed clear evidence that one of the two acridine residues intercalates between the C7-G16 and A8-T15 base pairs while the other acridine moiety inserts between the C7-G16 and C5-G18 base pairs (Fig. 11.13a).

Further evidence of **BisA** intercalation was given by the differences in chemical shifts of the aromatic protons in the free and bound molecules since on complexation most of the acridine protons resonances are shifted upfield up to 1.6 ppm. Intermolecular nuclear Overhauser effects (NOEs) were detected between one of the linkers and minor groove markers (A8-H1′, H4′, and C7-H1′) indicating that the two linker chains lie in opposite grooves. Intermolecular distance restraints determined from the NOESY spectra were used to perform a molecular modeling study in order to obtain a model of **BisA**-binding mode. The structure of the major complex proposed after calculations is represented in Fig. 11.13b. The macrocycle is inserted into the abasic pocket with one acridine unit replacing the missing base and the other one intercalated between the two adjacent base pairs. Dynamics calculations showed that the two acridine rings have quite different positions relative to their flanking base pairs. The long axis of the acridine moiety intercalated between C7-G16 and A8-T15 is oriented parallel to the long axis of the A8-T15 residues. This chromophore stacks partially over its adjacent C7 nucleotide while no stacking is observed with the G16 residue. By contrast, the major axis of the second

(a)

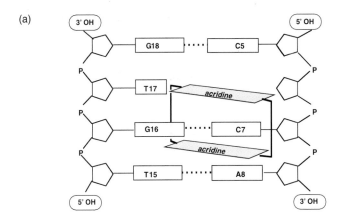

(b)

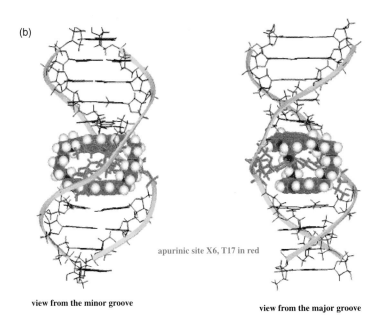

apurinic site X6, T17 in red

view from the minor groove

view from the major groove

Fig. 11.13. (a) Schematic representation of the major **BisA**–DNA complex. The chromophores thread through the helix with the linker chains lying in either groove. (b) Molecular modeling- based structure of the major **BisA**–DNA complex. The drug is drawn in CPK. Left: view from the major groove. Right: the same representation after 90°C angle rotation.

acridine moiety located in the abasic pocket forms approximately a 90° angle with the major axis of the C5-G18 base pair.

Actually, this chromophore stacks with the bases located on one strand (i.e. C5 and C7) and not with their complementary counterpart (G16 and G18). Similar partial intercalation overlap involving stacking of the drug with the two flanking bases on the same strand has only been described previously for threading intercalators such as nogalamycin. The two acridine moieties are oriented with the ring nitrogens pointing toward the major groove and the linking chains positioned in the middle of each groove. The abasic sugar is pushed out away from the minor groove, while the T17 residue is slightly shifted toward the major groove, stacking partially with G16 but the glycoside torsion angle remains roughly anti. The C7-G16 base pair that is sandwiched between the two acridine ring exhibits substantial buckling up to 36°C.

The existence of hydrogen bonds between the NH_2^+ groups of the linkers and hetero-atoms of the bases in both grooves are suggested from examination of the refined structure. In addition, other electrostatic interactions between the positively charged side chains and the negatively charged functional group of the nucleotides in both groove can contribute to the stabilization of the complex. The structure of minor complex was more difficult to determine. It seems that one acridine unit is inserted in the abasic pocket but the position of the second acridine moiety could not be defined. Formation of the minor complex could result from intercalation of the first acridine ring in the abasic pocket, the second acridine ring remaining outside the duplex. Opening of the C7-G16 would allow insertion of the second acridine ring between the two base pairs giving the major complex. This seems a reasonable hypothesis since the opening of a single base pair was previously demonstrated to be a rapid event. The presence of competing complexes and the few intermolecular NOEs available prevented complete quantitative analysis. Finally no complex formation was observed with the parent unmodified duplex TA in the same conditions.

For the first time, this study provides structural data on the interaction of **BisA** with oligonucleotides. It confirms the specific binding of the macrocycle to the abasic site and demonstrates clearly the lack of affinity for duplex DNA. Moreover a new binding mode was evidenced that could be defined as threading bisintercalation, which involves the opening of a base pair. The cost in energy of this event should be compensated by the high stability of the resulting complex formed with the macrocycle. Also it further opens the possibility of investigating the potential of **BisA** and other related macrocycles to act as reagents for the stabilization of short–lived single-stranded regions and for the detection of locally altered structures in DNA.

Quadruplex recognition   More recently, the interaction of **BisA** with DNA quadruplexes has been investigated [46]. From a structural point of view, intramolecular quadruplex structures formed from intramolecular folding of G-rich sequences exhibit hairpin loops that could constitute recognition motifs for cyclobisintercalands (Fig. 11.14a,b, left). From a biological point of view, the formation of quadruplexes in regulatory regions containing repetitive stretches of guanine is increasingly likely. Therefore their role in the regulation of DNA function, as well as their use as artificial tools, is the focus of intense research (see Chapter ▌).

Currently, one of the most concrete applications of G-quadruplexes is their

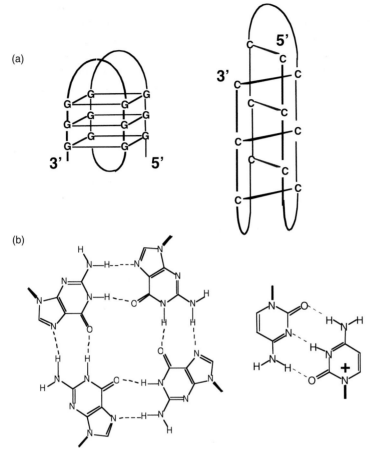

Fig. 11.14. (a) Possible folded structure of the G-rich (left) or C-rich (right) strand of human telomere. (b) Building blocks for quadruplex formation: the G-quartet (left) and the C.C+ hemiprotonated base pair (right).

capacity to inhibit telomerase. The single-stranded form of the telomere end is required for optimal telomerase activity and its folding into a G-quadruplex conformation has been shown to directly inhibit telomerase elongation. This enzyme, which is reactivated in most cancers, is an exciting target for designing new anticancer agents [47]. Therefore, a number of small ligands have been prepared to inhibit the function of telomerase by stabilizing G4-DNA structures. Moreover, telomeres are essential for DNA replication and for protecting chromosomes ends against double-strand breaks; they also participate in various aspects of the functional organization of the nucleus. All these functions might be altered by G-quadruplex-specific ligands independently of telomerase inhibition.

Thermal denaturation measurements have been carried out by fluorescence using doubly labeled oligonucleotides according to the FRET method developed by

Mergny *et al.* [48]. The oligonucleotide F21GT (5′-fluo-(GGGTTA)$_3$GGG-tamra 3′; fluo = fluorescein, tamra = tetramethylrhodamine)) was used; this models the human telomere repeats and forms an intramolecular quadruplex in the presence of KCl. Fluorescence melting curves showed that **BisA** is a good stabilizer of the G4-stranded structure ($\Delta T_m = +15°C$) whereas **MonoA** has little effect on the melting behavior of F21GT (Fig. 11.15a). A similar experiment was performed with the oligonucleotide F23CT, which corresponds to the complementary C-rich strand (5′ fluo-TA(CCCTAA)3CCC-tamra 3′) and forms a particular quadruplex called an i-motif (Fig. 11.14a,b, right). The melting profile revealed a strong stabilization by the macrocycle ($\Delta T_m = +33°C$) (Fig. 11.15b). The difference in the stabilizing activity of **BisA** towards the two quadruplexes should be interpreted in terms of kinetic effect of the folded forms rather than as a neat preference for the i-motif. However, **BisA** is one of the first i-DNA ligand to be discovered. A T_m experiment performed with the duplex formed from annealing F21GT and its complementary strand 21C revealed that **BisA** is able to prevent duplex formation (Fig. 11.16). This is attributed to the high affinity of **BisA** towards G- and C-tetraplexes that locks the oligonucleotides into a conformation that is no longer suitable for complementary hybridization. It is tempting to attribute the location of the macrocycle to the three loops which result from the folding of the C- and G-rich strands into quadruplexes. These results are fully in line with the helix-destabilizing properties of **BisA** described above.

The ability of **BisA** and **MonoA** to inhibit telomerase was determined using the TRAP assay (TRAP = telomerase repeat amplification protocol). **MonoA** was found to be completely inactive ($IC_{50} \gg 10$ μM, IC_{50} being the concentration of ligand required for a 50% decrease of the enzymatic activity). In contrast **BisA** is a potent inhibitor of telomerase ($IC_{50} = 0.75$ μM). This result confirms the correlation between the G-stabilizing effect and the inhibition of the activity of the telomerase, already observed for many G-quadruplex ligands [49].

The selectivity of **BisA** for various DNA conformations has been studied using a large panel of physico-chemical and biochemical methods. The results obtained on hairpins, helix destabilization, abasic sites, and quadruplexes all fit in with the general trend: **BisA** cannot bind DNA duplexes in a specific manner. Therefore, **BisA** preferentially recognizes any secondary structure with more accessible sites: single strands, hairpin loops, loop-containing quadruplexes, and abasic pockets. Quantitative data would be very informative in order to compare the affinities of **BisA** for these various structures as well as structural studies in order to characterize the interactions at the molecular level.

11.3.5
Miscellaneous

11.3.5.1 Naphthalene diimide derivatives
The macrocyclic bisintercaland **27**, composed of two naphthalenediimide subunits connected by two permethylated ammonium linkers, has been reported recently

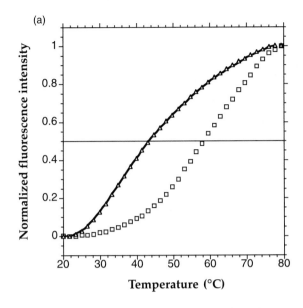

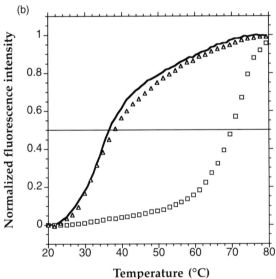

Fig. 11.15. FRET studies: fluorescence emission was monitored at 520 nm using a 470 nm excitation wavelength. (a) G-rich strand F21GT was chosen as the fluorescent probe for quadruplex formation. Solid line: oligonucleotide alone; triangles: +1 μM **MonoA**; squares: +1 μM **BisA**. Buffer conditions: 10 mM sodium cacodylate, 100 mM LiCl, pH 7.2. (b) C-rich strand F23CT was chosen as a fluorescent probe for i-DNA formation. Solid line: oligonucleotide alone; triangles: +1 μM **MonoA**; squares: +1 μM **BisA**. Buffer conditions: 10 mM cacodylate, pH 6.8 (heating curves).

(a)

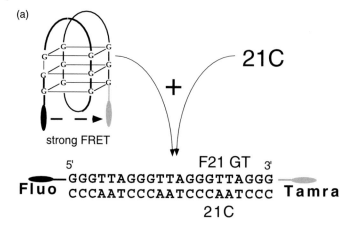

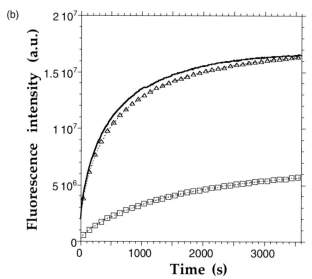

Fig. 11.16. (a) Design of the experiment: the G-rich strand is labeled with fluorescein (Fluo) and a rhodamine derivative (Tamra) at its 5′ and 3′ ends respectively. (b) Kinetic analysis of duplex formation: 21C (0.25 μM strand concen-tration) is added at 37°C to a preformed F21GT quadruplex alone (0.2 μM strand concentration, solid line) or in the presence of 0.5 μM **MonoA** (triangles) or **BisA** (squares), 100 mM KCl, 10 mM cacodylate buffer (pH 7.2).

[50]. Based on UV-vis measurements, a catenated structure was suggested for its complex with CpG. Likewise, it forms a strong complex with calf thymus DNA which dissociates 10^3-times more slowly than the corresponding monointercalator; this result also supports a catenated structure for the complex of **27** with dsDNA. However more experimental results would be necessary to assess the reality of this threading intercalation mode of binding.

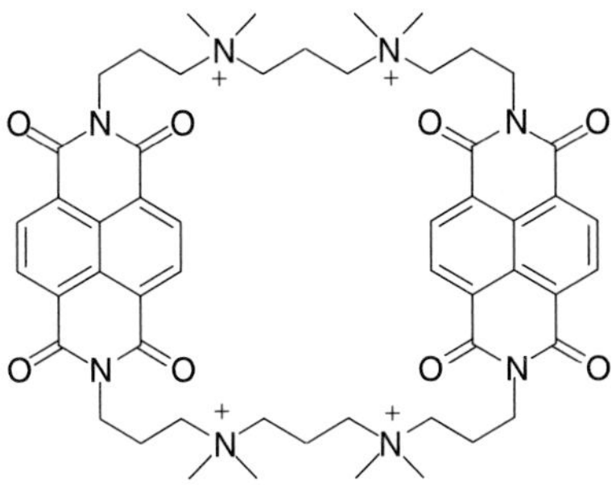

27

11.3.5.2 Phenazine derivatives

A tetracationic cyclointercaland made of two photosensitizing intercalating phenazine subunits linked by two viologen moieties, **28**, has been prepared [51]. Upon
irradiation this compound appears to strongly cleave supercoiled pBR322 plasmid,
generating mainly single-strand breaks.

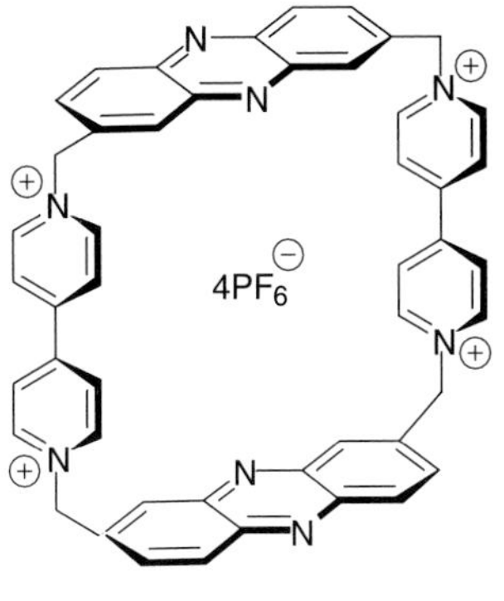

4PF$_6$

28

11.3.5.3 Aminocalixarenes and aminocyclodextrins

Since calixarenes and cyclodextrins can be considered to be special macrocycles,
the interactions of some of their amino derivatives with nucleotides and nucleic
acids are worth mentioning. It has been reported that cyclodextrins bearing amino
groups at their 6-positions bind nucleotides strongly and can discriminate between
isomeric 3'- and 5'-phosphates as well as between 2'-oxy and 2'-deoxy sugar phosphates [52]. Some permethylated aminocalixarenes, for their part, exhibit a remarkable preference for DNA in comparison to RNA [53]. In any case, these interactions are dominated by coulombic forces.

11.4
Conclusion and Perspectives

The synthetic macrocyles conceived by analogy with natural bisintercalators and on the basis of supramolecular concepts such as geometrical and functional complementary, show amazing and interesting binding behaviors towards DNA. The results reported throughout this chapter demonstrate that most macrocycles studied so far cannot form classical intercalation complexes with double-stranded DNA. More interestingly they display binding preferences for single-stranded secondary structures and some of them even exhibit the ability to destabilize DNA or RNA duplexes. Obviously, their particular behavior is determined by the steric clash expected from their cyclic framework, but the exhibited selectivities are also related to the chemical properties and to the topology of their constitutive subunits. As reported above, the combination of flat intercalative units and flexible poly-ammonium linkers led to the most selective and efficient binders (**BisA** and **SDM**). The flexibility of the latter could be the dominant structural feature contributing to fit the drug into its DNA target by allowing the optimization of specific polar and hydrophobic interactions.

Besides, the binding of macrocycle to DNA can result in strong distortion from normal double helix geometry. For instance, the 'intercalation platform' caused by **SDM** binding to duplexes DNA is a particularly intriguing phenomenon. It would be of great interest to know if these crosslinked species are biologically relevant and if they can be controlled for constructing well-defined multistranded DNA assemblies.

Similarly, the insertion of **BisA** into the abasic lesion through a threading bis-intercalation in violation of the neighbor-exclusion paradigm was difficult to predict. This behavior, which is unprecedented either for synthetic or natural ligands, demonstrates that cyclic ligands are compatible with dsDNA if a local pairing defect is present, allowing insertion. Moreover, this leads to remarkably stable associations resulting from interlocking of the two partners. This study offers the possibility of finding novel unusual binding modes which in turn should stimulate the conception of new multifunctional chemical architectures.

The above examples demonstrate that it is essential to develop structural analysis of the molecular interactions to shed light on the ability of DNA to accommodate complex compounds, thus providing new insight on the plasticity of this biopolymer. Furthermore, there is a growing need for real-time measurements of complexation events (i.e. quantitative access to kinetic parameters (K_{on}/K_{off})), especially when competitive binding of small molecules with proteins are contemplated. However, the use of NMR or of optical biosensors technology to monitor DNA/macrocycle interactions is systematically hampered by serious precipitation due to charge neutralization and/or crosslinking. Indeed, all the groups working with cationic macrocyclic compounds have reported that aggregation of DNA is frequently induced by ligand binding. This phenomenon also makes it difficult to use gel-mobility shift assays due to the inhibition of the migration of the complexes. The condensation of DNA with polyamines is a complicated process, which is not currently well understood. However, this phenomenon is related to fundamental biological problems such as gene vectorization and chromatin organization, and one can speculate that macrocycles might be potential candidates for studying or inducing condensation of DNA [54].

In most cases the molecular recognition of DNA by macrocycles has been investigated using model oligonucleotides or polynucleotides. The next step will be to examine the behavior of macrocycles with biologically relevant DNA targets, as has been done with **BisA** and G-quadruplexes. This could lead, for instance, to the use of macrocycles for the identification of structural-specific DNA elements either *in vitro* or *in vivo*; the detection could be achieved by using macrocycles containing fluorescent moieties or by grafting a fluorescent marker on the molecule. In addition, the trapping of nucleic bases by π-sandwiching between photo- or chemically active subunits could improve the redox-damaging activity. Such improvement could be expected from a longer residence time on the DNA or from a more efficient chemical reaction.

Another specific feature of macrocyclic structures possessing polyamino and heterocyclic moieties is their ability to form multinuclear coordination complexes with metallic cations (Cu^{2+}, Rh^{2+}, etc.). It would be of interest to take advantage of this property to induce redox damages at specific sites. The same rationale is valid for photocleavage, and this is currently under investigation with **BisQ** cyclobis-intercalands which possess quinacridine subunits able to damage DNA either when coordinated to Cu^{2+} [55] or via photoinduced electron transfer [56].

Finally, investigating more deeply the recognition of RNA by macrocyclic ligands could also be very fruitful. RNAs exhibit a large array of secondary and tertiary structures; among these a variety of bulge loops, hairpins, and pseudoknots present single-stranded regions which offer attractive specific binding sites for macrocyclic derivatives. Moreover the diversity of tertiary folding exhibited by RNAs can delineate clefts and pockets to accommodate macrocyclic compounds that could permit discrimination between DNA and RNA, which is still a challenge.

Altogether, the above observations may be seen as potent arguments for retaining consideration of macrocyclic structures as selective DNA ligands either in the drug discovery process or for use as biochemical tools.

Acknowledgements

The authors are grateful to all the collaborators who contributed to the works.

References

1 HOSSEINI, M. W., BLACKER, A. J., LEHN, J.-M. Multiple molecular recognition and catalysis. A multifunctional anion receptor bearing an anion binding site, an intercalative group, and a catalytic site for nucleotide binding and hydrolysis. *J. Am. Chem. Soc.* **1990**, *112*, 3896–3904 and references therein.

2 DIEDRICH, B., GUILHEM, J., LEHN, J.-M., PASCARD, C., SONVEAUX, E. Molecular recognition in anion coordination chemistry. Structure, binding constants and receptor-substrate complementarity of a series of anion cryptates of macrobicyclic receptor molecule. *Helv. Chim. Acta* **1984**, *67*, 91–104.

3 HOSSEINI, M. W., LEHN, J.-M. Binding of AMP, ADP, and ATP nucleotides by polyammonium macrocycles. *Helv. Chim. Acta* **1987**, *70*, 1312–1319.

4 HOSSEINI, M. W., BLACKER, A. J., LEHN, J.-M. Multiple molecular recognition and catalysis. Nucleotide binding and ATP hydrolysis by a receptor molecule bearing an anion binding site, an intercalative group, and a catalytic site. *J. Chem. Soc., Chem. Commun.* **1988**, 596–598.

5 FENNIRI, H., HOSSEINI, M. W, LEHN, J.-M. Molecular recognition of NADP(H) and ATP by macrocyclic polyamines bearing acridine groups. *Helv. Chim. Acta* **1997**, *80*, 786–803.

6 AGUILAR, J. A., GARCIA-ESPAÑA, E., GUERRERO, J. A. *et al.* Multifunctional molecular recognition of ATP, ADP and AMP nucleotides by the novel receptor 2,6,13,17,21-hexaaza[22] metacyclophane. *J. Chem. Soc., Chem. Commun.* **1995**, 2237–2239.

7 AGUILAR, J. A., CELDA, B., FUSI, V. *et al.* Structural characterization in solution of multifunctional nucleotide coordination systems. *J. Chem. Soc., Perkin Trans. 2* **2000**, 1323–1328.

8 RAGUNATHAN, K. G., SCHNEIDER, H.-J. Nucleotide complexes with azonia-cyclophanes containing phenyl-, biphenyl- or bipyridyl- units. *J. Chem. Soc., Perkin Trans. 2* **1996**, 2597–2600.

9 BAZZICALUPI, C., BENCINI, A., BIANCHI, A. *et al.* Basicity properties of two paracyclophane receptors. Their ability in ATP and ADP recognition in aqueous solution. *J. Chem. Soc., Perkin Trans. 2* **1997**, 775–781.

10 ODASHIMA, K., KOGA, K. in *Cyclophanes*, Vol. 2, eds P. M. Khun and S. M. Rosenfeld. Academic Press, New York, 1983, 629–677.

11 SOGA, T., ODASHIMA, K., KOGA, K. Modifications of hydrophobic cavity and their effects on the complex formation with a hydrophobic substrate. *Tetrahedron Lett.* **1980**, *21*, 4351–4354.

12 SCHNEIDER, H-J., BLATTER, T., PALM, B., PFINGSTAG, U., RÜDIGER, V., THEIS, I. Complexation of nucleosides, nucleotides, and analogs in an azoniacyclophane. van der Waals and electrostatic binding increments and NMR shielding effects. *J. Am. Chem. Soc.* **1992**, *114*, 7704–7708.

13 LEHN, J.-M., SCHMIDT, F. , VIGNERON, J.-P. Cyclointercalands. Incorporation of the phenazine group and of metal binding subunits into macrocyclic receptor molecules. *Tetrahedron Lett.* **1988**, *29*, 5255–5258.

14 CLAUDE, S., LEHN, J.-M., VIGNERON, J.-P. Bicyclo-bis-intercalands: synthesis of triply bridged bis-intercalands based on acridine subunits. *Tetrahedron Lett.* **1989**, *30*, 941–944.

15 CLAUDE, S., LEHN, J.-M., SCHMIDT, F., VIGNERON, J.-P. Binding of nucleosides, nucleotides and anionic planar substrates by bis-intercaland receptor molecules. *J. Chem. Soc., Chem. Commun.* **1991**, 1182–1185.

16 CLAUDE, S., LEHN, J.-M., PÉREZ DE VEGA, M.-J., VIGNERON, J.-P. Synthèse de bicyclo-bis-intercalants dérivés de l'acridine. *New J. Chem.* **1992**, *16*, 21–28.

17 CLAUDE, S. Etude de récepteurs bicyclo-bis-intercalants. PhD thesis, December 1990. Pierre & Marie Curie University, Paris.

18 CUDIC, P., ZINIC, M., TOMISIC, V., SIMEON, V., VIGNERON, J.-P., LEHN, J.-M. Binding of nucleotides in water by phenanthridinium bis(intercaland) receptor molecules. *J. Chem. Soc., Chem. Commun.* **1995**, 1073–1075.

19 CUDIC, P., ZINIC, M., SKARIC, V. *et al.* Synthesis of cyclo-bis-intercaland receptor molecules with phenan-thridinium units. *Croat. Chem. Acta* **1996**, *69*, 569–611.

20 PIANTANIDA, I., PALM, B. S., CUDIC, P., ZINIC, M., SCHNEIDER, H.-J. Phenanthridinium cyclobisinter-calands. Fluorescence sensing of AMP and selective binding to single-stranded nucleic acids. *Tetrahedron Lett.* **2001**, *42*, 6779–6783.

21 DHAENENS, M., LEHN, J.-M., VIGNERON, J.-P. Molecular recognition of nucleosides, nucleotides and anionic planar substrates by a water soluble bis-intercaland type receptor molecule. *J. Chem. Soc., Perkin Trans.* 2 **1993**, 1379–1381.

22 TEULADE-FICHOU, M.-P., VIGNERON, J.-P., LEHN, J.-M. Molecular recognition of nucleosides and nucleotides by a water soluble cyclo-bis-intercaland type receptor molecule based on acridine subunits. *J. Supramol. Chem.* **1995**, *5*, 139–147.

23 PARIS, T., VIGNERON, J.-P., LEHN, J.-M., CESARIO, M., GUILHEM, J., PASCARD, C. Molecular recognition of anionic substrates. Crystal structures of the supramolecular inclusion complexes of terephthalate and isophthalate dianions with a bis-intercaland receptor molecule. *J. Incl. Phenom.* **1999**, *33*, 191–202.

24 CUDIC, P., VIGNERON, J.-P., LEHN, J.-M., CESARIO, M., PRANGÉ, T. Molecular recognition of azobenzene dicarboxylates by acridine-based receptor molecules. Crystal structure of the supramolecular inclusion complex of *trans*-3,3′-azobenzene dicarboxylate with a cyclo-bis-intercaland receptor. *Eur. J. Org. Chem.* **1999**, 2479–2484.

25 BAUDOIN, O., TEULADE-FICHOU, M.-P., VIGNERON, J.-P., LEHN, J.-M. Cyclobisintercaland macrocycles: synthesis and physicochemical properties of macrocyclic polyamines containing two crescent-shaped dibenzophenanthroline subunits. *J. Org. Chem.* **1997**, *62*, 5458–5470.

26 BAUDOIN, O., GONNET, F., TEULADE-FICHOU, M.-P., VIGNERON, J.-P., TABET, J.-C., LEHN, J.-M. Molecular recognition of nucleotide pairs by a cyclo-bis-intercaland-type receptor molecule: a spectrophotometric and electrospray mass spectrometry study. *Chem. Eur. J.* **1999**, *5*, 2762–2771.

27 SCHNEIDER, H.-J., BLATTER, T. Interactions between acyclic and cyclic peralkylammonium compounds and DNA. *Angew. Chem. Int. Ed. Engl.* **1992**, *31*, 1207–1208.

28 FERNANDEZ-SAIZ, M., SCHNEIDER, H.-J., SARTORIUS, J., WILSON, W. D. Cationic cyclophane that forms a base-pair open complex with RNA duplexes. *J. Am. Chem. Soc.* **1996**, *118*, 4739–4745.

29 FERNANDEZ-SAIZ, M., RIGL, C. T., KUMAR, A. *et al.* Design and analysis of molecular motifs for specific recognition of RNA. *Bioorg. Med. Chem.* **1997**, *5*, 1157–1172.

30 HAMILTON, A. D., LEHN, J.-M., SESSLER, J. L. Mixed substrate supermolecules: Binding of organic substrates and of metal ions to heterotopic coreceptors containing porphyrin subunits. *J. Chem. Soc., Chem. Commun.* **1984**, 311–313.

31 SLAMA-SCHWOK, A., LEHN, J.-M. Interaction of a porphyrin-containing macrotetracyclic receptor molecule with single-stranded and double-stranded polynucleotides. A photo-physical study. *Biochemistry* **1990**, *29*, 7895–7903.

32 WILSON, W. D., WANG, Y. H., KRISHNAMOORTHY, C.R., SMITH, J. C. Poly(dA).poly(dT) exists in an unusual conformation under physiological conditions: propidium binding to poly(dA).poly(dT) and poly[d(A-T)].poly[d(A-T)]. *Biochemistry* **1985**, *24*, 3991–3999.

33 ZIMMERMAN, S. C., LAMBERSON, C. R., CORY, M., FAIRLEY, T. A. Topologically constrained bifunctional intercalators: DNA intercalation by a macrocyclic bisacridine. *J. Am. Chem. Soc.* **1989**, *111*, 6805–6809.

34 VEAL, J. M., LI, Y., ZIMMERMAN, S. C. *et al.* Interaction of a macrocyclic bisacridine with DNA. *Biochemistry* **1990**, *29*, 10918–10927.

35 TANIOUS, F., YEN, S., WILSON, W. D. Kinetic and equilibrium analysis of a threading intercalation mode: DNA sequence and ions effects. *Biochemistry* **1991**, *30*, 1813–1819.

36 CROTHERS, D. M. Calculation of binding isotherms for heterogeneous polymers. *Biopolymers* **1968**, *6*, 575–584.

37 YANG, X., ROBINSON, H., GAO, Y.-G., WANG, H.-J. Binding of a macrocyclic bisacridine and ametantrone to CGTACG involves similar unusual intercalation platforms. *Biochemistry,* **2000,** *39,* 10950–10957.

38 SLAMA-SCHWOK, A., TEULADE-FICHOU, M.-P., VIGNERON, J.-P., TAILLANDIER, E., LEHN J.-M. Selective binding of a macrocyclic bisacridine to DNA Hairpins. *J. Am. Chem. Soc.* **1995,** *117,* 6822–6830.

39 SLAMA-SCHWOK, A., PERONNET, F., HANTZ-BRACHET, E. *et al.* A macrocyclic bisacridine shifts the equilibrium from duplexes towards DNA hairpins. *Nucleic Acids Res.* **1997,** *25,* 2574–2581.

40 MONAGHAN, A., WEBSTER, A., HAY, R. T. Adenovirus DNA binding protein: helix destabilizing properties. *Nucleic Acids Res.* **1994,** *22,* 742–748.

41 TEULADE-FICHOU, M.-P., FAUQUET, M., BAUDOIN, O., VIGNERON, J.-P., LEHN, J.-M. DNA double helix destabilizing properties of cyclobisintercaland compounds and competition with a single strand binding protein. *Bioorg. Med. Chem.* **2000,** *8,* 215–222.

42 BLACKER, A. J., TEULADE-FICHOU, M.-P., VIGNERON, J.-P., FAUQUET, M., LEHN J.-M. Selective photocleavage of single stranded nucleic acids by cyclobisintercaland molecules. *Bioorg. Med. Chem. Lett.* **1998,** *8,* 601–606.

43 LOEB, L. A., PRESTON, B. D. Mutagenesis by apurinic/apyrimidinic sites. *Annu. Rev. Genet.* **1986,** *20,* 201–230.

44 BERTHET, N., MICHON, J., LHOMME, J., TEULADE-FICHOU, M.-P., VIGNERON, J.-P., LEHN, J.-M. Recognition of abasic sites in DNA by a cylobisacridine molecule. *Chem. Eur. J.* **1999,** *5,* 3625–3630.

45 JOURDAN, M., GARCIA, J., LHOMME, J., TEULADE-FICHOU, M.-P., VIGNERON, J.-P., LEHN, J.-M. Threading bis-intercalation of a macrocyclic bisacridine at abasic sites in DNA: ^{1}H NMR and molecular modeling study. *Biochemistry* **1999,** *38,* 14205–14213.

46 ALBERTI, P., REN, J., TEULADE-FICHOU, M.-P. *et al.* Interaction of an acridine dimer with DNA quadruplex structures. *J. Biomol. Struct. Dyn.* **2001,** *19,* 505–513.

47 PERRY, P. J., ARNOLD, J. R. P., JENKINS, T. C. Telomerase inhibitors for the treatment of cancer: the current perspective. *Exp. Opin. Invest. drug* **2001,** *10,* 2141–2156.

48 MERGNY, J.-L., MAURIZOT, J. C. Fluorescence resonance energy transfer as a probe for G-quartet formation by a telomeric repeat. *Chembiochem.* **2001,** *2,* 124–132.

49 MERGNY, J.-L., LACROIX, L., TEULADE-FICHOU, M.-P. *et al.* Discovery of new G4-based telomerase inhibitors by fluorescence resonance energy transfer. *Proc. Natl Acad. Sci. USA* **2001,** *98,* 3062–3067.

50 TAGAGI, M., YOKOYAMA, H., TAKENAKA, S., YOKOYAMA, M., KONDO, H. Poly-intercalators carrying threading intercalator moieties as novel DNA targeting ligands. *J. Incl. Phenom.* **1998,** *32,* 375–383.

51 LORENTE, A., FERNANDEZ-SAIZ, M., HERRAIZ, F., LEHN, J.-M., VIGNERON, J.-P. Photocleavage of DNA by tetracationic intercalands containing phenazine and viologen subunits. *Tetrahedron Lett.* **1999,** *40,* 5901–5904.

52 ELISEEV, A. V., SCHNEIDER, H.-J. Aminocyclodextrins as selective hosts with several binding sites for nucleotides. *Angew. Chem. Int. Ed. Engl.* **1993,** *32,* 1331–1333.

53 SHY, Y., SCHNEIDER, H.-J. Interactions between aminocalixarenes and nucleotides or nucleic acids. *J. Chem. Soc., Perkin Trans. 2* **1999,** 1797–1803.

54 SUKHANOVA, A., DÊVY, J., PLUOT, M. *et al.* Human DNA topoisomerase I inhibitory activities of synthetic polyamines: relation to DNA aggregation. *Bioorg. Med. Chem.* **2001,** *9,* 1255–1268.

55 BAUDOIN, O., TEULADE-FICHOU, M.-P., VIGNERON, J.-P., LEHN, J.-M. Efficient copper (II)-mediated nuclease activity of *ortho*-quinacridines. *J. Chem. Soc. Chem. Commun.* **1998,** 2349–2350.

56 TEULADE-FICHOU M.-P., PERRIN, D., BOUTORINE, A. *et al.* Direct photocleavage of HIV-DNA by quinacridine derivatives triggered by triplex formation. *J. Am. Chem. Soc.,* **2001,** *123,* 9283–9292.

12
Triplex- versus Quadruplex-specific Ligands and Telomerase Inhibition

Patrizia Alberti, Magali Hoarau, Lionel Guittat, Masashi Takasugi, Paola B. Arimondo, Laurent Lacroix, Martin Mills, Marie-Paule Teulade-Fichou, Jean-Pierre Vigneron, Jean-Marie Lehn, Patrick Mailliet, and Jean-Louis Mergny

12.1
Introduction

A number of alternative DNA structures have been described to date. As stressed previously, DNA comes in many forms [1]. It can exist in a variety of double-, triple-, and quadruple-stranded forms. The list of alternative conformations that can be adopted by this molecule is still growing with the recent addition of DNA pentaplexes [2, 3]. Multistranded structures rely on the formation of specific base pairs, triplets, or quartets (Fig. 12.1).

G-quadruplexes are a family of secondary DNA structures formed in the presence of monovalent cations that consist of four-stranded structures stabilized by G-quartets (Fig. 12.1a) [4]. There is a renewed interest in quadruplex structures due to their putative biological regulatory function. Optimal telomerase activity requires the non-folded single-stranded form of the primer, and G-quartet formation has been shown to directly inhibit telomerase elongation *in vitro* [5]. Therefore, ligands that selectively bind to G-quadruplex structures may modulate telomerase activity [6, 7].

Other multistranded nucleic acid structures may also be observed. Cytosine-rich oligomers may also form a quadruplex called i-DNA [8–10], based on the formation of hemiprotonated base pairs (Fig. 12.1b). Triplexes are obtained when a third strand binds to the major groove of double-stranded nucleic acids [11, 12]. Triple helices were first observed in 1957 formed by polyribonucleotides [13]. At least three structural classes of triple helix exist that differ in the base composition of the third strand, the relative orientation of the phosphate-deoxyribose backbone and the thermodynamical parameters [14]. In the first class, the pyrimidine motif (TC motif), the third strand binds parallel to the purine strand of the duplex by Hoogsteen hydrogen bonds, forming T.A*T and C.G*C+ triplets (Fig. 12.1c, left). This class of triple helix is stabilized by an acidic pH. In the second class, the GA motif, oligonucleotides containing guanines and adenines bind in antiparallel ori-

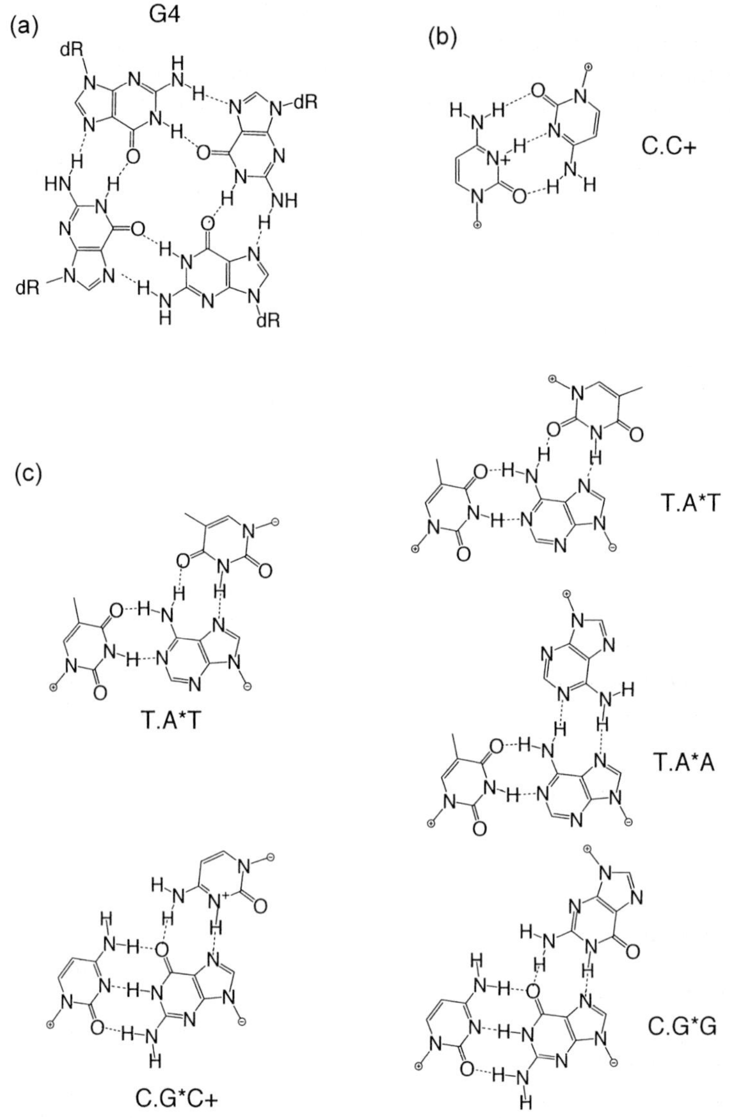

(a)

G4

(b)

C.C+

(c)

T.A*T

T.A*T

T.A*A

C.G*C+

C.G*G

Fig. 12.1. Base pairing schemes. (a) A G-quartet involving four guanines. (b) A C.C+ base pair found in i-DNA. (c) Base triplets. Left: Hoosteen pairing. Right: reverse Hoogsteen. TC triplexes are composed of T.A*T and C.G*C+ base triplets; GA and GT triplexes are composed of T.A*A and C.G*G or T.A*T and C.G*G base triplets, respectively. Watson–Crick base pair is indicated by a dot, Hoogsteen or reverse-Hoogsteen pairing by *. The base belonging to the third strand is indicated last. Strand orientation is indicated by − and +, except for the G-quartet, where it may vary. Note that the formation of the C.C+ base pair and of the C.G*C+ base triplet requires protonation of one cytosine at its N3 position.

entation to the purine strand of the duplex by reverse Hoogsteen hydrogen bonds, forming C.G*G and T.A*A base triplets (Fig. 12.1c, right). In the third class, the GT motif, oligonucleotides containing thymines and guanines bind either parallel or antiparallel to the purine strand of the duplex depending on the base sequence [15].

Significant progress has been made over the past few years in studies of drug–DNA interactions. Alternative DNA structures offer significant differences in terms of electrostatics, shape, and rigidity compared with single- or double-stranded DNA. Therefore, specific recognition of unusual DNA structures by small molecules should be possible. This assumption was shown first to be correct for triplexes [16], then for quadruplexes [6]. A drug that stabilizes quadruplexes could interfere with telomerase and telomere replication [17, 18]. A range of molecules containing tricyclic, tetracyclic, and pentacyclic aromatic chromophores has been shown to inhibit telomerase [6, 19–24]. Nevertheless, recent reports have shown that some molecules thought to be quadruplex-specific are rather triplex-specific [25], arguing for a systematic binding evaluation to a variety of nucleic acids structures. The sequence and structural selectivity of various different DNA-binding agents has been previously explored using a thermodynamically rigorous competition dialysis procedure [25–27]. In this competition dialysis method, different nucleic acid structures are dialyzed against a common ligand solution. More ligand accumulates in the dialysis tube containing the structural form with the highest ligand-binding affinity. We have decided to modify the experimental format chosen by Chaires and colleagues in order to accommodate several new triplex structures, as well as a few other DNA conformations. We will first present the nucleic acids structures chosen for this assay before analyzing the selectivity of some new ligands towards these conformations.

12.2
Nucleic Acids Samples

The low specificity of some compounds towards telomerase inhibition prompted us to evaluate the specificity of these compounds for quadruplexes with a competitive dialysis experiment. In such an experiment, different nucleic acids structures are dialysed against a common ligand solution [25]. More ligand accumulates in the dialysis tube containing the structural form with the highest ligand binding affinity. Because our laboratory is interested in the design of triplex- as well as quadruplex-specific ligands, we decided to select a slightly different set of structures (8 different, 10 identical) than previously described [26]. Formation of some of these structural motifs requires a slightly acidic pH and/or magnesium. As a consequence, the buffer chosen for all experiments is somewhat different (pH 6.5, with 10 mM $MgCl_2$), keeping the same relatively high monocation concentration (0.185 M NaCl) in order to limit electrostatic interactions between positively charged dyes and nucleic acids.

Tab. 12.1. Nucleic acid structures.

Number	Name	Type[a] (length)	Structure	T_m (°C)[d]
1	TC triplex	oligos (30 + 13)	Triplex	38
2	GA triplex	oligos (30 + 13)	Triplex	53
3	GT triplex	oligos (30 + 13)	Triplex	53
4	poly(dA)-2poly(dT)	poly	Triplex	71
5	GA duplex	oligo (24)	"Duplex"[b]	37
6	ps duplex	oligos (24 + 24)	"Duplex"[b]	39
7	24CTG	oligo (24)	"Duplex"[b]	64
8	poly(dA-T)	poly	Duplex	66
9	poly(dG-C)	poly	Duplex	>90
10	CT DNA	poly	Duplex	86
11	ds26	oligo (26)	Duplex	75
12	poly(dC)	poly	i-DNA	51[e]
13	22CT	oligo (22)	ss/i-DNA[c]	13[e]
14	22AG	oligo (22)	G4	62
15	24G20	oligo (24)	G4	>90
16	poly(dT)	poly	single-str	–[f]
17	poly(dA)	poly	single-str	–[f]
18	poly(rU)	poly	single-str	–[f]

All polynucleotides were ordered from Amersham-Pharmacia. Oligonucleotides were synthesized by Eurogentec, Belgium on the 1 μmol scale, checked by denaturing gel electrophoresis and used without further purification. Eighteen different nucleic acids structures are used (samples labeled 1–18, left column).

The TC, GA and GT triplexes result from the association of two strands of different lengths (13 and 30 nucleotides) [14]: 5′-GAAAGAGAGGAGG and 5′-CCTCCTCTCTTTCCCTTCTTTCTCTCCTCC (TC triplex, sample 1); 5′-CCTCCTCTCTTTC and 5′-GAAAGAGAGGAGGCCTTGGAGGAGAGAAAG (GA triplex, sample 2); 5′-CCTCCTCTCTTTC and 5′-GAAAGAGAGGAGGCCTTGGTGGTGTGTTTG (GT triplex, sample 3).

The GA "duplex" (sample 5) results from the self-association of the 24 GA oligonucleotide (5′-GAGAGAGAGAGAGAGAGAGAGAGA) which probably leads to the formation of a parallel-stranded duplex [28, 29].

The parallel-stranded duplex (sample 6) results from the association of two AT strands [30]: 5′-AAAAAAAAAATAATTTTAAATATT and 5′-TTTTTTTTTTTATTAAAATTTATAA.

The 24CTG (sample 7) mimics eight repeats of the trinucleotide unit: 5′ CTGCTGCTGCTGCTGCTGCTGCTG.

ds26 (sample 11) is a 26-base-long duplex formed with the self-complementary oligonucleotide 5′-CAATCGGATCGAATTCGATCCGATTG.

22CT (sample 13) is an oligonucleotide that mimics the cytosine-rich strand of human telomeres: 5′-CCCTAACCCTAACCCTAACCCT [31], whereas 22AG (sample 14) is an oligonucleotide that mimics the guanine-rich strand of human telomeres: 5′-AGGGTTAGGGTTAGGGTTAGGG [32]. 24G20 ($T_2G_{20}T_2$, sample 15) may form an intermolecular quadruplex 5′-(TTGGGGGGGGGGGGGGGGGGGGTT)₄ [26]. Due to the

The 18 samples chosen for analysis and the sequence of oligonucleotides are listed in Tab. 12.1. Structural forms included in the assay range from single strands (poly(dA), poly(dT), and poly(rU)), through a variety of duplexes, to four different triplexes (TC, GA, and GT as well as poly(dA)-2poly(dT)), as well as tetra-plex forms. Among these samples, two correspond to G-quartet-containing motifs: 24G20 ($T_2G_{20}T_2$), which forms a parallel-stranded intermolecular quadruplex and 22AG which forms a folded antiparallel quartet structure. 22CT and poly(dC) are prone to i-DNA formation.

The stability of the various structures may be analysed by thermal denaturation experiments. Absorbance versus temperature is recorded for each sample at two different wavelengths (260 and 295 nm) and the results are shown in Figs 12.2 and 12.3. Absorbance at 260 nm was chosen as it corresponds to the absorbance maxi-mum of nucleic acids of mixed purine–pyrimidine content. Absorbance at 295 nm has been previously shown to be useful as the melting of some DNA structures is easily evidenced at this wavelength [35]. Data for poly(dT), poly(dA), poly(rU) (which fail to form any structure between 0 and 90°C), poly(dA-T), poly(dG-C), and calf thymus DNA (which melt at 66, >90 and 86°C, respectively) is omitted for clarity. All the other 12 melting profiles are shown. The two i-DNA-forming oligo-nucleotides exhibit a non-reversible melting behavior: one cannot superimpose the heating and cooling profiles (Fig. 12.3c,d). At this pH, the 22CT structure is rela-tively unstable and this oligonucleotide is mostly single stranded at 20°C (Fig. 12.3d). All other structures are formed at room temperature as their T_m is $\geq$ 37°C. The melting of the pyrimidine triplex is biphasic at pH 6.5: the triplex dissociates at a lower temperature than the duplex (Fig. 12.2a). Only the melting of the pyr-imidine triplex gives a clear transition at 295 nm (Fig. 12.2a) as a result of cytosine deprotonation upon triplex unfolding [36–39]. Melting of the GA and GT triplexes is monophasic (Fig. 12.2b,c) and the transition corresponds to a triplex-to-single-strand phenomenon [14]. As expected, the stability of the poly(dA)-2poly(dT) tri-plex is much higher.

Three unusual duplexes were then studied. The GA duplex (Fig. 12.2e), resulting from the self-association of a (GA)-repeat and the parallel-stranded DNA duplex (Fig. 12.2f) have relatively similar stabilities with T_m values of 37 and 39°C, re-

presence of magnesium in the incubation buffer which favors ribonucleotide degradation, only one RNA sample is present (poly rU) among the 18 samples. This oligomer was aliquoted in the absence of $MgCl_2$.

[a] poly, polynucleotide; oligo, oligonucleotide; oligos, structure formed by the association of two different oligonucleotides. The length of the oligomer(s) is shown within parenthesis. Polynucleotides are >100 bases long.

[b] These three duplexes are rather unusual as they involve the formation of non-classical base pairs.

[c] 22CT may form an i-DNA structure but is mainly single stranded at room temperature.

[d] Obtained in an buffer identical to the equilibrium dialysis protocol.

[e] Hysteresis. T_m obtained while heating.

[f] No transition.

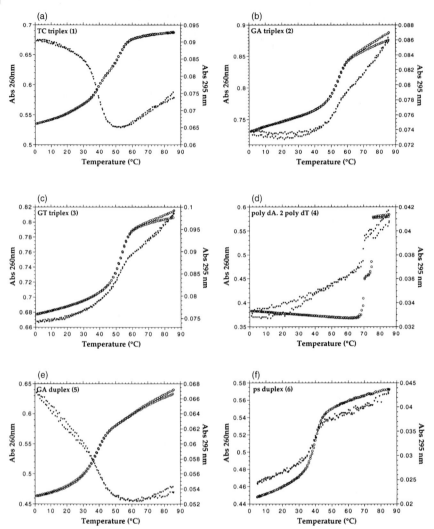

Fig. 12.2. Stability of some DNA structures: T_m analysis (1). The experimental conditions are: 0.185 M NaCl, 15 mM Na cacodylate pH 6.5, MgCl$_2$ 10 mM. All T_m values were recorded starting from high temperature (90°C) cooling at 0.2°C min^{-1} to 0°C followed by a heating cycle at 0.2°C min^{-1} [14]. T_m values were obtained at 2–5 μM strand concentration. The profiles were recorded at 260 nm (open circles) and 295 nm (filled triangles). Note the separate y-axis for these two wavelengths. All melting profiles are kinetically reversible, as shown by the superimposition of the heating and cooling curves. (a) TC triplex (1); (b) GA triplex (2); (c) GT triplex (3); (d) poly(dA)-2poly(dT) triplex (4); (e) 24 GA duplex (5); (f) parallel-stranded AT duplex (6).

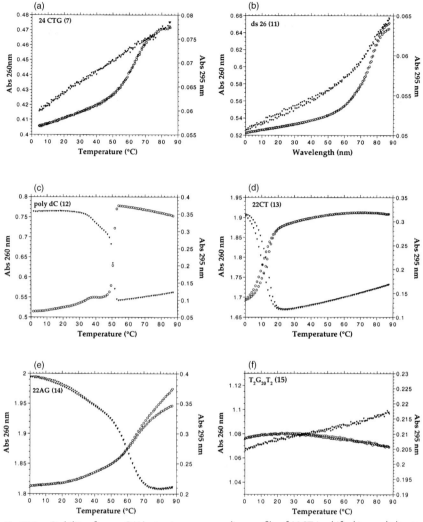

Fig. 12.3. Stability of some DNA structures under dialysis conditions: T_m analysis (2). Same conditions as in Fig. 12.2. All melting profiles were recorded at 260 nm (open circles) and 295 nm (filled triangles). Note the separate *Y*-axis for these two wavelengths. All melting profiles except c and d are kinetically reversible, as shown by the superimposition of the heating and cooling curves. i-DNA formation at near neutral pH is slow, leading to an hysteresis phenomenon. As a result the cooling profile of 22CT is shifted towards lower temperatures. For poly(dC), the effect is even more pronounced, and i-DNA formation is extremely slow. Only the denaturation profile is shown. (a) 24 CTG trinucleotide repeat (7); (b) short 26-bp duplex (11); (c) poly(dC) (12); (d) i-DNA forming oligonucleotide 22CT (13); (e) intramolecular G-quadruplex 22AG (14); (f) parallel G-quadruplex 24G20 (15) ($T_m >$ 90°C, no fusion observed).

spectively. The 24-base-long oligonucleotide that mimics a trinucleotide repeat, $(CTG)_8$, forms a highly stable structure with a T_m of 64°C (Fig. 12.3a). This T_m value is close to the melting temperature (75°C) of a perfectly matched 26-base-long Watson–Crick autocomplementary oligonucleotide ds26 (Fig. 12.3b). Among the two i-DNA-forming samples only poly(dC) gives a relatively high T_m (51°C) at pH 6.5 (Fig. 12.3c,d). Finally, an oligonucleotide mimicking the G-rich strand of human telomeres has a T_m of 62°C under these conditions (Fig. 12.3e), whereas the parallel-stranded quadruplex involving 20 contiguous guanines cannot be unfolded even at 90°C (Fig. 12.3f).

The nature of the folded form may also be analyzed by recording absorbance spectra below and above the melting temperature. The resulting differential absorbance spectrum shown in Fig. 12.4 may be seen as specific signatures for each sample. For example, the relatively broad peak between 245 and 270 nm (Fig. 12.4a) and the small negative peak at 295 nm is specific for a pyrimidine triplex requiring the protonation of the cytosines in the third strand in order to form C.G*C+ triplets. On the other hand, it is difficult to distinguish between the melting of a GT triplex (Fig. 12.4b) or a GA triplex (not shown) and the melting of a classical Watson–Crick duplex [14]. Both unusual duplexes have specific, differential absorbance signatures. The shoulder shown in Fig. 12.4c between 285 and 300 nm is absolutely specific of a parallel duplex, whereas the trinucleotide repeats (Fig. 12.4d) gives a maximum differential at an unusually high wavelength (277 nm).

Concerning quadruplexes, both types of structures give highly specific differential absorbances. Melting of an i-DNA structure (Fig. 12.4e) is associated with a positive peak at very short wavelengths (238 nm) and a large negative peak at 295 nm resulting from cytosine deprotonation [37]. On the other hand, melting of a G-quadruplex gives three reproducible peaks a 242, 255, and 271 nm and a negative peak at 295 nm (Fig. 12.4f) [35]. This figure illustrates the practical use of recording absorbances at several wavelengths to study nucleic acids structures.

12.3
Dialysis Results

We have recently identified several families of quadruplex ligands [23, 24, 34, 40], and we decided to analyse their binding specificity in the dialysis assay. The formulas of some of these molecules are shown in Fig. 12.5. Results are shown in Figs 12.6–12.7 for eight different compounds. Figure 12.6a shows the dialysis profile for ethidium bromide, a well-known intercalator. Its profile has previously been determined by Ren *et al.*, and this compound was therefore chosen as the reference. This dye interacts preferentially with regular duplexes such as poly(dA-T), poly(dG-C), and calf thymus DNA. Surprisingly, its preferred substrate is the unusual CTG repeat duplex. A detailed analysis demonstrates that this dye weakly interacts with triplexes [41], with a slight preference for poly(dA)-2poly(dT) [42].

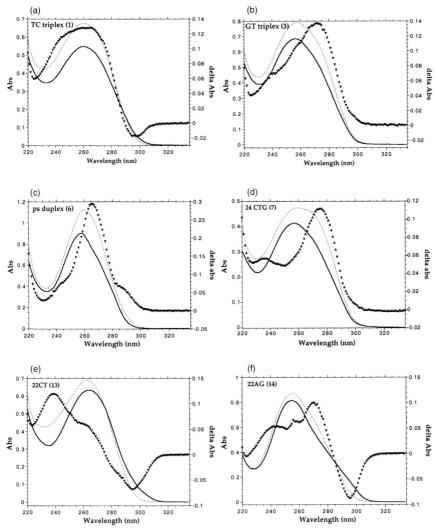

Fig. 12.4. Absorbance of some DNA structures under dialysis conditions at low and high temperature. Experimental conditions: 0.185 M NaCl, 15 mM Na cacodylate pH 6.5, MgCl$_2$ 10 mM. The UV-scan was recorded at 0° (full line) and 85° (dotted line). The absorbance values are shown on the left y-axis. The difference between high and low temperature absorbance is shown by black triangles (right scale). A negative ΔAbs means that the absorbance at 85°C is lower than the absorbance at 0°C. (a) TC triplex (1); (b) GT triplex (3); (c) parallel-stranded AT duplex (6); (d) 24 CTG trinucleotide repeat (7); (e) i-DNA forming oligonucleotide 22CT (13); (f) intramolecular G-quadruplex 22AG (14).

Binding to quadruplexes is confirmed [43], although the amount of dye bound to G-quadruplexes is low [23]. On the other hand, **9944**, an ethidium derivative that efficiently inhibits telomerase, shows a marked preference for G-quadruplexes and antiparallel triplexes (Fig. 12.6b).

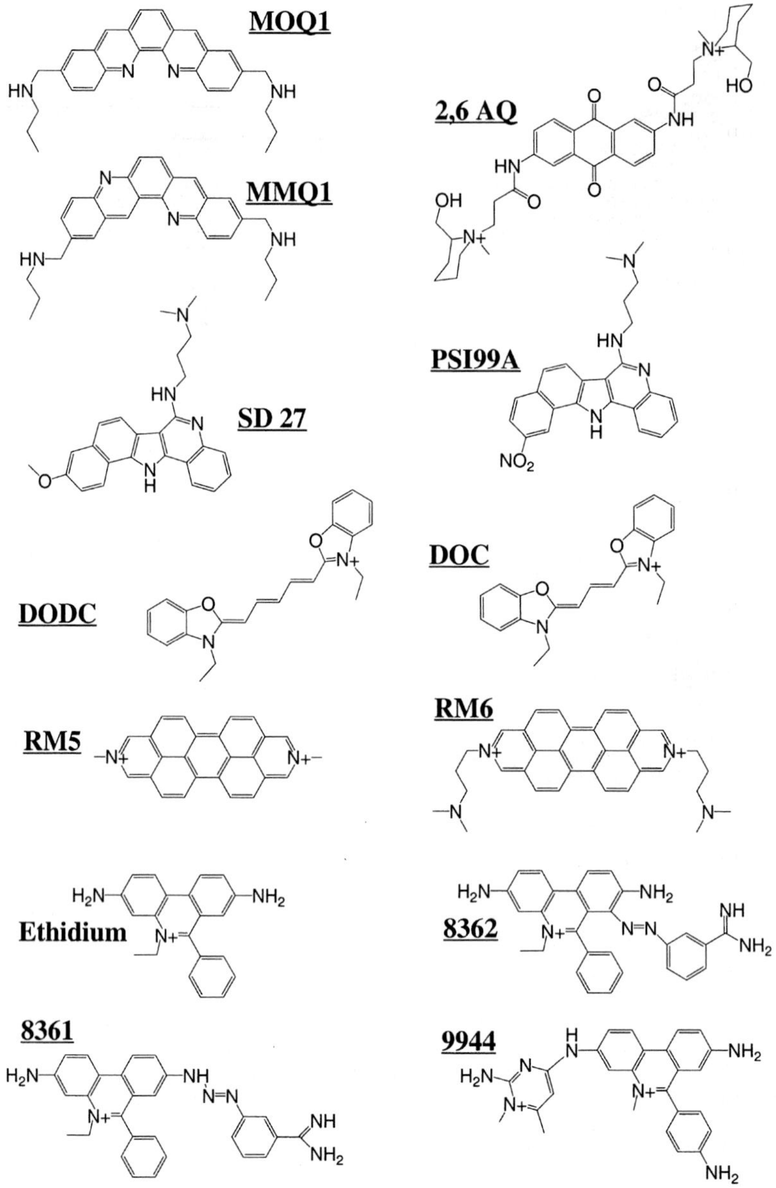

Fig. 12.5. Formulas of the compounds. **2,6 AQ** stands for 2,6-disubstituted amidoanthraquinone (or BSU 1051). **DODC** is the abbreviation of 3,3'-diethyloxadicarbocyanine. **DOC** is the abbreviation of 3,3'-diethyloxacarbocyanine. The synthesis of ethidium derivatives [23], dibenzophenanthrolines [24], **RM5** and **RM6** [33] and benzoindoloquinolines [34] has been described previously. **DODC** and **DOC** were obtained from Sigma-Aldrich. **2,6 AQ** (or BSU 1051) was synthesized and purified according to a published protocol [6].

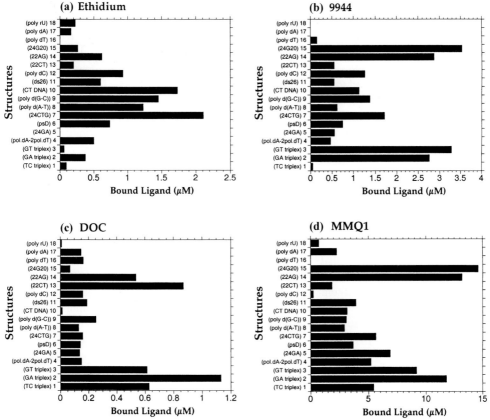

Fig. 12.6. Dialysis results (part 1). All measurements were performed using a methodology adapted from Chaires and colleagues [25–27]. Different nucleic acids were tested against a 1 μM dye solution in 500 mL of dialysis buffer (0.185 M NaCl, 15 mM Na cacodylate pH 6.5, 10 mM MgCl$_2$). A volume of 200 μL (at 75 μM monomeric unit: nucleotide, base pair, base triplet, or quartet) of each of the DNA samples listed in Table 12.1 was pipeted into a separate dialyzer unit (Pierce). All 18 dialysis units were then placed in the beaker containing the dialysate solution. The beaker was covered with Parafilm and wrapped in foil, and its contents were allowed to equilibrate with continuous stirring at room temperature (20–22°C) overnight. At the end of the equilibration period, DNA samples were carefully removed to microfuge tubes, and treated as described previously [26]. The ligand concentration in each sample was determined by absorbance or fluorescence. T$_2$G$_{20}$T$_2$ and 22AG form G-quadruplexes. Note the differences in x-axis limits between various compounds, which reflect large differences in binding affinities. (a) **Ethidium**; (b) **9944**; (c) **DOC**; (d) **MMQ1**.

This compound illustrates the structural difference between the different classes of triplexes: binding to the pyrimidine triplex is approximately 30 times weaker than to the GA and GT parallel triplexes. Such differences could be the result of the positive charges present on each C.G*C+ triplet in the TC triplex [41] or of differences in strand orientation and stacking interactions. On the other hand, there is little difference between the classes of quadruplexes (antiparallel, no. 14 and parallel no. 15). Binding to single strands is very low, whereas binding to duplexes is intermediate. Within the duplex set, a slight preference for the CTG repeat is obtained.

Other classes of DNA ligands were then analysed. Carbocyanines (**DODC**) have been reported to interact with triplexes [25] and quadruplexes [44, 45]. We chose to analyse **DOC**, a close relative of **DODC**. As shown in Fig. 12.6c, **DOC** binds preferentially (but weakly, note x-axis values) to triplexes and some quadruplexes. The marked **DODC** preference for triplexes is lowered for **DOC**, showing that the net distance between the cyanine groups could fine-tune the structural selectivity. Surprisingly, binding to the 22CT oligonucleotide is much higher than to the other i-DNA prone sample (poly(dC)). **DOC** is also the compound for which the relative difference between antiparallel (no. 14) and parallel (no. 15) quadruplexes is the largest. Binding to all other structures is very low (<0.2 µM), especially when considering that the uncertainty of the measurement is ~ 0.1 µM for that compound.

MMQ1 is a G-quadruplex ligand that efficiently inhibits telomerase [24]. Figure 12.6d confirms that this molecule interacts preferentially with G-quadruplexes, but does not distinguish between antiparallel and parallel structures. This molecule also strongly interacts with triplexes, especially the GA and, to a lesser extent, GT parallel triplexes. On the other hand, binding to single strands is almost negligible. Binding to duplexes is lower than to multistranded structures. Within the duplex class, very little difference is found between the different classical Watson–Crick duplexes (samples 8–11). The affinity for the three unusual duplexes (samples 5–7) is at least as high as for the regular duplexes.

13H-benzo-[6,7]-indolo[3,2-c]quinolines (latter referred to as benzoindoloquinolines) are pentacyclic aromatic molecules that carry a positively charged chain [46] and recognize quadruplexes [34]. Two benzoindoloquinolines (**PSI99A** and **SD27**) have been tested. As shown in Fig. 12.7a, **PSI99A** preferentially interacts with quadruplex-forming DNA samples. It also binds strongly to the GA triplex. Binding to duplexes is moderately strong, whereas binding to single strands is very low. On the other hand, **SD27** exhibits a limited structural specificity (Fig. 12.7b).

Another ligand, PIPER, was shown to interact with quadruplexes [47, 48]. Its shape is similar to dimethyldiazapyrenium dications, which interact with nucleic acids [33, 49]. We therefore tested the affinity of two derivatives, **RM5** and **RM6**, for various DNA structures. As shown in Fig. 12.7c,d, both molecules interact with most types of nucleic acids. This observation allows us to conclude that most planar polyaromatic cationic molecules interact with quadruplexes, but not necessarily in a specific manner. Compounds with a bend shape might be more selective for quadruplexes and/or triplexes.

All these profiles confirm that equilibrium dialysis is a useful method to test the specificity of G4-based telomerase inhibitors [50]. Some, but not all, of the compounds presented here exhibit a significant preference for quadruplexes as compared to duplexes or single strands. However, it has been relatively difficult to dissociate triplex and quadruplex affinity. Most of the quadruplex ligands exhibited similar and sometimes higher affinity for antiparallel triplexes. Another potential

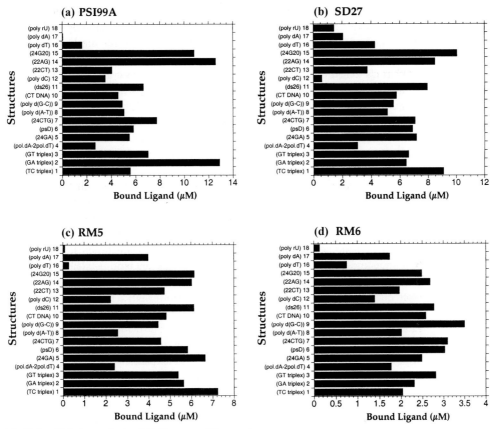

Fig. 12.7. Dialysis results (part 2). Same conditions as in Fig. 12.6. (a) **PSI 99A**; (b) **SD27**; (c) **RM5**; (d) **RM6**.

problem, when *in vivo* applications are contemplated, comes from the fact that very few molecules, with the possible exception of **DOC**, distinguish between the various classes of G-quadruplexes (intra- or intermolecular, parallel or antiparallel). In order to achieve specific telomerase inhibition, it might be interesting to identify compounds that would only bind to the antiparallel telomeric quadruplex (sample 14). As none of the ligands fulfill this condition, non-specific and undesired cellular effects are expected to occur [50].

12.4
Induction of Quadruplex Structures

In all previous experiments, binding of ligands to preformed DNA structures (and G-quadruplexes in particular) was measured. Another point would be to demonstrate that these molecules may induce the formation of the multistranded structures starting from single strands. The formation of some bimolecular quadruplex structures may be a very slow process. Small ligands such as perylene diimide [48] or ethidium derivatives [23] may behave as drivers for the assembly of nucleic acid secondary structures. It has been demonstrated by a gel-shift experiment that PIPER can dramatically accelerate the association of a DNA oligomer (Tr2) containing two tandem repeats of the human telomeric sequence (TTAGGG) into di- and tetrameric G-quadruplexes [48]. By analogy, eight different quadruplex ligands were tested for their ability to promote two different quadruplex-folded oligonucleotides, and the results are shown in Figs 12.8 and 12.9.

As expected, ethidium is a poor G-quadruplex ligand, it fails to induce significant amount of quadruplex (Fig. 12.8a). On the other hand, **8361**, **8362**, and **9944** efficiently promote quadruplex formation in a dose-dependent manner. Among these ethidium derivatives, only **8361** (Fig. 12.8b) seems to distinguish between the two quadruplexes: Tr2 (human telomeric quadruplex, crosses) is induced at a lower concentration than Ox-1T (*Oxytricha* telomeric repeat, circles). One should remember that this accelerated assembly is measured at a relatively high dye concentration and that other factors such as dye aggregation, binding stoichiometry, or dimerization could also play a role. The four other G-quadruplex ligands also promote quadruplex formation, but with a slightly lower efficiency (Fig. 12.9). **SD27** and **2,6 AQ** seem to prefer the *Oxytricha* over the human telomeric motif (Fig. 12.9a,b).

The situation is more complex for the dibenzophenanthrolines **MMQ1** and **MOQ1**. At low compound concentrations, Ox-1T is partially converted to a quadruplex species, whereas Tr2 remains single-stranded. However, with 20 μM **MMQ1** or **MOQ1**, G-quadruplex formation is favored with the Tr2 sequence. The following compounds were also tested: **DODC** is a very poor quadruplex inducer (<5% quadruplex at 20 μM dye concentration) whereas **RM5** induces a significant conversion to a multistranded species (data not shown).

12.5
Triplex versus Quadruplex Stabilization

Our laboratory has been involved in the search of triplex ligands for a decade. Various methods have been implemented to evidence triplex stabilization, such as UV-melting [16, 51], footprinting, and gel retardation experiments [52]. Of particular interest is the observation that some parallel triplexes may be formed only in the presence of these ligands [53] (Takasugi, unpublished results). We then reasoned that if a triplex stabilizer exhibits some affinity for other nucleic acid structures, adding increasing amounts of such competing structures should trap the triplex ligand and destabilize the triplex. This experiment was implemented using non-denaturing gel electrophoresis. Formation of a 16-base-long triplex is evidenced by a retarded band when the target duplex (40RY) is radiolabeled. Under these experimental conditions, no triplex is obtained in the absence of **SD27**,

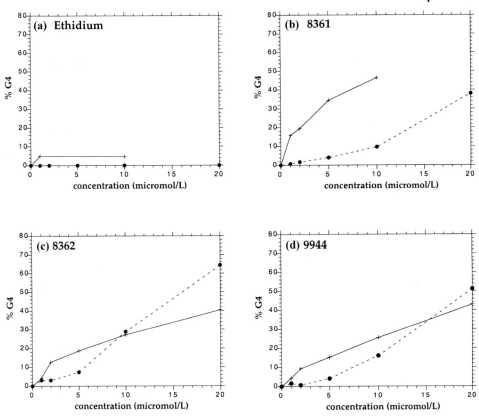

Fig. 12.8. G4 induction by ethidium derivatives. The purified Tr2 (5′ TACAGATAGTTAGGGTTAGGGTTA) and Ox-1T (5′ ACTGTCGTACTTGATAGGGGTTTTGGGGAATGTGA) oligodeoxynucleotides were 5′-end labeled and used at a final strand concentration of 8 μM. The solution was heated to 95°C for 5 min in a TE buffer containing 0.1 M KCl and slowly cooled to room temperature. Ligands were added (0–20 μM), and reaction mixtures were then incubated for 1 h at room temperature, then loaded on a native 12% acrylamide vertical gel in a 0.5× TBE buffer supplemented with 20 mM KCl. The gel was run at 4°C for 6 h, dried and analyzed with a phosphorimager. Quadruplex formation is efficiently promoted by some derivatives as shown by the appearance of a new band of retarded mobility. The dye concentration is indicated in micromolars. %G4 represents the relative amount of radioactivity found in the band corresponding to a quadruplex species. Crosses: Tr2 oligonucleotide; filled circles: Ox-1T. (a) **Ethidium**; (b) **8361**; (c) **8362**; (d) **9944**.

whereas a complete conversion to the triplex form is observed in the presence of 5 μM **SD27** (Fig. 12.10a, lane 1 and 2). Adding increasing amounts of a non-radiolabeled quadruplex leads to the disappearance of the triplex band (Fig. 12.10, lane 3–5). Adding an excess of the ligand (50 μM) restores the triplex band, showing that the trapping phenomenon is saturable and therefore does not result

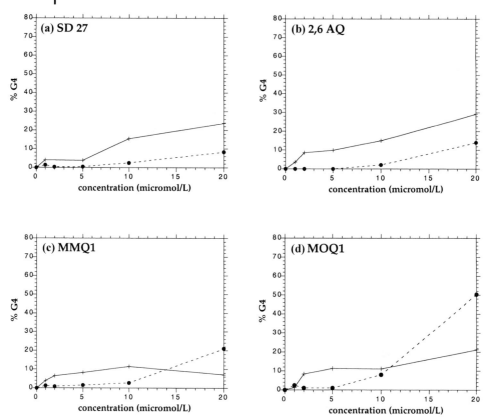

Fig. 12.9. G4 induction by other compounds. Conditions as in Fig. 12.8. %G4 represents the relative amount of radioactivity found in the band corresponding to a quadruplex species. Crosses: Tr2 oligonucleotide; filled circles: Ox-1T. (a) **SD27**; (b) **2,6 AQ**; (c) **MMQ1**; (d) **MOQ1**.

from an interaction between the competitor nucleic acid and the third strand of the duplex (16GT). Identical results are obtained in the presence of an excess of some, but not all, nucleic acid species. For example, single-stranded poly(dA) is an inefficient competitor, whereas an unrelated triplex leads to the rapid disappearance of the retarded band (Fig. 12.10a, lanes 7–9, 11–13).

The results of this method are quantified on Fig. 12.10b. The structures that inhibit triplex formation at low concentrations are expected to strongly bind **SD27**. This ranking is in agreement with the equilibrium dialysis result for most of the competitors structures: **SD27** has the best affinity for triplexes and G-quadruplexes structures. Nevertheless, some discrepancies are found: poly(dA)-2poly(dT) which is the best competitor according to this method has a relatively low affinity for **SD27** in the equilibrium dialysis experiment.

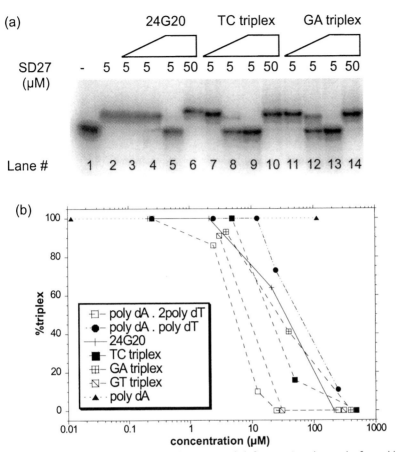

Fig. 12.10. Competition for triplex stabilization: a gel-shift assay. A triplex can be formed by the interaction of a 40-mer duplex (40R + 40Y) and a GT third strand (16GT) in the presence of ligand (**SD27**, see Fig. 12.5).

40R   5′-CATTCGTACGTTAGGAGGAAAAGGAGGATGCAAGTAATAC-3′

40Y   5′-GTATTACTTGCATCCTCCTTTTCCTCCTAACGTACGAATG-3′

16GT 5′-TGGTGGTTTTGGTGGT-3′

The triplex radiolabeled on the 40Y strand mixture was incubated in the absence or in the presence of **SD27** (5 or 50 μM) in a 50 mM HEPES (pH 7.2) buffer containing 140 mM KCl. This triplex mixture was also incubated in the presence of several non-radiolabeled nucleic acids prone to form different structures (single strands, duplexes, triplexes, and quadruplexes). For each competitor, the incubation at the higher competitor concentration was also performed with a 10-fold excess of **SD27** (50 μM). Duplex (40R + 40Y) and triplex (40R + 40Y + 16GT) species were separated by non-denaturing gel electrophoresis (12% acrylamide (acrylamide:bisacrylamide 19:1), HEPES 50 mM, pH 7.2) at 4°C. The gels were then dried and analyzed with a Phosphorimager SP instrument (Molecular Dynamics). Incubation 3 days at 4°C (HEPES 50 mM pH 7.2, KCl 140 mM, sucrose 10%, duplex 40R + 40Y* 10 nM (* = 5′ ^{32}P labeled), 16GT 10 μM). (a) Gel shift: Lane 1: nothing added; lanes 2–5, 7–9, 11–13: **SD27** 5 μM; lanes 6, 10, and 14: **SD27** 50 μM. Competitor added (concentrations are expressed in base triplets or quadruplets according to the competing structure): lanes 3–6 24G20 (2.1, 21, 210, 210 μM); lanes 7–10 TC triplex (4.9, 48.9, 489, 489 μM); lanes 11–14 GA triplex (3.9, 39.3, 393, 393 μM). Electrophoresis at 4°C in a 12% non-denaturing polyacrylamide gel (50 mM HEPES pH 7.2). (b) Quantification of the triplex signal detected in the gel-shift assay for various competitors.

12.6
Conclusion and Further Developments

Equilibrium dialysis is a useful method for the analysis of the structural selectivity of nucleic acids ligands. The data presented here, as well as previous observations, confirm that a significant overlap exists between triplex and quadruplex recognition [6, 19, 55, 56]. Within a given family of molecules, the detailed positioning of key substituents modulates triplex versus quadruplex preference. However, it is relatively difficult to completely prevent binding to one form without losing affinity for the other multistranded structure. A number of small ligands have been discovered to inhibit the function of telomerase by stabilizing G4-DNA structures. Evidence for the recognition of G4-DNA may also be provided by absorbance and fluorescence spectroscopy [23, 24, 34]. These ligands lock a telomeric G-overhang into a folded conformation which cannot be extended by telomerase. Once G4 stabilization and induction had been established, it is interesting to test whether these molecules inhibit telomerase as measured by a modified TRAP assay [23, 54]. Most G-quadruplex ligands described in this chapter inhibit telomerase with a submicromolar IC_{50} and may therefore be considered as potent inhibitors [23, 24, 34, 40].

Even more important is the compared affinity with duplexes [50]. Although some evidence has been presented to demonstrate that some drugs prefer to bind tetraplexes rather than duplexes, an exquisite specificity would be required to target this particular structure in the presence of a massive excess of double-stranded DNA [57]. Such preference is difficult to obtain as only a 2- to 10-fold preference for a tetraplex over a duplex has been demonstrated [58]. One trisubstituted acridine has a better specificity (40-fold) [59]. Most quadruplex ligands are polyaromatic molecules bearing one or more positive charge (see Fig. 12.5). A notable exception to that rule is NMM [60], an anionic porphyrin, which binds rather exclusively to quadruplexes. Equilibrium dialysis assay leads to the conclusion that this derivative, perhaps thanks to its negative charge and despite a relatively low affinity, is the most selective quadruplex ligand studied so far [26]. The race for a high-affinity highly specific G-quadruplex ligand has not ended yet!

12.7
Summary

Specific recognition of a given DNA conformation by small molecules may be demonstrated by equilibrium dialysis using a specific set of nucleic acid structures. We have designed a set of 18 different nucleic acid structures, ranging from single strands to G- or C-quadruplexes, and analyzed the binding of several ligands. Some of the molecules that bind to preformed G-quadruplexes were also shown to induce G-quadruplexes in a gel-shift assay. These families of derivatives were shown to inhibit telomerase *in vitro*.

Acknowledgments

We thank J. F. Riou (Reims, France), J. B. Chaires, J. Ren (Jackson, Mississippi) F. Gallaire, T. Garestier and C. Hélène (MNHN, Paris, France) for helpful discussions. This work was supported by an ARC grant (no. 4321), a French-South African grant and an Aventis research grant (to J-L. M.).

References

1 RICH, A. DNA Comes in many forms. *Gene* **1993**; *135*, 99–109.

2 CHAPUT, J. C., SWITZER, C. A DNA pentaplex incorporating nucleobase quintets. *Proc. Natl Acad. Sci. USA* **1999**; *96*, 10614–10619.

3 SEELA, F., KROSCHEL, R. Quadruplex and pentaplex self-assemblies of oligonucleotides containing short runs of 8-aza-7-deaza-2′-deoxyisoguanosine or 2′-deoxyisoguanosine. *Bioconjug. Chem.* **2001**; *12*, 1043–1050.

4 SEN, D., GILBERT, W. Formation of parallel four-stranded complexes by guanine-rich motifs in DNA and its applications for meiosis. *Nature* **1988**; *334*, 364–366.

5 ZAHLER, A. M., WILLIAMSON, J. R., CECH, T. R., PRESCOTT, D. M. Inhibition of telomerase by G-quartet DNA structures. *Nature* **1991**; *350*, 718–720.

6 SUN, D., THOMPSON, B., CATHERS, B. E. *et al.* Inhibition of human telomerase by a G-quadruplex-interactive compound. *J. Med. Chem.* **1997**; *40*, 2113–2116.

7 MERGNY, J. L., HÉLÈNE, C. G-quadruplex DNA: A target for drug design. *Nature Med.* **1998**; *4*, 1366–1367.

8 GEHRING, K., LEROY, J. L., GUÉRON, M. A tetrameric structure with protonated cytosine.cytosine base pairs. *Nature* **1993**; *363*, 561–565.

9 LEROY, J. L., GEHRING, K., KETTANI, A., GUÉRON, M. Acid multimers of oligodeoxycytidine strands: stoichiometry, base pair characterization and proton exchange properties. *Biochemistry* **1993**; *32*, 6019–6031.

10 KANG, C. H., BERGER, I., LOCKSHIN, C., RADLIFF, R., MOYZIS, R., RICH, A. Crystal structure of intercalated four stranded d(C3T) at 1.4 Å resolution. *Proc. Natl Acad. Sci. USA* **1994**; *91*, 11636–11640.

11 LE DOAN, T., PERROUAULT, L., PRASEUTH, D. *et al.* Sequence specific recognition, photocrosslinking and cleavage of the DNA double helix by an oligo α thymidylate covalently linked to an azidoproflavine derivative. *Nucleic Acids Res.* **1987**; *15*, 7749–7760.

12 MOSER, H. E., DERVAN, P. B. Sequence specific cleavage of double helical DNA by triple helix formation. *Science* **1987**; *238*, 645–650.

13 FELSENFELD, G., DAVIES, D. R., RICH, A. Formation of a three-stranded polynucleotide molecule. *J. Am. Chem. Soc.* **1957**; *79*, 2023.

14 MILLS, M., ARIMONDO, P., LACROIX, L. *et al.* Energetics of strand displacement reactions in triple helices: a spectroscopic study. *J. Mol. Biol.*, **1999**; *291*, 1035–1054.

15 SUN, J. S., DE BIZEMONT, T., DUVAL-VALENTIN, G., MONTENAY-GARESTIER, T., HÉLÈNE, C. Extension of the range of recognition sequences for triple helix formation by oligonucleotides containing guanines and thymines. *CR. Acad. Sci. Paris (III)* **1991**; *313*, 585–590.

16 MERGNY, J. L., DUVAL-VALENTIN, G., NGUYEN, C. H. *et al.* Triple helix specific ligands. *Science* **1992**; *256*, 1691–1694.

17 KERWIN, S. M. G-quadruplex DNA as a target for drug design. *Curr. Pharm. Design* **2000**; *6*, 441–471.

18 PERRY, P. J., ARNOLD, J. R. P., JENKINS, T. C. Telomerase inhibitors for the treatment of cancer: the current perspective. *Expert. Opin. Invest. Drugs* 2001; *10*, 2141–2156.

19 PERRY, P. J., RESZKA, A. P., WOOD, A. A. *et al.* Human telomerase inhibition by regioisomeric disubstituted amidoanthracene-9,10-diones. *J. Med. Chem.* 1998; *41*, 4873–4884.

20 PERRY, P. J., GOWAN, S. M., READ, M. A., KELLAND, L. R., NEIDLE, S. Design, synthesis and evaluation of human telomerase inhibitors based upon a tetracyclic structural motif. *Anti Cancer Drug Des.* 1999; *14*, 373–382.

21 NEIDLE, S., HARRISON, R. J., RESZKA, A. P., READ, M. A. Structure-activity relationships among guanine-quadruplex telomerase inhibitors. *Pharmacol. Ther.* 2000; *85*, 133–139.

22 CAPRIO, V., GUYEN, B., OPOKUBOAHEN, Y. *et al.* A novel inhibitor of human telomerase derived from 10H-indolo[3,2-b]quinoline. *Bioorg. Med. Chem. Lett.* 2000; *10*, 2063–2066.

23 KOEPPEL, F., RIOU, J. F., LAOUI, A. *et al.* Ethidium derivatives bind to G-quartets, inhibit telomerase and act as fluorescent probes for quadruplexes. *Nucleic Acids Res.* 2001; *29*, 1087–1096.

24 MERGNY, J. L., LACROIX, L., TEULADE-FICHOU, M. P. *et al.* Telomerase inhibitors based on quadruplex ligands selected by a fluorescent assay. *Proc. Natl Acad. Sci. USA* 2001; *98*, 3062–3067.

25 REN, J. S., CHAIRES, J. B. Preferential binding of 3,3'-diethyloxadicarbocyanine to triplex DNA. *J. Am. Chem. Soc.* 2000; *122*, 424–425.

26 REN, J. S., CHAIRES, J. B. Sequence and structural selectivity of nucleic acid binding ligands. *Biochemistry* 1999; *38*, 16067–16075.

27 REN, J. S., BAILLY, C., CHAIRES, J. B. NB-506, an indolocarbazole topoisomerase I inhibitor, binds preferentially to triplex DNA. *FEBS Lett.* 2000; *470*, 355–359.

28 RIPPE, K., FRITSCH, V., WESTHOF, E., JOVIN, T. M. Alternating d(G-A) sequences form a parallel-stranded DNA homoduplex. *EMBO J.* 1992; *11*, 3777–3786.

29 EVERTSZ, E. M., RIPPE, K., JOVIN, T. M. Parallel-stranded duplex DNA containing blocks of trans purine-purine and purine-pyrimidine base pairs. *Nucleic Acids Res.* 1994; *22*, 3293–3303.

30 RIPPE, K., JOVIN, T. M. Substrate properties of 25-nt parallel-stranded linear duplexes. *Biochemistry* 1989; *28*, 9542–9549.

31 PHAN, A. T., GUERON, M., LEROY, J. L. The solution structure and internal motions of a fragment of the cytidine-rich strand of the human telomere. *J. Mol. Biol.* 2000; *299*, 123–144.

32 WANG, Y., PATEL, D. J. Solution structure of the human telomeric repeat d[AG3(T2AG3))3] G-tetraplex. *Structure* 1993; *1*, 263–282.

33 SLAMA-SCHWOCK, A., JAZWINSKI, J., BERE, A. *et al.* Interaction of dimethyldiazaperopyrenium dication with nucleic acids: 1. Binding to nucleic acids components and to single-stranded polynucleotides and photocleavage of single-stranded oligonucleotides. *Biochemistry* 1989; *28*, 3227–3234.

34 ALBERTI, P., SCHMIDT, P., NGUYEN, C. H., HOARAU, M., GRIERSON, D., MERGNY, J. L. Benzoindoloquinolines interact with DNA quadruplexes and inhibit telomerase. *Bioorg. Med. Chem. Lett.* 2002 *12*, 1071–1074.

35 MERGNY, J. L., PHAN, A. T., LACROIX, L. Following G-quartet formation by UV-spectroscopy. *FEBS Lett.* 1998; *435*, 74–78.

36 XODO, L. E., MANZINI, G., QUADRIFOGLIO, F. Spectroscopic and calorimetric investigation on the DNA triplex formed by d(CTCTTCTTTCTTTTCTTTCTTCTC) and d(GAGAAGAAAGA) at acidic pH. *Nucleic Acids Res.* 1990; *18*, 3557–3564.

37 MERGNY, J. L., LACROIX, L., HAN, X., LEROY, J. L., HÉLÈNE, C. Intramolecular folding of pyrimidine oligodeoxynucleotides into an i-DNA motif. *J. Am. Chem. Soc.* 1995; *117*, 8887–8898.

38 LAVELLE, L., FRESCO, J. R. UV spectroscopic identification and thermodynamic analysis of protonated third strand deoxycytidine residues at neutrality in the triplex d(C+-T):[d(A-G)6.d(C-T)6]; evidence for a proton switch. *Nucleic Acids Res.* **1995**; *23*, 2692–2705.

39 LACROIX, L., MERGNY, J. L., LEROY, J. L., HÉLÈNE, C. Inability of RNA to form the i-motif: Implications for triplex formation. *Biochemistry* **1996**; *35*, 8715–8722.

40 ALBERTI, P., REN, J., TEULADE-FICHOU, M. P. et al. Interaction of an acridine dimer with DNA quadruplex structures. *J. Biomol. Struct. Dyn.* **2001**; *19*, 505–513.

41 MERGNY, J. L., COLLIER, D., ROUGÉE, M., MONTENAY-GARESTIER, T., HÉLÈNE, C. Intercalation of ethidium bromide in a triple-stranded oligonucleotide. *Nucleic Acids Res.* **1991**; *19*, 1521–1526.

42 SCARIA, P. V., SHAFER, R. H. Binding of ethidium bromide to a DNA triple helix. *J. Biol. Chem.* **1991**; *266*, 5417–5423.

43 GUO, Q., LU, M., MARKY, L. A., KALLENBACH, N. R. Interaction of the dye ethidium bromide with DNA containing guanine repeats. *Biochemistry* **1992**; *31*, 2451–2455.

44 CHEN, Q., KUNTZ, I. D., SHAFER, R. H. Spectroscopic recognition of guanine dimeric hairpin quadruplexes by a carbocyanine dye. *Proc. Natl Acad. Sci. USA* **1996**; *93*, 2635–2639.

45 KERWIN, S. M., SUN, D., KERN, J. T., RANGAN, A., THOMAS, P. W. G-quadruplex DNA binding by a series of carbocyanine dyes. *Bioorg. Med. Chem. Lett.* **2001**; *11*, 2411–2414.

46 NGUYEN, C. H., MARCHAND, C., DELAGE, S. et al. Synthesis of 13H-benzo[6,7]- and 13H-benzo[4,5]indolo[3,2-c]quinolines: a new series of potent specific ligands for triplex DNA. *J. Am. Chem. Soc.* **1998**; *120*, 2501–2507.

47 FEDOROFF, O. Y., SALAZAR, M., HAN, H., CHEMERIS, V. V., KERWIN, S. M., HURLEY, L. H. NMR-based model of a telomerase inhibiting compound bound to G-quadruplex DNA. *Biochemistry* **1998**; *37*, 12367–12374.

48 HAN, H. Y., CLIFF, C. L., HURLEY, L. H. Accelerated assembly of G-quadruplex structures by a small molecule. *Biochemistry* **1999**; *38*, 6981–6986.

49 SLAMA-SCHWOCK, A., ROUGÉE, M., IBANEZ, V. et al. Interaction of dimethyldiazapyrenium dication with nucleic acids: 2. Binding to double-stranded polynucleotides. *Biochemistry* **1989**; *28*, 3234–3242.

50 PERRY, P. J., JENKINS, T. C. DNA tetraplex binding drugs: structure-selective targeting is critical for antitumour telomerase inhibition. *Mini Rev. Med. Chem.* **2001**; *1*, 31–41.

51 ESCUDÉ, C., NGUYEN, C. H., KUKRETI, S. et al. Rational design of a triple helix-specific intercalating ligand. *Proc. Natl Acad. Sci. USA* **1998**; *95*, 3591–3596.

52 ESCUDÉ, C., SUN, J. S., NGUYEN, C. H., BISAGNI, E., GARESTIER, T., HÉLÈNE, C. Ligand-induced formation of triple helices with antiparallel third strands containing G and T. *Biochemistry* **1996**; *35*, 5735–5740.

53 ROULON, T., HÉLÈNE, C., ESCUDÉ, C. A ligand-modulated padlock oligonucleotide for supercoiled plasmids. *Angew. Chem. Int. Ed.* **2001**; *40*, 1523–1526.

54 KRUPP, G., KUHNE, K., TAMM, S. et al. Molecular basis of artifacts in the detection of telomerase activity and a modified primer for a more robust 'TRAP' assay. *Nucleic Acids Res.* **1997**; *25*, 919–921.

55 HAQ, I., LADBURY, J. E., CHOWDHRY, B. Z., JENKINS, T. C. Molecular anchoring of duplex and triplex DNA by disubstituted anthracene-9,10-diones: calorimetric, UV melting, and competition dialysis studies. *J. Am. Chem. Soc.*, **1996**; *118*, 10693–10701.

56 BAUDOIN, O., MARCHAND, C., TEULADE-FICHOU, M. P. et al. Stabilization of DNA-triple-helices by crescent-shaped dibenzo-phenanthrolines. *Chemistry Eur. J.* **1998**; *4*, 1504–1508.

57 MERGNY, J. L., MAILLIET, P., LAVELLE, F., RIOU, J. F., LAOUI, A., HÉLÈNE, C. The development of telomerase inhibitors: the G-quartet approach. *Anti Cancer Drug Des.* **1999**; *14*, 327–339.

58 ANANTHA, N. V., AZAM, M., SHEARDY, R. D. Porphyrin binding to quadrupled T4G4. *Biochemistry* **1998**; *37*, 2709–2714.

59 READ, M., HARRISON, R. J., ROMAGNOLI, B. *et al.* Structure-based design of selective and potent G quadruplex-mediated telomerase inhibitors. *Proc. Natl Acad. Sci. USA* **2001**; *98*, 4844–4849.

60 ARTHANARI, H., BASU, S., KAWANO, T. L., BOLTON, P. H. Fluorescent dyes specific for quadruplex DNA. *Nucleic Acids Res.* **1998**; *26*, 3724–3728.